Dynamic RAM
Technology Advancements

Dynamic RAM
Technology Advancements

Muzaffer A. Siddiqi

CRC Press
Taylor & Francis Group
Boca Raton London New York

CRC Press is an imprint of the
Taylor & Francis Group, an **informa** business

CRC Press
Taylor & Francis Group
6000 Broken Sound Parkway NW, Suite 300
Boca Raton, FL 33487-2742

First issued in paperback 2017

Version Date: 20120822

ISBN 13: 978-1-138-07705-8 (pbk)
ISBN 13: 978-1-4398-9373-9 (hbk)

This book contains information obtained from authentic and highly regarded sources. Reasonable efforts have been made to publish reliable data and information, but the author and publisher cannot assume responsibility for the validity of all materials or the consequences of their use. The authors and publishers have attempted to trace the copyright holders of all material reproduced in this publication and apologize to copyright holders if permission to publish in this form has not been obtained. If any copyright material has not been acknowledged please write and let us know so we may rectify in any future reprint.

Library of Congress Cataloging-in-Publication Data

Siddiqi, Muzaffer A.
    Dynamic RAM : technology advancements / Muzaffer A. Siddiqi.
        p. cm.
    Includes bibliographical references and index.
    ISBN 978-1-4398-9373-9 (hardback)
    1. Semiconductor storage devices. 2. Random access memory. I. Title.

TK7895.M4S53 2012
004.5'3--dc23   2012030359

Visit the Taylor & Francis Web site at
http://www.taylorandfrancis.com

and the CRC Press Web site at
http://www.crcpress.com

Contents

Preface

Semiconductor memories were introduced in the late 1960s. Increased storage density, speed, and reduced cost were their main advantages. In the basic form, semiconductor random access memory (RAM) information is stored in a cell that is replicated for each bit. RAMs are broadly classified as nonvolatile, such as flash memories, or volatile, such as static RAM (SRAM) and dynamic RAM (DRAM). No single memory satisfies all system needs. However, the highest density and lowest cost per bit of DRAMs have given them a prominent role, whether in mainframe computers and PCs, telecommunications, or so many other high-tech applications such as avionics and space. Hence, during the last four decades the number of DRAM bits/chip has increased four times every three years and the cost/bit ratio has declined nearly by the same order. Communication systems, graphic subsystems, and the extremely fast expanding mobile industry became catalysts for the rapid growth in the demand for large-volume memories. The present work is an attempt to describe the technological developments that allowed the design and optimization of high-density and cost-efficient DRAMs.

The one-device DRAM cell was conceptualized and then manufactured at the 1 Kbit level in planar form. Rapid advancements have presently enabled the one-device cell as three-dimensional entity with 2 Gbit DRAMs (DDR3) firmly established in the market and 4 Gbit DRAMs on the threshold. This book tries to look into the breakthroughs that contributed to the fast evolution of DRAMs.

A DRAM cell consists of a storage capacitor and a select transistor with cells arranged in arrays of rows (word lines or WLs) and columns (bit lines or BLs). For higher density DRAMs, multiple subarrays replace a single large array to shorten the WLs and the BLs and thereby make the cell faster. Once a transistor selects a cell, it can be *read* or *written*. Storage capacitor changes bit line potential and a sense amplifier detects this small signal change. With continuous decrease in cell size and working voltage, storage capacitance value must be maintained above a critical value and a large amount of research is going on to maintain the critical value of the charge on it without increasing (rather decreasing) its projected chip area. Advancements in the sensing techniques for this critical charge and hence, design of sense amplifiers, assume great importance. At the same time development of storage capacitance with thinly layered high dielectric constant materials compatible with other materials gained extreme significance.

The reading process of DRAM disturbs stored information; hence, it is to be restored after each *read*. In addition, cell capacitor charge continuously leaks and thus needs continuous replenishment, requiring additional power to retain the correct data. Different types of leakages, and the circuits and technologies employed as the remedial measures, have become extremely important, especially with increasing DRAM density.

The topic of semiconductor memories does find a regular place in undergraduate technology curricula and elective courses are given at institutes at the graduate level; research work has been limited *mainly* to the big industries involved in DRAM manufacture. For a large number of people to be involved, which is essential for rapid developments, technological advancement needs to be presented to generate wider interest. It is a fair assumption that more books must be made available as references and in textbook form on the topic of semiconductor memory to make it popular as a separate subject in academic circles. Only a few books are available on semiconductor memories in general and fewer on DRAMs. It is expected that the present work will be a worthwhile addition for enhancing interest in the study of semiconductor memories. It is also likely to be useful to people in semiconductor manufacturing and the electronic industry as expert groups involved in a different nature of activities might like to look for related topics. Academicians and researchers interested in state-of-the-art work on semiconductor memories may also find it useful as parts of the book can be included in a graduate-level course.

Although a great amount of effort and attention has been paid in the preparation of the manuscript, the book may still have some shortcomings and mistakes. The author would greatly appreciate any correction or suggestion, whether with regard to the method of presentation, technical flaws, or content.

The book has been divided into eight chapters. The first chapter includes basic construction, working and simple modes of operation of DRAM, and its evolution from the six-transistor SRAM. Some important developments of the 1970s such as the introduction of sense amplifiers in DRAMs and a discussion on open and folded bit line structures are added. The rest of the chapter is devoted to some other emerging competitors of DRAM, such as capacitor-less DRAM and potential nonvolatile memories, namely, ferroelectric RAM (FRAM), magnetoresistive RAM (MRAM), including STT MRAM, and phase-change RAM.

It is not only that the size of the cell is to be reduced with increasing DRAM density but the way the two basic components, the transistor and the storage capacitor, are realized and positioned physically with respect to each other, and how the word and bit line are fabricated are also important for its functional and economic feasibility. Early- and medium-stage DRAM cells are discussed in Chapter 2. It includes planar DRAM cells used for 1–4 Mbit density, in which a metal-oxide-semiconductor field-effect transistor (MOSFET) and a capacitor were fabricated side by side. As it was physically not possible to accommodate higher-density DRAMs in a two-dimensional small area, capacitor was located below the substrate surface in *trench capacitor cells* or the capacitor above the surface of the substrate in *stacked capacitor cells*. Types of first-generation trench capacitor cells in which the storage node was outside the dielectrically lined trench were the initial stage three-dimensional cells. With the main intention of decreasing cell leakage and improving soft error rate, inverted trench cells were fabricated in which the storage electrode was inside the isolated trench. Further increases in the density of DRAMs brought cells very close to each other, increasing leakage. Moreover, to get sufficient capacitance, the trench was required to be deeper. Facing a bottleneck in technology advancements, stacked capacitors, with horizontal and vertical fin–structured capacitor cells, were found to be more suitable. With improved technology and better fabrication processes, higher value dielectric-based capacitors realized higher-density DRAM cells.

In the planar DRAM cell, a thoroughly studied planar MOSFET was used. However, 1 Gbit and beyond DRAMs found the drivability of the conventional transistor impractical because of its small channel width. Cell array voltage was also coming down and problems of short channel effect, leakage currents, hot electron generation, and so on, forced researchers to pursue different/advanced transistors. Recessed types of transistors, delta-structured transistors that later took the form of Fin field effect transistors (FinFETs), replaced the conventional transistors. Along with a discussion on these transistors, a few other transistors like body-tied MOSFET bulk FinFET and saddle MOSFETs are included in Chapter 4. A major advancement came in the form of the surrounding gate transistor (SGT) and stacked SGT, with the help of which cell size could be highly reduced. Vertical cells and some advanced cells that combined the better features of earlier developed transistors have also been used. Buried strap (BEST) and vertical BEST (VERIBEST) DRAM cells are examples of cells with advanced trench capacitors.

Fabrication of storage capacitance in an ever decreasing available chip area is a critical issue for DRAMs. Considerable attention has been paid to the use of a variety of dielectrics, materials for the capacitor electrodes, and the physical formation of the capacitor. Obviously planar transistors with polysilicon electrodes and SiO_2 as dielectric had a limit, even when dielectric thickness was reduced to its minimum permissible value of 3–4 nm. Use of textured storage nodes has been shown effective in increasing realized capacitance not only in planar capacitors but also in capacitors with other complex shapes. However, major advancement behind capacitance enhancement came with the use of metal-insulator-metal (MIM) capacitors with high dielectric constants value insulators like Ta_2O_5 and barium strontium titanate (BST).

An increase in DRAM density and improvement in its performance could become possible with improvements in fabrication processes, circuit innovations, and technological advances. A cornerstone in DRAM development is the reduction in minimum feature size F or technology scaling. Technological developments since the early stage to around 1 Gbit DRAM density are included in Chapter 3, and advanced developments are included in Chapter 6 after going through the study of advanced cell transistors and storage capacitor enhancement techniques in Chapters 4 and 5, as the technology is strongly connected with the two basic components of the DRAM. Change over to complementary MOSFET (CMOS) technology was a major event that occurred mainly to reduce power consumption. An important connected circuit innovation was ($V_{DD}/2$) precharging of the bit line before the read operation, which changed many design rules in a positive way. Stacked capacitor developments especially in crown shape with Ta_2O_5 as dielectric-related technological developments were very significant. Another class of stacked capacitor arrangement in which the realized capacitor enclosed the bit line has continued its usage as it helped in minimizing bit-line noise considerably.

Advancements in the DRAM technologies are focused mainly on fabricating a high-performance transistor and a compact capacitor in a minimum-sized cell and interconnections with a minimum of parasitic capacitance and resistance with the help of advanced lithographic technologies. Obviously every generation brought new challenges. Initial-stage capacitor-over-bit-line (COB) cells, discussed in Chapter 3, were improved to remove some drawbacks of difficulty in making connection with the buried bit line. The advancements in technology were gradual in a sense that at every stage, some new technologies were merged with the older ones; however, on the two sides of 100 nm nodes, technology perspectives were significantly different. The height of the storage node now needed to be so high that it needed mechanical stability. Two prominent technologies for the purpose discussed in Section 6.3 are leaning exterminated ring-type insulator (LERI) and mechanically enhanced storage node for unlimited height (MESH). Most of the advanced transistors and capacitors were already discussed in Chapters 4 and 5, and advantages obtained through them were integrated with other technologies such as developments in lithography and resolution enhancement technologies (RETs), without which ever-decreasing features could not be fabricated. Other important support technologies are in the form of isolation techniques such as moving from local oxidation of silicon (LOCOS) to shallow trench isolation (STI) and the filling of dense oxide in trenches and gaps. Formation of bit lines, word lines, and their connection with source/drain and gate, which assumed significance and performance, had to be improved through the use of low-resistance materials with a minimum of parasitic capacitance.

With decreasing dimension, fringe capacitance dominates and bit-line capacitance does not come down. Use of Si_3N_4 capping of bit and word lines also adds to the parasitic capacitance. As the word line is to be connected to the transistor gate, while remaining isolated from source/drain, and the storage capacitor node is to be connected with the source, cell connections assume significance; use of materials like Ti and W has found favor. Lithographic error tolerance reduction has to be attended to as well. Smaller width of connecting wires, word lines, and bit lines results in stress-induced fabrication failures, and use of aluminum was constrained. Multilayer connection wire and other metals like Cu became necessary. At the same time as all the wires were surrounded by dielectrics, which resulted in interwire capacitance, use of a low-dielectric environment along with a high-conductivity wire strip became a topic of intense research.

After a brief survey of the technologies deployed at various stages in DRAM development, technology flow is described in Section 6.11. It is hoped that the problems faced and the solutions employed at the 0.18 µm stage, 100 nm to 50 nm, and sub-50 nm levels will help readers form their views about the technological steps. Selection of dielectrics, formation of capacitor electrodes, filling of gaps without or with a minimum of voids, formation of contact holes, planarization, and self-aligned contact (SAC) change their implementation and the direction of the technology progress. Evolution of photolithography while using KrF and ArF has to be paid maximum attention, without which any conceived design would not be born.

Going below 50 nm using stacked capacitor technology presented new challenges, which are being faced through the use of MIM capacitors with dielectric materials like $HfO_2/Al2O_3$ stacks and ZrO_2 films in conjunction with advanced transistors such as recessed transistors, body tied FinFETs, vertical pillar transistors, and saddle transistors. The buried word-line scheme, which uses extended U-shaped devices with metal gates, is another important technology. Use of high-k materials has also been employed in deep trench technology in sub-100 nm levels. A checkerboard layout (CKB), having highly symmetrical lines and spaces, is helpful in realizing smaller technology nodes. One of the technological solutions for faster data transfer with DRAM was developed in the form of embedded DRAM (eDRAM) in which good points of high-density DRAM and logic modules were combined to lead toward realizing a system-on-chip (SOC). Some contradictory requirements while fabricating high-density DRAM and the logic modules on the same chip are discussed at the end of Chapter 6, and the technologies adopted to overcome the problem faced are described. Fabrication of eDRAM was mostly done on bulk silicon up to 65 nm technologies, but beyond that silicon-on-insulator (SOI) substrate was used in a simplified deep trench process. SOI logic has been realized in 45 nm and 32 nm node technology for advance applications. Stacked capacitor cells, which were not preferred earlier because of the high-temperature processes involved, were also modified to give full-metal eDRAM using MIM capacitor.

Power dissipation in the ever-increasing transistor count on a chip has always been one of the most important issues. One major successful remedy is the reduction in the working voltage, which was otherwise also critical in the highly expanding mobile market. Therefore, low-voltage low-power memories/logic and systems are receiving big attention. Different kinds of leakage currents, which are mainly responsible for power dissipation in DRAMs, are discussed in Chapter 7; their effect on cell signal charge and data retention is studied. Depending upon applications, power dissipation in active mode, sleep mode, or both modes is critical and needs to be minimized. Several methods are now available, which are applicable in general or specific in nature for the active or sleep mode. Another very important factor is the study of power consumption during refreshing of DRAMs and the frequency with which refreshing has to be done. Technological developments have greatly affected this area. A considerable amount of research work is going on looking into different aspects of the DRAM's power consumption reduction.

Peripherals greatly affect the cost and speed of the DRAM. Though the basics of all the peripherals have remained same, their construction has changed considerably with the increasing complexity of DRAMs. Hence, only the simplest versions of row and column decoders have been included; actual architecture does depend on the way the high-density DRAM is divided in blocks and sub-blocks and the way the bit lines and/or word lines are segmented/divided/shared. The sense amplifier is one very important component and has undergone many changes, despite the basic gated flip-flop remaining firmly placed. Keeping the cost of the DRAM competitive is one major aim; therefore, the use of redundancy techniques and error correction coding (ECC) is also very important. Combinations of redundancy and ECC for high-density DRAMs are also studied briefly in the last chapter on DRAM peripherals.

Acknowledgments

I must acknowledge Dr. Ashok K. Sharma, author of a beautiful book, *Semiconductor Memories—Technology, Testing and Reliability* (IEEE Press), as his book provided me encouragement to introduce a graduate-level course on semiconductor memories at Z. H. College of Engineering and Technology, Aligarh Muslim University, Aligarh, India. After teaching this course for a few years and testing the endurance of my students, encouragement from colleagues enabled me to take up this project. All the department colleagues shared my added responsibility of the chairmanship, which was a great help. I am indebted to my wife, Shagufta Siddiqi; my children, Zeba Syed, Subuhi Riaz, Soofia Ghufran, Bushra Shahbaz; and my close friends, who were tolerant enough of my dry attitude toward them in the recent past. Their consistent support kept me afloat.

Continuous help from Dr. Gagandeep Singh of CRC Press and the suggestions of learned referees have gone a long way toward completing the book and improving the contents. Crisp and prompt notes from Laurie Schlags and the confidence shown by the editorial board of Taylor & Francis are acknowledged with thanks. I shall be failing if I don't thank Nadeem Nami, who did all the word processing, Mohammad Sulaiman, who drew most of the figures, and my son, Belal, who made all the corrections in the manuscript while proofreading.

Muzaffer A. Siddiqi
Aligarh, India

1

Random Access Memories

1.1 Introduction

The last few decades have seen a tremendous increase in usage of semiconductor memories, and there has been no looking back. Digital circuits and systems are using semiconductor memories in ever-increasing proportion. Advances in technology and fabrication processes have resulted in a high rate of continuous increase in the memory density. Performance has also been improving, which has opened new application areas considered unreachable. A broad categorization of semiconductor memories is in terms of their ability to retain stored data when supply is stopped; volatile memories lose their data whereas nonvolatile memories retain it. Be it volatile or nonvolatile, in most of the semiconductor memories information can be stored or retrieved from any location, hence the term *random access memories* (RAMs). The basic arrangement of storage of data/information is done either in a bistable flip-flop called static RAM (SRAM) or through charging a capacitor in dynamic RAM (DRAM). Both SRAM and DRAM are volatile memories.

In the last four decades SRAM and DRAM have seen tremendous growth. Both have been developed, improved, and used in large quantities. However, in terms of volume, DRAMs have remained on top as main memory in large systems, apart from other widespread applications, because of their higher density and low cost per bit of information stored. At the same technology generation level, SRAM occupies nearly four times the chip area of DRAM, but it is faster and has low power consumption. Therefore, DRAM designers continuously try to make it faster as well as better in terms of power consumption.

The basis entity in a semiconductor memory is a *cell* in which a bit of information is stored. SRAM employs six or four transistors in each cell and the DRAM uses a single transistor, called *access transistor,* and a small value capacitor, called *storage capacitor.* Since the cell has to be replicated as many times as the density of the RAM, all efforts are concentrated toward the improvements in the formation of cell, or on the access transistor and the storage capacitor. This book is intended to be a study on important technological aspects of the single-transistor single-capacitor (1T1C) DRAM. As the DRAM evolved from the six-transistor SRAM through reduction in the use of number of transistors, Chapter 1 begins with a brief introduction of the construction and working of SRAM. Changeover to three-transistor (3T) DRAM, along with its basic operation and construction, is taken next and then its conversion to the (1T1C) DRAM, which enabled it to be produced in 4 Mbit and 16 Mbit density levels. Early-stage DRAM cell was a simple entity with a planar metal-oxide-semiconductor field-effect transistor (MOSFET) and a planar capacitor fabricated side by side. Its major difference with the 3T DRAM was that the retrieving of information or read operation became destructive in the DRAM. It became a must to restore the information in each cell of the DRAM, which was done through a *sense amplifier*. It was and shall remain one of the most important peripheral circuits for the DRAM cell operation. While using the sense amplifier, reduction in the generated noise was to be minimized and one of the early methods, which is continuously in use, is *folded bit-line connection* in place of *open bit-line configuration*. Advantages and limitations of both the structures, which also decide the location of the sense amplifiers, are briefly discussed. Other than the advancements in DRAM technology for improving its performance, several modes of operation were used. A major change came in the form of conversion from asynchronous to synchronous architectures. All the modes of operation, in either configuration, improved the rate of transfer of information from the DRAM to the outside world. Study of architecture and enhancement of the rate of transfer of information has become extremely important. However, here the modes of operations have been mentioned only in brief.

Most of the early and middle-age advancements in DRAM fabrication were done on bulk silicon substrate. With the reduction in minimum feature size, cost of silicon on insulator (SOI) technology became comparable to the bulk silicon technology, with the added advantage of lesser parasitic capacitance. SOI technology was also applied to the conventional DRAM structure for performance enhancement. In addition, property of storing charge in the body of SOI substrate was recently used to fabricate DRAM cell without external capacitor in order to achieve even smaller size cell. Basic concept of such capacitor-less DRAM with some examples is included.

Nonvolatile memories also find widespread applications, where the data is written permanently or many times after erasing it, but the information is not destroyed once the supply is off. As a result of continuous research and developments a number of nonvolatile memories are emerging and claiming to give tough competition to the DRAMs. One such structure is in the form of *flash memories* in which either complete data or big blocks of stored data can be erased electrically, instead of unreliable and time-consuming methods. Flash memories use either NOR or NAND structure to get random access and faster writing (programming). Nonvolatile memories generally have used properties of materials, which change in digital form on the application of proper bias condition. Ferroelectric materials have been used as dielectric materials for the storage electrodes for the realization of nonvolatile ferroelectric RAMs (FRAMs). Polarization of dielectric film in opposite direction corresponds to the two digital states and removal of applied voltage does not change the stored state. Change of magnetoresistance in the presence of magnetic fields has been used in the integration of another nonvolatile memory-MRAM. There are two main memory elements used in it, which are giant magnetoresistance and magnetic tunnel junction. MRAM got a boost in the form of spin-torque-transfer (STT) MRAM, in which injection of polarized electrons was used to change the direction of magnetic layer, instead of using an external magnetic field. Distinct values of resistance in different directions of magnetic field resulted in two clear states for data storage. The third nonvolatile memory discussed at the end of Chapter 1 is the phase-change RAM (PRAM) in which chalcogenide alloy–based memory element was used, which changed its state on the application of controlled amount of heat. Resistance of the material is changed on the *set* and *reset* conditions, which provides sufficient read signal margins. A comparison between the three leading nonvolatile memories is also included.

1.2 Static Random Access Memory

The basic six-transistor CMOS SRAM cell consisting of two cross-coupled inverters M_1-M_2 and M_3-M_4 and two access transistors M_5 and M_6 is shown in Figure 1.1. Access to the cell is enabled by the word line, which controls the

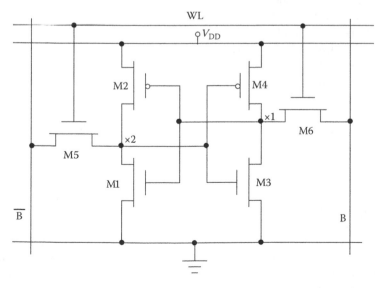

FIGURE 1.1
Six-transistor CMOS SRAM cell.

transistors M_5 and M_6. Two bit lines are required for writing (storing) in the cell as well as for reading the stored (written) signal. The cell has two stable logic states, a 1 (high) and a 0 (low). If in the logic state 1 (0) node x_1 (x_2) is high, then node x_2 (x_1) shall have to be low, and transistor M_1 and M_4 shall be on (off) whereas transistor M_2 and M_3 shall be off (on). During its read/ write operation, the selected word line is made high through a row decoder (not shown in the figure) making the access transistors M_5 and M_6 conduct. The column address decoder can select the bit line B and inverse bit line $\overline{B}$ of any cell in the addressed column.

The first step in the read operation is to make bit lines B and $\overline{B}$ high and to select a word line (WL). Enabling of the word line connects outputs of the cross-coupled inverter nodes x_1 and x_2 to B and $\overline{B}$ respectively; depending on the data 1 or 0 stored, one of the bit-line voltages is pulled low by a small amount. The differential signal on B and $\overline{B}$ lines is detected, is amplified to full logic value, and is made available at the output.

Assume that a high is stored at x_1 and that both bit lines have been pre-charged to V_{DD}. As the WL has been selected, it enables both the access transistor M_5 and M_6 after the initial WL delay. Data values stored at x_1 and x_2 are transferred to the bit lines by leaving B at its precharged value and $\overline{B}$ discharging through transistors M_1–M_5.

For a cell to be of small size, it is essential to keep the size of all transistors as close to minimum as possible, though it makes their on-resistance large. Combined with larger value of bit-line capacitance in large memories, on-resistance makes the rate of drop in one of the bit-line voltage very slow. However, as soon as the difference of potential between B and $\overline{B}$ reaches a small but critical value, the sense amplifier is activated to accelerate the change in the bit-line voltage. If the sense amplifier is not used, it will take too long to reach its final value so as to correctly read the data.

In the meantime, there is a problem to be sorted out during this read process, as M_1 and M_5 form a potential divider section. To prevent a substantial current from flowing through the M_3-M_4 inverter, it is necessary to keep the resistance of transistor M_5 larger than that of transistor M_1.

The boundary constraints on the device size (resistance) are derived by solving current equations at the maximum allowable value of the potential difference between B and $\overline{B}$. It is observed that even in the worst case the only requirement is that transistor M_5 has to be made weaker by increasing its length, thereby making the basic cell a bit larger. However, in practice, this is not needed as the second bit line B, which is precharged to V_{DD} and now clamped to node x_1 at $V_{DD,}$ makes the read operation safe.

With transistor M_3-M_4 being of minimum size and M_5 and M_6 slightly weaker for read constraints, node x_2 has to be below nearly 0.4 V (assuming V_{DD} as 2.5 V), and a high cannot be written as such. Writing high can be ensured only if node x_1 can be pulled below the threshold value of transistor M_1. Fortunately, this constraint is met easily when both the transistors M_4 and M_6 are minimum-size devices [1].

As the bit lines are precharged to V_{DD}, the PMOS transistors of 6T SRAM cell are not involved in the pull-up process. The requirement through this path is only to maintain the state of the cell by compensating for the small leakage currents, which is on the order of 10^{-15} A/cell [2]. This job can be done in different ways, and one such method is to use very large value resistors in place of the PMOS transistors M_2 and M_4 of Figure 1.1. Use of high-value resistors minimizes the static power consumption in addition to reducing the SRAM cell size by nearly one-third. Such large-value resistors in the teraohm range are fabricated in a compact form by using undoped poly-silicon through a specialized fabrication process, reducing area and power consumption. However, the majority of SRAM designs use the conventional six-transistor configuration, because of constraints in applying special fabrication techniques for the resistors.

1.3 Dynamic Random Access Memories: Basics

A conventional SRAM cell used six transistors and a number of connect lines for storing one bit of information. The cell consumed large chip area, hence, raising cost per bit. As the two load transistors (M_3 and M_4 in Figure 1.1) were used only to provide a path for supplying leaked charge, they could be eliminated, resulting in a four-transistor memory cell as shown in Figure 1.2(a). Its operation was similar to the 6-T SRAM; however, it is required to be *refreshed* periodically, that is, the charge that was stored at the parasitic and gate capacitance of a node connected with high voltage bit line is required to be replenished.

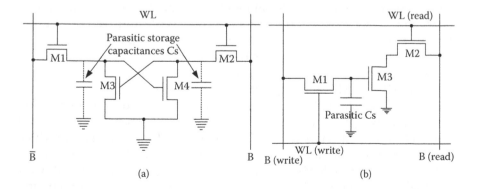

FIGURE 1.2
(a) Four-transistor DRAM cell with two storage nodes. (b) Three-transistor DRAM cell with two lines and two word lines. (Modified from *CMOS Digital Integrated Circuits—Analysis and Design*, S. M. Kang and Y. Leblebici, Tata McGraw Hill, 2003.)

To further reduce the complexity and size of the cell, the three-transistor DRAM cell was evolved from the four-transistor cell in the early 1970s, by eliminating one of the storage transistors, M_4. As shown in Figure 1.2(b), transistor M_3 remains the single storage device and it turns on or off depending on the presence or absence of the charge stored at its gate capacitances. One transistor each is used for the read and write access; M_1 for read and M_2 for write. The read operation of the three-transistor DRAM cell is nondestructive, but it still requires four connecting lines, two each for bit lines and word lines along with their contacts and vias.

1.3.1 Three-Transistor DRAM Cell

Elimination of the load transistors/resistors in a six-transistor SRAM cell and one driver transistor led to the three-transistor cell, which became the first popular dynamic memory cell as the stored data depended on charge stored at a capacitor located at a high-impedance node. A commercial dynamic RAM (DRAM) chip of 1 Kbit density was produced by Intel in 1970 using the 3T cell. It was then followed by 4 Kbit DRAM in a single-polysilicon, single-metal in 10 µm technology. Initially non-multiplexed addressing was used; however, soon address-multiplexed DRAMs were made available. At that stage sense amplifiers were simply made up of two static inverters and two switches with a bootstrapped output stage for fast charging of load capacitances.

1.3.1.1 Construction and Operation

A circuit diagram of a typical three-transistor dynamic RAM cell and some peripherals is shown in Figure 1.3. The transistors M_1, M_2, and M_3 are made small to minimize the cell area. The data is stored on parasitic capacitance C_1 in the form of presence/absence of charge on it. The storage transistor M_2 is turned on or off depending on the amount of charge stored in C_1 and transistors M_1 and M_3 act as access transistors for data write and read operations. There are two separate word lines for write and read. Separate data lines for read and write are also necessary as the stored charge of C_1 would be lost if M_1 was turned on during read. Necessary peripherals include two precharge transistors MP and MP' and the column read/write circuitry.

The three-transistor DRAM cell and its peripheral circuitry operation are based on the two-phase nonoverlapping clock (CLK) scheme. The first half of each read and write cycle is a precharge phase (CLK being high) in which columns data input (write), DW, and data output (read), DR, are charged to valid high level through MP and MP'. As precharge is initiated with CLK going high, the column pull-up transistors became activated and the column capacitances C_B and C'_B are charged up to logic-high level.

All read and write operations are performed during the second phase, that is, when CLK is low. For the write high operation, $\overline{DATA}$ is made low,

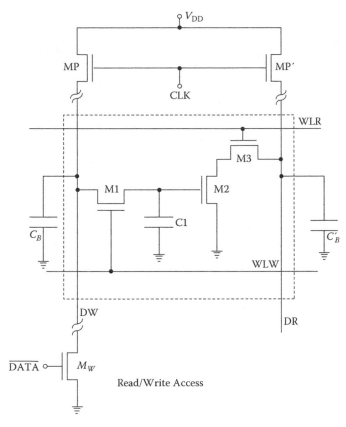

FIGURE 1.3
Three-transistor DRAM cell with the pull-up transistors. (Modified from *CMOS Digital Integrated Circuits—Analysis and Design*, S. M. Kang and Y. Leblebici, Tata McGraw Hill, 2003.)

turning the data write transistor M_w off, leaving the voltage level at line DW high after the precharge phase. Once the signal WLW is turned high during the second phase of ($\overline{CLK}$), the write access transistor M_1 is turned on and the charge on C_B is shared with C_1 as shown in Figure 1.4(a). However, C_B being very large in comparison to C_1, charge transferred from C_B is very small and voltage level of the column capacitance C_B as well as that of storage capacitance is almost the same at logic-high level. When a high has been written, transistor M_1 is turned off. Now transistor M_2 remains turned on due to the storage capacitance C_1 charged to a logic high.

To read a stored high, the read-select WLR is pulled high (once again with phase ($\overline{CLK}$) low, following a precharge phase) which turns access read transistor M_3 on, and conducting M_2 and M_3 create a conducting path between the column capacitance C'_B and ground. The capacitance C'_B discharges to ground and lowering of DR column voltage is detected by data read circuitry

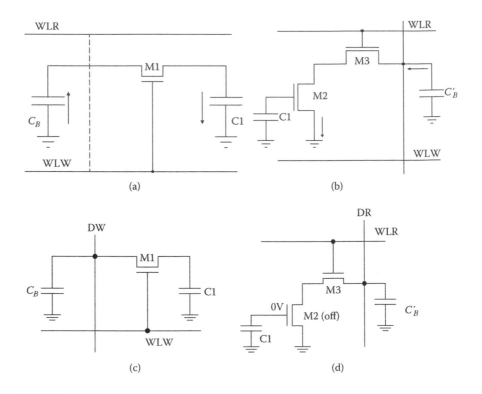

FIGURE 1.4
(a) Charge sharing while writing "1". (b) The column capacitance C'_B is discharged during the read "1" operation. (c) Both C_1 and C_B are discharged via M_1 during the write "0". (d) The bit line capacitance $C_{B'}$ cannot discharge during the read "0" cycle. (Modified from *CMOS Digital Integrated Circuits—Analysis and Design*, S. M. Kang and Y. Leblebici, Tata McGraw Hill, 2003.)

as stored logic 1 as shown in Figure 1.4(b). It is important to note that the read process is not destructive and the read operation can be performed repeatedly.

For the write low operation, $\overline{\text{DATA}}$ is made high, turning the write transistor M_w on, which pulls down the logic level of line DW low after the precharge phase (CLK). The signal WLW is turned high, during the second phase ($\overline{\text{CLK}}$); hence access transistor M_1 is turned on. The voltage level on C_B, which is at logic low, pulls the voltage level of C_1 to low through M_1, as shown in Figure 1.4(c). At the end of write low, the storage capacitance C_1 is left with only a small charge, resulting in low gate voltage for M_2, turning the transistor off.

To read the stored low, read select WLR is pulled high with ($\overline{\text{CLK}}$). The read access transistor M_3 turns on; however, C'_B cannot discharge because transistor M_2 is off and no conducting path is available between DR line and the ground as shown in Figure 1.4(d). High level is retained on the DR column, which is interpreted by the data read circuitry as a low bit.

It is to be noted that repeated read operations can be performed because the read operation is nondestructive. However, charge stored on the storage capacitor C_1 cannot be retained for long for various reasons of leakage, the main culprit being the drain junction leakage current of the write access transistor M_1. Hence the leakage charge has to be restored at regular intervals. This restoration of charge or data, commonly known as refreshing the data, has to be done every 2 to 4 ms. For refreshing the data, it must be read, inverted, and then written back in the cell. Inversion of the read data is necessary during refreshing since the data output level reflects the inverse of the stored data. A number of refresh methods are used, the most common being one in which all the cells in a row are refreshed simultaneously.

An extra inverter is used in either the read or the write data path to get same logic in the memory data input and output. Simply put, refreshing does require some overheads; even then, cost per bit of DRAM is much less than for SRAMs. A major advantage of 3T DRAM cells is the absence of any static power dissipation during data storage, because of the fact that there is no continuous current path between V_{DD} and ground. In addition, dynamic power dissipation is also reduced due to the use of precharge cycles instead of static pull-up. If static pull-up devices are used at the data output line DR, a higher average drain current would be needed after reading a high. Similarly, if static drivers are used at the data input line DW, fast changes in logic levels would require excessive power [3].

1.4 One-Transistor DRAM Cell

Dramatic reduction in cell size and complexity was achieved in a one-transistor DRAM cell, though at the expense of certain cell properties. One of the important constructional differences from 3T cell is shown in the one-transistor DRAM cell of Figure 1.5, which has an explicit storage capacitor C_s. Instead of using the parasitic gate and diffusion capacitances of the transistor M_1 for data storage, a separate capacitor C_s must be fabricated. In the write operation, after the word line is enabled, the data is written into the cell through the transistor M_1 and stored at the storage capacitor C_s. In the read operation, charge stored on the storage capacitor is shared with the bit line capacitance and hence the charge is changed significantly, making the read process destructive. For large memories, bit-line capacitance (column capacitance C_B) is much larger than the storage capacitor; hence, only a small voltage change takes place at the bit line. It is therefore a must to detect this small change in voltage, amplify it quickly to full logic level, and rewrite (refresh) the data into the cell.

With only one transistor, one capacitor, and one line each for bit and word line, the DRAM cell consumes the smallest silicon area of all the static and

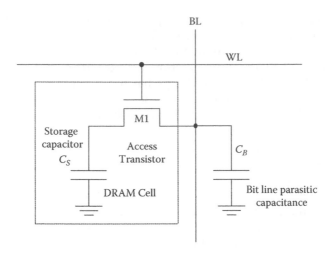

FIGURE 1.5
A one-transistor DRAM cell with its access transistor and data line. (Modified from *CMOS Digital Integrated Circuits—Analysis and Design*, S. M. Kang and Y. Leblebici, Tata McGraw Hill, 2003.)

dynamic memory cells. However, a minimum size storage capacitance is essential to develop small but sufficient voltage difference (Δv) on the bit line, which is to be detectable for correct read operation. The amount of Δv depends on the operating voltage (with a decreasing trend) and the ratio of the total bit line capacitance (which is increasing with increase in the size of the DRAM density) to the storage cell capacitance. DRAM designers and manufacturers have to make a continuous effort to develop and realize an effective minimum value storage capacitor that consumes as little chip area as possible.

Major reduction in cell area and complexity in comparison to a 3T DRAM cell was achieved in a single-transistor (1T) DRAM cell of Figure 1.5, which forced considerable difference in their operation. It happens because read-select and write-select lines are merged in single word line (WL) and data input and data output lines are also merged in single bit line (BL). Its basic operations are very simple. For both read and write, transistor M_1 is turned on with the single WL. During the write cycle, data is written into the cell by placing data on the bit or column line. Depending on the data value, storage cell capacitance C_s is ether charged or discharged. However, before starting to read, BL is precharged to a voltage V_{PRE}. Once WL is selected (high), charge distribution takes place between the BL capacitance C_B and the storage capacitance C_s through the transistor M_1. As a result, a small change takes place on the BL voltage of magnitude (Δv) as mentioned before, depending on the capacitance values (charge transfer ratio) C_B and C_s, V_{PRE}, and data voltage to be written. The direction of Δv depends on the logic state stored in the cell. The small voltage charge has to be converted to its final logic value, high or low, for completing the functionality of the (1T) DRAM cell. Thus, presence of a sense amplifier is a must for each bit line, not only for speeding up the read operation, as in the case of SRAM or 3T DRAM, but also for completing its function. Moreover, the sense amplifier design now becomes complicated due to the single-ended nature of 1T DRAM cell in comparison to the other cases where both data and its complement are available on two bit lines.

During the read operation the amount of charge on C_s is modified and remains so. Therefore, after reading, its original value must be restored or refreshed, so output of the sense amplifier is fed again onto the BL during the read process. Transistor M_1 should remain on by keeping the WL raised during the refresh.

Another important point is the requirement of an explicitly fabricated capacitor C_s instead of using the parasitic gate and diffusion capacitance of the transistor M_1. For reliable operation of the cell, the charge transfer ratio cannot be very small; hence, C_s must have a value not less than (say) 30 fF to have sufficient charge on it. Though small in absolute terms, realization of C_s in the submicron range is a tough problem. Some of the methods for realizing C_s, especially without consuming a bigger area, shall be discussed later in detail in Chapter 5.

1.4.1 One-Transistor DRAM Structures and Switching Waveforms—Review

In the early stages of the one-transistor DRAM cell, electrodes of the transistor and the storage capacitance were fabricated separately in a single polystructure [4]. With the introduction of double polysilicon structure, polysilicon electrode of the transistor and the polysilicon electrode of the capacitance were formed in different layers in an overlapping form, thus reducing the cell area. Further reduction in the cell area was achieved when triple polysilicon structures were developed. In this scheme bit line was fabricated as third polysilicon layers over the capacitor electrode. Reduction in the planar DRAM cell size resulted in considerable increase in the DRAM density. However, this arrangement in planar cell proved to be inadequate beyond 1–4 Mbit DRAM density. Several innovations and structures were developed for obtaining 30–40 fF storage capacitors without any increase in the cell area.

At initial stages, selection of a cell was done by the application of (N/2) address inputs each to both the row and column decoders in a (2^N) capacity DRAM. For reading the data at the selected cell location, the chip is put in the read mode by asserting $\overline{R/W}$ and enabling the chip through lowering the chip enable ($\overline{CE}$) pin of Figure 1.6(a). Figure 1.6(b) shows the timing relation between non-multiplexed address inputs and data available at the output pin D_{out}. Important timing specification for the DRAM are Read Cycle Time (t_{RC}) and Read Access Time (t_{RAC}), where t_{RC} represents the minimum time required between any two read operations and t_{RAC} is the duration between

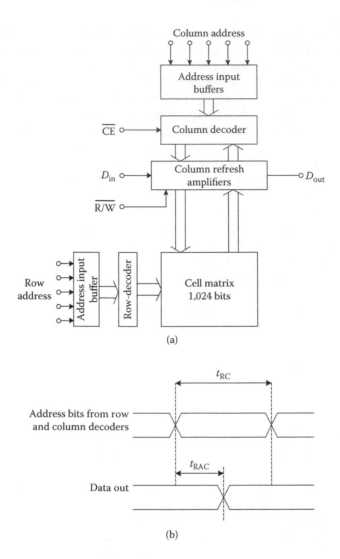

(a)

(b)

FIGURE 1.6
(a) Typical functional diagram of a 1kbit DRAM. (b) Typical DRAM Read cycle. (c) Typical DRAM Write cycle. (Modified from *CMOS Digital Integrated Circuits—Analysis and Design*, S. M. Kang and Y. Leblebici, Tata McGraw Hill, 2003.)

time instants when address is applied and the valid data becomes available at the output.

For writing data to a DRAM location, $\overline{R/W}$ is made low with valid data available at the data input pin D_{in}. As shown in Figure 1.6(c), Write Cycle Time (t_{WC}) is the minimum time required between any two write operations and it normally equals t_{RC}, and the Write Access Time t_{WAC} is the duration between time instants when $\overline{R/W}$ is made low or address is applied and

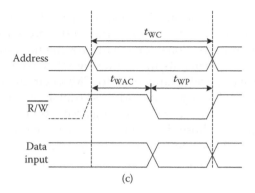

FIGURE 1.6 (continued)

when valid data is written. In addition, Write Pulse Width t_{WP} is the duration for which input must remain present and then $\overline{R/W}$ can change to high for another read/write.

For refreshing the DRAM, all the cells are accessed periodically through selecting every row one by one using the falling edge of the $\overline{RAS}$ strobe. All the cells in a row are read and written back.

1.5 Initial-Stage DRAM Technology Developments

In spite of increased complexity especially due to refreshing, low cost per bit of DRAM has made it the memory of choice for large-volume commercial applications. Both 3T and 1T DRAM configurations became standard for various positive reasons. Even in 1973, one-transistor DRAM was easily available for 4 Kbit density. However, as DRAM's read operation is destructive, sufficient value cell capacitance C_s was essential to get adequate value of voltage differential on bit line, especially as its capacitance increases for large-density memories. A number of technologies contributed in the progress of the DRAM developments—for example, use of high-permittivity dielectric material and advanced structures for the storage capacitance, without increasing effective occupied chip area; which was crucial and shall be studied separately in Chapters 3 and 5. Other key factors included advanced processing techniques including (but not limited to) photolithographic and etching techniques, without which it was not possible to continuously reduce DRAM size per bit. Improvement in array architecture like folded data line arrangement for noise reduction [5] and $\frac{1}{2}V_{DD}$ data-line precharge [6] were also extremely important factors.

Advantages of precharging to $\frac{1}{2}V_{DD}$ are manifold; for example, it improves noise immunity, reduces power consumption of bit line, and reduces electric field intensity across the capacitor C_s. Bit and sensing line equalizers are used to precharge the bit lines and sensing nodes to the precharge level of $\frac{1}{2}V_{DD}$ before any operation starts. Considerable improvement in supporting circuit technologies like improvements in bit line signal sensing, use of dynamic amplifiers and drivers, and changeover to CMOS circuit, and so on, were other very important developments. Address transition detection (ATD) circuits were found to be quite useful in saving access time by performing useful circuit functions for those durations which were earlier wasted because of RC delay in word lines. Some of the technical developments shall be discussed in this chapter and others shall be taken at later stages in Chapters 3, 4, 5, and 6.

1.5.1 Sense Amplifiers

A DRAM cell uses a single bit line through which data is written into as well as read from the charge storage capacitor. When storage capacitor is connected to the bit line, charge transfer takes place between the storage capacitor C_s and the bit line capacitance C_B. An important figure of merit is the charge transfer ratio, given as $C_s/(C_s + C_B)$. Value of C_s has to be kept small so that it occupies less chip area where as C_B goes on increasing with the increase in the size of DRAM. Thus an amplifier commonly known as sense amplifier must be available, for the reason already mentioned. The sense amplifiers used with DRAMs are generally different from sense amplifiers used with SRAMs and ROMs. DRAM sense amplifiers are discussed in more detail in Chapter 8.

In SRAM sense signal is available differentially on the two bit lines, where as DRAM is essentially a single bit line structure. Arrangement has to be made of a reference bit line against which differential voltage change is detected. Since storage capacitance charge is disturbed in the reading process and it also loses charge due to leakage, its original state cannot be restored/refreshed unless done intentionally through the sense amplifier. Once sense signal is amplified to its full logic value, it remains stored in the sense amplifier output until the cell is precharged for the new read operation. Since one sense amplifier is connected with one row, it effectively acts as and is referred to as row buffer also. Correct detection of small column signal or bit line charge is one of the most difficult parts of the sense amplifier design for 1T DRAM cell.

A read and refresh schematic for a 1T DRAM is shown in Figure 1.7 in a simplified form. In normal course, small voltage change in single bit is difficult to detect, so the bit lines are divided in two halves. Consequently bit-line capacitance is also reduced by half, doubling the charge transfer ratio. The regenerative switching of dynamic flip-flop, working as a sense amplifier, detects the small deviation in bit line signal and restores/refreshes the high or low signal level [7].

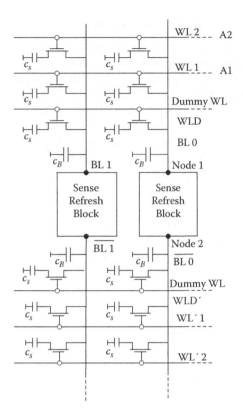

FIGURE 1.7

Sense/refresh arrangement with dummy and storage cells. ("Storage Array and Sense/Refresh Circuit for Single Transistor Memory Cell," K.U. Stein, A. Sihling and E. Doering, IEEE Int. S.S. Circuits Conf., pp. 56–57, 1972.)

To read from 1T DRAM array, column lines (bit lines) are precharged to V_{PRE}. Dummy cells on both sides of the sense amplifier are also turned on, charging the dummy cell capacitor to V_{PRE}. One of the word lines, say WL1 is made high so that first set of cells on the top side is selected. Simultaneously, the dummy cell, WLD' located on the opposite side of the sense amplifier is selected. No change of voltage takes place on the bit line $\overline{BL0}$ as its voltage level is same as that of the voltage of WLD'. However, the voltage level on the top side of BL0 will change in value toward higher or lower side of V_{PRE} depending on whether a high or a low is stored in the cell. It is important to select dummy cell on the other side of the sense amplifier with respect to the selected cell to balance the common-mode noise. The direction of the resulting small differential voltage difference between the selected cell and the dummy cell on the opposite half determines the final data state to be sent to the output buffer.

1.5.2 Open and Folded Bit-Line Structures

While differential sensing is the preferred approach, some variants, like charge-redistribution amplifiers schemes, were used in DRAMs, because of their inherent single-ended nature. However, this scheme works with a very small noise margin and needs a very careful design. At the same time with increase in DRAM density, noise components start increasing and single ended sensing does not remain a practical solution. Circuit modifications are done so that a differential sensing becomes possible, like that shown in Figure 1.7 where a differential amplifier is placed in the middle of the bit line. This scheme of having the bit line is called *open bit-line architecture* (OBLA). Advantage of the OBLA is that every crossing point of word line and bit line contains a storage cell. One of the important advantages of the OBLA is that one transistor–one capacitor DRAM cell and its interconnections with WL and BL can be fabricated in $4\ F^2$ chip area, where F is the minimum feature size during the fabrication process, independent of the working technology. With continued demand on reducing the cell size, F become further small, bringing word lines and bit lines closer, storage cell smaller; and size of the transistors used in the cell as well as in the sense amplifiers also has to go down. However, it has limitations in terms of noise minimization. Coupling noise between word line and bit line increases considerably especially when word line is selected and the bit line is low. Parameters of the transistors in sense amplifiers like threshold voltage become mismatched and produce noise; two halves of the OBLA also do not match exactly. All these and some other noise sources combine together to make this architecture less robust. Many corrective measures have been suggested to reduce noise components; however, changeover to folded bit-line architecture (FBLA), as shown in Figure 1.8, is extremely useful in reducing the differential noises [8]. In this arrangement sense amplifiers are at one end, instead of being in the middle, and column decoders are at the other end. As the bit line is folded back on itself, noise generated on the column decoders appears as common mode and gets canceled. As seen from the diagram, cells are not placed at every cross section of word line and bit line. Minimum possible cell area in this architecture is $8\ F^2$.

1.6 DRAM Operating Modes

DRAM operating modes can be classified as (1) asynchronous and (2) synchronous. In asynchronous mode all control and operational signals depend on $\overline{\text{RAS}}$ and $\overline{\text{CAS}}$ strobes whereas in synchronous mode, these depend on a system clock.

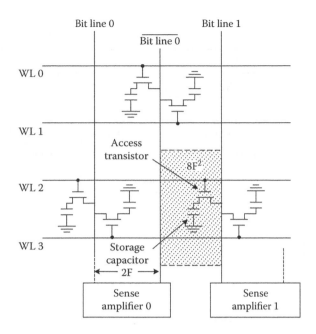

FIGURE 1.8
Memory architecture for a typical DRAM with folded bit-lines.

1.6.1 Asynchronous Modes

First $\overline{\text{RAS}}$ is made low for any DRAM operation, which, along with activating $\overline{\text{CAS}}$, starts operation with row address buffers, row decoders, word line drivers, bit-line sense amplifiers, and so on. Depending on the address supplied by the selected combination of row and column address in a multiplexed address scheme, bit location is selected on the falling edge of $\overline{\text{RAS}}$ and $\overline{\text{CAS}}$, respectively. Some precharge time is needed by the $\overline{\text{RAS}}$ and $\overline{\text{CAS}}$ strobes before locating a new address. Time duration between two $\overline{\text{RAS}}$ falling edges is called read/write cycle time as mentioned in Section 1.4.1. Figure 1.9(a) shows simple single-bit access asynchronous operation mode. Typical operating frequency in this mode is in 20–30 MHz range. Other modes of asynchronous DRAM operation are page mode, nibble mode, static column mode, fast page mode (FPM), and extended data out (EDO).

Figure 1.9(b) shows the timing diagram of a page mode read and writes. With $\overline{\text{RAS}}$ going low, selected word line is made high. Prior to the row selection it is precharged to a voltage V_{PRE} (usually $V_{\text{CC}}/2$) and the input/ output (I/O) pin is in the high-z state. Next $\overline{\text{CAS}}$ is made low and column address is decoded which selects addressed bit line, and data on the select bit location is read and goes out through (I/O) pin. When $\overline{\text{CAS}}$ goes high, I/O pin goes back again to the high-z state. Another bit line can be accessed

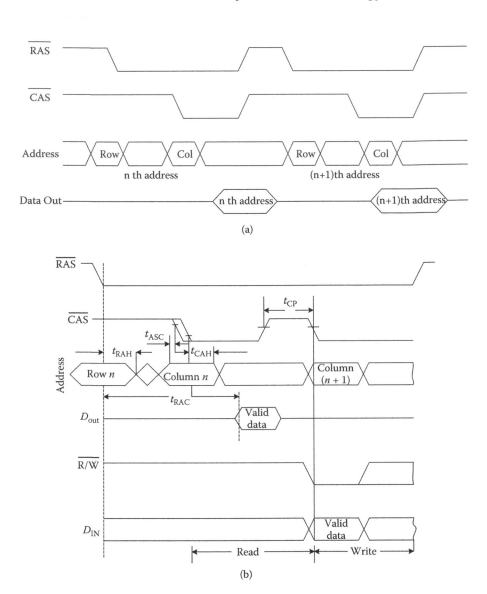

FIGURE 1.9
(a) Single bit read in an asynchronous DRAM. (b) Page mode operation.

through the column address decoder, from the same selected word line to
either read or write. The entire bit in a selected word line can be accessed by
raising all bit lines one-by-one and total bits in a word line form one *page*.
In fact, for bigger-size memories when the cell array is horizontally divided
in N blocks, generally one word line address opens one word line in each
of the block; in that case size of the page is increased by a multiplicative
factor N.

In the fast page mode (FPM), column address is applied early, that is, even when $\overline{CAS}$ is high. Some of the time taken in propagating the column address and high $\overline{CAS}$ overlaps, this makes the mode of operation fast. Extended data output (EDO) is simply an FPM scheme in which the output is not forced to go to high-z state when $\overline{CAS}$ goes high. In the *nibble mode* operation, column address changes by strobing $\overline{CAS}$ through an internal presettable counter. Column address is loaded into the counter when the $\overline{CAS}$ goes low first time (after the first $\overline{RAS}$ changing to low). With $\overline{CAS}$ toggling, next cells are automatically selected for reading or writing without providing column address. In the nibble mode only four cells are selected in this way—hence the name *nibble mode*.

In the *static column mode* operation flow-through latches are used in the column address path. With $\overline{CAS}$ low, when the column address is changed, it directly goes to the column address decoder; it does not wait for the falling edge of the $\overline{CAS}$. Output of the DRAM does not go to the high-z state and DRAM speed is increased. From the above discussion it is obvious that DRAM is being made fast by keeping a row selected, thus saving time.

1.6.2 Control Logic

Figure 1.10 shows a simplified structure of a 64 Mbit FPM DRAM with number of rows, columns, signal lines, and so on, also mentioned alongside [9]. Captions inside the blocks in the figure state their use. DRAMs based on

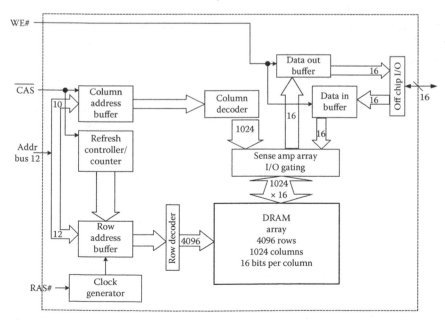

FIGURE 1.10
A 64 Mbit fast page mode DRAM simplified architecture.

other modes of operation have similar structures with subtle variations. All DRAM structures contain a logic circuitry in addition to the function blocks, like decoders, data buffers, sense amplifier array, and core array. The logic circuitry controls and assigns direction to all signals inside the DRAM chip, and in addition, it sends and accepts the signals from outside. In Figure 1.10, control logic receives, $\overline{RAS}$, $\overline{CAS}$, and $\overline{WE}$ signals as shown, as well as data-in at the I/O module. The rest of the signals inside the chip are generated and directed by the control logic. Important to note is that all movement of data is asynchronous in nature as it is not controlled by an external clock. The asynchronous nature of DRAM working has the advantage that different memory controllers can be designed at different frequencies depending on the internal clock generators. However, asynchronous nature restricts the use of pipelining, which is an important tool for speeding up processes in digital systems.

1.6.3 Synchronous Modes

Up to 16 Mbit density level, basic structure of Figure 1.10 was used, though other improvements were done for performance enhancements. However, the structure had major limitation in terms of only small improvement in DRAM's access time. In 1993, JEDEC standards committee decided a new standard for synchronous operation of DRAMs in which all signals were to operate on a single clock. All control logic on the DRAM chip was to depend on this external clock. This brought a remarkable change in the DRAM complexity and the way bits moved around, and possibility of further increasing DRAM speed became a reality. A synchronous interface is added to the basic DRAM core so that all commands and operations are dependent on the rising edge of the master clock. Figure 1.11 shows a block diagram of a DDR synchronous DRAM (SDRAM) comprising of basic DRAM array with 32 bit I/O, command interface on the left and a DDR SDRAM interface on the right [10]. All previous generation signals like $\overline{CS}$, $\overline{WE}$, $\overline{CAS}$, and $\overline{RAS}$ continued to do the same operation. Fundamental operations remained as before; however, because of modified structure of DRAM having multiple banks, necessary changes were to occur in the selection of row, column, and particular bit(s). Usage of number of banks enhances the speed of the DRAM, partly because of reduced bit line length for any selected bit(s), but a major contribution toward speed enhancement is because of the usage of pipelining. While one bank may be in precharging state, reading may be taking place in other bank and so on. $\overline{CAS}$ latency and burst length and burst type are stored in a mode register; and stored latency field value decides the number of cycles between output and assertion of column read command. Burst type value determines the ordering of the data availability and the length gives the number of columns that the SDRAM will return to the memory controller as a result of one column read command. The mode

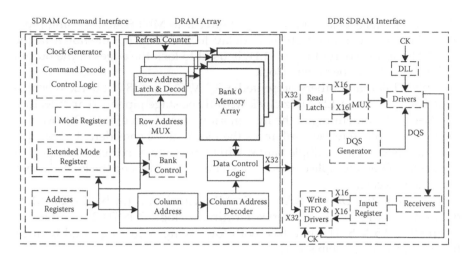

FIGURE 1.11
Block diagram of a DDR SDRAM showing the underlying SDRAM and DRAM. ("Application Specific DRAMs Today," B. Prince, Proc. Int. Workshop on Memory Technology, Design and Testing (MTDT), 2003.)

registers go on becoming more complex for advanced synchronous DRAMs, commonly known as DDRx SDRAMs.

1.6.4 Double Data Rate SDRAM

DRAMs, which have been used as main memory devices, have quadrupled in size with every generation, but with limited speed improvements. Asynchronous DRAMs, be that EDO or Burst EDO, have not exceeded the bus speed beyond 66 MHz. The asynchronous interface and internal logic cannot achieve the page mode accesses required for bursts at over 66 MHz. High speed SRAM caches usually filled the bandwidth (= bus width × cycle frequency) gap between CPU and DRAMs which heavily relied on low miss rate to avoid main memory access. However, SRAM cache could not satisfy the requirements of software applications like multimedia, so miss rate increased. Hence improvement in bandwidth was possible with increase in bus transfer rate. In an intuitive evolution from the single data rate SDRAM, data transfer was doubled in double data rate (DDR) SDRAM. The data was transferred on both the rising and falling edges of the clock, though all commands and operations took place only on the rising edge of the clock [11].

In DDR SDRAM, a wide internal bus prefetches two words simultaneously, and they are parted as two separate words of equal width. Prefetch doubles the data rate with an additional advantage of lowering the array power as the array runs at half speed, compared to an array with higher data rate without prefetch. As an example, 32 bits are fetched from the array and converted into two 16-bit words as shown in the right-hand side of Figure 1.11, the DDR

SDRAM interface. Choi and others [12] gave a 16 Mbit SDRAM using pipe-lined 2-bit prefetch architecture. A delay locked loop (DLL) shown in the SDRAM interface part of Figure 1.11, is used to lock the edges of the output and data strobe to the system clock.

In 1993, Takai and others presented a 3.3 V, 512 k × 18 × 2 bank synchronous DRAM using a 125 MHz clock with three stage pipelined architecture [13]. Use of the pipelining of the information, in and out of the SDRAM having synchronous interface, enhances the clock speed considerably. Speed is also increased using multiple banks, which are shown in the middle part of Figure 1.11. Even multiple banks can also be left open using the loaded sense amplifiers as caches.

The commodity DDR SDRAM is 256 Mb, with four array banks, 266 MHz data rate having 2.5 V stub series terminated logic 2 (SSTL2) interface. A terminated 2.5 V SSTL2 interface with low swing reduces ground bounce, hence improving system speed.

The DDR2 SDRAM simplifies the DDR SDRAM operation. Elpida Corporation gave a DDR2 SDRAM in 2002 having 256 Mbit densities, operating at 1.8 V and contained four array banks [14]. It used a clock speed of 200 MHz and provided data rate of 400 Mb/sec/pin in a large system using DIMMs and a data-rate of 600-Mb/sec/pin in a point-to-point system. A four word prefetch with two way data interleave was used to improve speed. In 2004, Fujisawa and others [15] gave a 1.8 V, 400 MHz clock, 800 Mbps/pin, 1 Gbit DDR2 SDRAM. It used a drivability-adjustable off-chip driver for DDR1 operation and an adjustable on-die termination register in DDR2 operation. A dual clock input latch scheme helped in reducing the cycle time from 3.0 ns to 2.15 ns. Several performing schemes were given in subsequent years for DDR2 and then for DDR3. One of the given configurations, SDRAM was a 1.5 V, 1.6 G bps/pin 1 Gbit DDR3 SDRAM, which used an 80 nm triple-metal dual-gate poly CMOS process [17].

Variants of DDR SDRAM have been reported in large numbers. Since the scope of discussion here is limited, the topic is not treated in depth. As mentioned before, specific applications required further higher bandwidths and graphic memories has been one such area. To achieve required bandwidth novel I/O technologies have been adopted and a whole new area of advanced research has been opened. Graphic DDR4 SDRAM at 4 Gb/s pin speed and GDDR5 SDRAM at 7 Gb/s pin bandwidth have been realized [18] and further improvements are ongoing.

1.7 Silicon-on-Insulator Technology

With other fabrication processes being almost the same as in the bulk complementary MOS (CMOS) process, fabrication in silicon-on-insulator (SOI) technology is performed in a thin layer of silicon, which is deposited on a thick layer of SiO_2. A number of techniques are available for producing starting wafer which is the basic difference with the bulk CMOS. SIMOX (separation by implanted oxygen) is the most commonly used technique in which a high dose of oxygen (O^-) ions prepares a silicon wafer with a high quality layer of SiO_2. Energy level of the implant is so selected that the peak of the implant is nearly 0.3–0.5 µm deep in the silicon. Crystallinity of the silicon surface is restored by heating it at 400°C during implantation, and then post-implant annealing is performed. Crystalline silicon surface layer is usually 100–300 µm thick [18] and if necessary epitaxial silicon is grown to increase its thickness.

Along with the use of lateral isolation techniques, completely latch up-free circuits are realizable in the SOI technology. Reduced parasitic and better on-off characteristic of transistors are its basic advantages. Other advantages include reduced short channel effect (SCE), reduced hot electron generation and sharper sub threshold slope. Usage of SOI chips remained restricted in high-performance, high-cost applications for long because of economic reasons. However, with continued reduction in minimum feature size in the bulk silicon technology, cost differential has come down considerably along with its improved performance in terms of reduced parasitic capacitance, lower threshold voltage, and body charging effects. Due to its floating substrate, *kink effect*, a dip in current-voltage relation, is observed in SOI MOSFETs, but it has been virtually eliminated in fully depleted SOI MOSFETs.

In another method wafers are produced in so-called *bonded SOI wafer production* in which ion implantation is not used. First two silicon wafers are heated so that oxide is formed on their surface. Next, the two oxide surfaces are bonded together to form the buried oxide, and one side of the silicon is partially removed. The SOI wafer is annealed and polished finally.

Figure 1.12 shows the cross-section of a partially depleted (PD) SOI device, in which every fabrication is the same as that in bulk CMOS except that

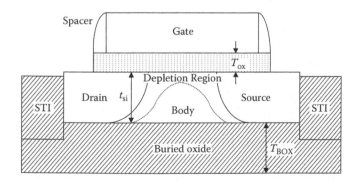

FIGURE 1.12
Cross-section of a partially depleted SOI FET. (Modified from *SOI Circuit Design Concepts*, K. Bernstein and N. J. Rohrer, Springer Science, 2007.)

buried oxide was already present. Silicon body having thickness t_{si} is completely surrounded on three sides and partially opens at the top. For a PDSOI device depletion region extends into the body of FET, as shown in the figure and does not deplete all the charge in the body. Remaining charge in the body is mobile and variable as it depends on the terminal voltages. This mobility of charge makes PDSOI device unique and is responsible for quite different characteristics than a bulk MOSFET.

In PDSOI-based MOSFETs, current-voltage characteristics are not constant as they depend on the amount of charge in the body. This prominent property is known as the *history effect*. Total charge in the body is the algebraic sum of incoming and outgoing charges due to various reasons like impact ionization which generates electron-hole pairs, junction leakage, source-body n-p junction, and so on. All of these depend on the gate, drain, and source potential.

Impact ionization is an important contribution of charge to the body, and it is observed that its peak occurs when the gate voltage is half of the supply voltage. So, any operation in which gate voltage remains close to $V_{DD}/2$ for long duration, increases the body potential and threshold voltage variability. There are many more attributes and characteristics different from bulk CMOS and texts like [19] can give an excellent exposure.

DRAMs on thick SOI substrate were fabricated as early as 1994 at 256 Mbit density levels [20] and soon SOI DRAMs for multi-gigabit density were proposed [21]. For increased density on SOI substrate one transistor-one capacitor DRAM quite a few schemes were given, however a capacitor-less single transistor DRAM on SOI substrate was given in 2001 [22]. The idea of not using a capacitor was with the purpose of saving chip area and the complex fabrication steps involved in three-dimensional capacitors. Few such capacitor-less DRAMs are briefly discussed in the following section.

1.7.1 Capacitor-Less DRAMs

It was expected in 2007 that by 2011 high-end processors would be using 83% of chip area for the memory [23]. Hence, the designer shall be forced to accommodate continuously increasing amount of memory in as small chip area as possible. Though 1T1C DRAMs are continuously increasing density through many technological improvements, circuit innovations, and advanced transistors and capacitors, use of external capacitor was completely avoided in a slightly different idea in *capacitor-less DRAMs*. While pursuing the idea it was important not to add new complexities and materials in the already existing procedures, especially at sub-100 nm technology level. The first such attempt was made in 2001 [22] and in the following years [24] in which capacitor-less 1T DRAM was proposed using SOI substrate. Floating body of a partially depleted (PD) SOI MOSFET was used as a charge storage node, which served/replaced the purpose of the physical capacitor of a conventional 1T1C DRAM.

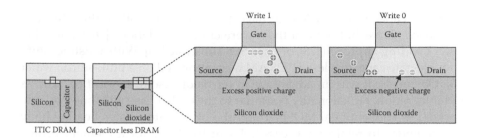

FIGURE 1.13
Changeover to capacitor-less DRAM. Presence or absence of excess positive charge decides its state. (Redrawn from "Masters of Memory," S.K. Moore, *IEEE Spectrum*, pp. 45–49, 2007.)

A capacitor-less SOI DRAM cell have been given much earlier, using two transistors, but it was operationally complex and occupied larger chip area [24], whereas the aim was to get a DRAM cell in an area of 4 F^2. Such a simple structure, which used three signal lines and a single channel, was given [23] which used the property of body-charging effect in PD SOI MOSFET, an effect considered problematic in normal SOI MOSFET working. Utility of the concept was shown on MOSFETs at 0.25 μm and 0.13 μm SOI CMOS technology. Positive substrate charge was created through impact ionization near the drain by applying a positive drain voltage pulse, which increased the source current and it corresponded to a *high*. A high can also be written using source side impact ionization by negatively biasing the drain and gate. For a 0.25 μm NMOS, writing operation took 3 ns. A *low* state was written, again in 3 ns, through removing holes from the substrate at the source or drain side. Figure 1.13 shows the basic storage of high and low. Data stored was read by comparing the source current with that of a reference cell using a current sense amplifier. Reading was done at a low drain voltage to make it nondestructive.

At reduced feature size, cell retention time becomes extremely important. Moreover, sense margin obtained in early stage of capacitor-less SOI based DRAM was low. A double-gate fully depleted FinFET DRAM at gate length less than 50 nm [25] was used to try to overcome the limitation.

Excess holes can be generated not only by the impact ionization but also by gate-induced drain leakage (GIDL); due to band-to-band tunneling, GIDL current can flow under both negative gate bias and positive drain bias. One-transistor DRAM using GIDL is able to operate at high speed and with lower power consumption. Double-gate depleted FinFET DRAM was used to overcome limitations caused by the reduction of minimum feature size. The front part of such a FinFET functioned as a conventional transistor, whereas the back gate MOS structure was used to store excess holes in the body with proper biasing condition. Since the double-gate Fin DRAM uses very low doping concentration in the floating body, junction leakage current is very low and hence retention time is improved.

Several capacitor-less DRAMs have been reported other than the one mentioned above, but some of these were either simulations [26,27] or using wafer bonding approach [28], or used SOI substrate [29]. With a distinct shift, a double-gate capacitor-less 1T DRAM was fabricated on a bulk-silicon wafer [30]. Obviously, it was an advantage in terms of ease in fabrication process and integration. Functionality is similar to the earlier case with one gate side working as conventional MOSFET and the other gate creates floating-body storage node. To retain the excess holes in the body for a memory action, the second gate is to be reverse-biased. Because of the vertical orientation of source and gate, the two gates are physically isolated and there is no lithographic constraint on the channel length. As before, use of low doping concentration, junction leakage is low resulting in higher retention time. Impact ionization generates holes at the drain side and a quasi-floating-body is created in a few nanoseconds at the back (second) gate interface. Scaling of such a DRAM was achievable even up to 22-nm technology node. Apart from being called capacitor-less DRAM [26,28,30,31] it has been given some other names like zero (ZRAM) [22], unified (URAM) RAM [32,33], or simply 1T DRAM [34]. In the URAMs, functions of DRAM and nonvolatile memory are combined. It is expected that more efforts shall be made to make this kind of memory more acceptable for manufacturing.

1.8 Advanced Nonvolatile Memories

A large number of semiconductor nonvolatile (NV) memories, such as ROMs in which data is written permanently, PROMs in which data is entered by the user only once, and floating-gate technology EPROMs and EEPROMs in which data can be written and erased with endurance cycle in the range of 10^4–10^6 have been widely used. Special NV memories, which use static RAM array and backup array of EEPROM as well as those based on DRAMs and MNOS [35], were also in development. Memories in which all contents or big blocks of data can be erased electrically called flash memories in NOR and NAND structures are giving hard competition to the DRAMs. Requirements from the NV memories are very similar to those of other semiconductor memories, like low cost per bit, high scalability, fast access and read/write operation, and low power consumption. In addition its endurance cycle should be as high as possible. None of the NV RAMs has all the qualities and continuous efforts strive to find better NV technologies.

One such successful attempt was in the development of thin film ferroelectric technology in which lead zirconium titanate $Pb\,Zr_4Ti_{1-x}O_3$ (PZT) (and few other dielectrics) was used as nonvolatile capacitive storage element to give ferroelectric random access memories (FRAMs) [36]. FRAMs were found to have higher endurance cycle and low cost per bit in comparison to other NV RAMs in the 1990s.

Another technological development is in the form of magnetoresistive (MR) memories (MRAMs), which are based on possible change in a material's magnetoresistance in the presence of a magnetic field. The MR effect has been effectively utilized in forming storage elements comprised of a pair of magnetic film strips with a dielectric in between [37]. MRAMs have been fabricated in Mbit density. Spin-torque transfer (STT) magnetizing switching has further given a significant technological advancement, in which external magnetic field is not necessary that was required in a conventional MRAM. Read operation in MRAMs is non-destructive, and it possesses considerably higher endurance cycle in comparison to FRAMs.

A high-performance NV memory technology has been developed in the form of phase-change RAM (PCRAM), which uses GST (GeSbTe) chalcogenide alloys as memory elements [38]. The GST alloy is set to crystallize and reset or amorphize in a reversible process by the application of proper heating. Effect of the transition between set and reset is that resistance ratio between the two is in the range of 10 to 10^4, which is mainly responsible for giving large signal sensing margins. Theoretically PCRAMs have almost unlimited scalability. However a major limitation of PCRAMs is the requirement of larger current during reset process of GST, which in turn, requires large size of cell transistor. Efforts have continued to overcome the limitation.

1.8.1 Flash Memories

Be it EPROM or EEPROM, memory is programmed using alteration in the threshold voltage of floating gate transistors. The programmed transistor retains modified threshold even when the supply is turned off. For reprogramming, an erase procedure was required, and it is this erase procedure which was mainly deciding the type of NV memory, its advantages, its limitations, and its applicability. By 1992 conventional EPROMs and EEPROMs lost their space to flash memories because of their larger cell size and the inconvenience in their use, especially for the erase procedure.

Flash memories have combined better properties of EPROM and EEPROM of higher density and faster programming capability, respectively. In flash devices content of the memory cells is erasable through the use of an electrical erase signal as a whole or in terms of big blocks. Most of the flash memories use avalanche hot-electron injection for programming of the cells, wherein a high-voltage application between source and gate-drain terminals converts electrons as *hot* with sufficient energy to cross thin gate oxide and get trapped on the floating gate, thus changing threshold voltage of the transistor. Erasure mechanism is through Fowler-Nordheim tunneling, as in the EEPROM cells.

A number of similar flash cells are available. One of these shown in Figure 1.14(a) is the ETOX flash cell given by Intel [39] which is similar

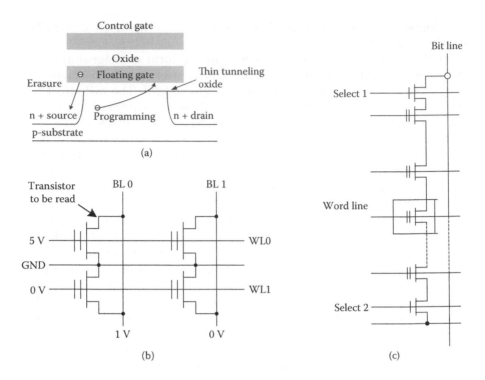

FIGURE 1.14
(a) ETOX device as used in flash EEPROM memories. (b) Basic read operation in a NOR flash memory. (c) NAND-based flash DRAM module. (Modified from *Digital Integrated Circuits: A Design Perspective*, J.M. Rabaey, A. Chandrakasan and B. Nikolic, Pearson, Prentice Hall, 2003.)

to a FAMOS gate, except that tunneling gate oxide is very thin (~10 nm). Programming is performed through the right-hand area as shown, by applying 12 V on gate and drain with source grounded. Erasing is performed using left-hand area with gate grounded and the source at +12 V. Flash cells can be connected in NOR ROM or NAND ROM. For a NOR ROM, the programming cycle starts with complete erasure along with a monitoring procedure of the threshold voltage of all the transistors for reliability's sake; writing is then performed for the selected cells. The memory is then ready for read operation(s) as shown in Figure 1.14(b). NOR ROM architecture has fast random access but writing the cell is slow on account of monitoring of the threshold voltage of the transistors. If faster random access is relaxed, cheaper, denser, and fast erasing and programming with fast access is obtained in NAND ROM where 8 to 16 floating-gate transistors are connected in series, as shown in Figure 1.14(c). Chain of the transistors is sandwiched by bit line and source line in series with two select transistors 1 and 2 [40]. Here select 1 ensures selectivity and select 2 ensures no current flow during programming of a cell. The cell can be erased as well as programmed using Fowler-Nordheim tunneling. Program cycle consists of (1) initial program by applying 20 V to the selected control gate (with 10 V applied to the rest of the gates) and grounding the bit lines for writing 1. For writing 0, bit lines are raised to 10 V. (2) For reading, control gate of the selected cell is grounded and 5 V is applied to the gate of the rest of the cells, with both select transistors 1 and 2 enabled.

Many more architectures for flash memories are available, though the basic principle remains same. An excellent review is available in [41]. It is observable that all early flash memories used high voltages (>10 V) for programming and erasure. For those memories requiring smaller currents on chip charge pumps could be used. However, larger current memories needed external supply, increasing complexity of the system.

A unique advantage of flash memories is multibit storage in a single cell which enables high densities. However, there are limitations in scaling the technology node. Since NOR flash memories need higher drain voltage for hot electron generation and their injection into the floating gate, non-scalable Si-SiO$_2$ energy barrier height becomes a basic scaling limiting factor [42,43]. This high drain voltage can create intercell interference. Furthermore, scaling down of transistor has short channel effect (SCE). To minimize these problems NOR flash technology adopted use of high dielectric material with low energy barrier height like HfO$_2$ [44] and advanced transistor like FinFET [45]. It was expected that sub-50 nm technology node could be deployed for NOR flash. NAND flash structure has more scaling capability because of its architecture and device physics [46] but cell-to-cell interference is a major limitation for scaling. Closely placed cells have capacitive coupling which makes it probable to over program or under-erase failures [46]. As the NAND flash technology moves toward 50 nm sensing signal margin is reduced in conventional-type transistors. To overcome the scaling problem, once again use of low dielectric material and FinFET based SONOS (silicon-oxide-nitride-oxide-semiconductor) cell structures [47] are very effective.

Simplest cell structure and smallest cell size per bit make the NAND flash suitable for highest density. Packing density is further accelerated by 3-bit per cell and 64-bit cell string structures [48]. Kurata et al. [49] gave a 1 Gbit multilevel cell (MLC) flash memory with precise control of $V_{th.}$ In 2007, Shibata et al. [50] presented 16 Gbit flash memory using 16-level cell at 70 nm technology. Another important flash cell development revival was in the form of a MONOS (metal-oxide-nitride-semiconductor) cell first given by Nozaki [51] in 1990 for embedded flash in microcontrollers, workable at 5 V. Depending on the speed of evolution of NAND DRAM, it was expected that 32–64 Gbit flash memories would be available by 2010 in sub-40 nm technology [52]. However, it required overcoming the scaling challenges in the form of lithography constraints and device characteristics. One of the promising techniques is the self-aligned double patterning [53], which has been shown to be feasible at 30 nm node. Main challenges in the device characteristics are coupling ratio, and previously mentioned cell-to-cell interference and SCE which becomes acute at smaller technology level. One of the solutions for reducing cell-to-cell interference comes in the form of charge trap NAND flash memory [54,55] and it also meets the coupling ratio requirement. Requirement of sufficient I_{on}/I_{off} ratio in transistors beyond 40 nm is difficult to solve as it needs an increase in channel doping, which in turn increases depletion capacitance, degrading V_{pass} window; additionally, the problem of SCE is to be solved. Advanced transistors have been used for this purpose; a three dimensional transistor hemi-cylindrical FET (HCFET) appears to be a suitable candidate [56]. Beyond the 20 nm node, the most effective method of increasing NAND flash density is the vertical cell stacking [57,58]. It is expected that it will reduce the mentioned constraints of lithography, cell-to-cell interference, SCE, coupling ratio, and so on.

1.8.2 Ferroelectric RAM (FRAM)

A FRAM cell is similar to a DRAM cell, as it comprises an access transistor, a word line, a bit line, and a storage capacitor, which is realized using two metal electrodes and dielectric film in between. Similarity ends here as the dielectric used is a thin film of ferroelectric material. A number of such materials are available; however, lead zirconium titanate (PZT) is most commonly used, as it can store large charge on the order of 10 $\mu C/cm^2$ (nearly 100 times that for SiO_2). In a ferroelectric capacitor, when a positive voltage more than a critical voltage called *coercive voltage* is applied, the dielectric film is polarized in the positive direction and reaches a saturation level. If the applied voltage is removed, the film still remains positively polarized to a slightly lesser level than that of saturation level, and the level of polarization is called *remnant polarization* value. This polarization level is normally defined as *low* state. Changeover of applied voltage in the negative direction converts the polarization of the film in the opposite direction, reaches saturation, and once again rests at the opposite remnant value when applied voltage is removed. This value of polarization now corresponds to *high* state. The net result of the operation is that two digital states are clearly defined as the difference between remnant polarizations and the polarization condition remains in either state without application of external electric field or current. Hence storage of data does not require power for data retaining [59].

FRAM cell comprising one transistor and one capacitor operates similar to DRAM cell; however, the dual-element cell shown in Figure 1.15(a) is more common for reliability reasons as common-mode variations in the ferroelectric capacitor get canceled. In fact, one of the earliest FRAMs used such architecture to realize an experimental 512 Mbit NV memory. Here, a word line controls two access transistors, bit (BL) and bit bar ($\overline{BL}$) collect charge from the capacitor, and an additional component is a common drive line which actively drives the capacitors. Figure 1.15(b) show FRAM memory in its basic form [36]. For the write operation, desired data is supplied to the sense amplifier, which in turn sets BL and $\overline{BL}$ in such a way that when the

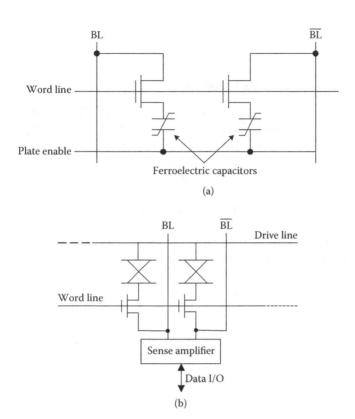

FIGURE 1.15
(a) A dual-element memory FRAM cell. (b) Read/write basic arrangement in FRAM. (Redrawn from *Semiconductor Memories, Technology, Testing and Reliability*, Ashok K. Sharma, IEEE Press, 1997.)

drive line is pulsed high, grounded BL writes a *low*. When drive line returns to ground level, a high is written in the second capacitor connected to $\overline{BL}$. Data written on the BL represent the stored data. During the read operation of FRAM, the sense amplifier is turned off with bit lines floating, and the drive line is pulsed. Because of opposite polarization in the differential cell BL and $\overline{BL}$ combination provides signal differential of the two.

For the purpose of scaling of FRAM, it is observed that limiting factor is the realization of capacitor. Very obviously, sensing signal in FRAM depends on the capacitor area and the remnant polarization charge density mostly depending on the thin ferroelectric film. Beyond a certain FRAM density level, instead of planar ferroelectric capacitor, a three-dimensional structure has to be employed to increase its area, which may bring down the cell size in the range of $6\,F^2$ to $8\,F^2$. Thickness of the ferroelectric film is also reduced to get same remnant polarization at smaller cell size, but it also has a practical limit, possibly a few nanometers. At thin ferroelectric layer, its degradation reduced the remnant polarization and data retention time, which is greatly improved by better film growth techniques like MOCVD and optimization of seed layer.

1.8.3 Magnetoresistive RAM (MRAM)

Magnetoresistive random access memory (MRAM) is based on the integration of magnetic memory elements with silicon CMOS technology. Memory elements are based on the basic principle that the presence of magnetic field will change magnetoresistance of certain materials. Mainly two types of memory elements, namely giant magnetoresistance (GMR) [60,61] and magnetic tunnel junction (MTJ) [62,63], were studied and employed extensively. Current-in-plane (CIP) GMR structure of pseudo spin valve (PSV) type structure consisted of two magnetic layers of different thickness with copper in between as shown in Figure 1.16(a). At submicron dimensions, due to the shape anisotropy, magnetic layers having different thicknesses have

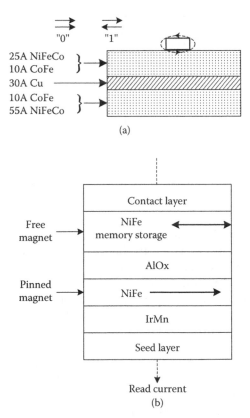

(a)

(b)

FIGURE 1.16
(a) GMR based material stack structure. (b) Typical MTJ material stack for a memory bit. (Redrawn from "Progress and Outlook for MRAM Technology," S. Tehrani et al., IEEE Trans. Magnetic, Vol. 35, pp. 2814–2819, 1999.)

different switching fields [64]. The magnetic moments of the two layers can be parallel or antiparallel and that makes resistance of the film low or high, respectively. Magnetoresistance ratio (MR), defined as the ratio between the difference of resistance in antiparallel (R_{AP}) to parallel (R_P) layers with R_p, that is, $(R_{AP} - R_P)/R_p$, ranges between 6% and 9% for the GMR structure. As the GMR sheet resistance is low, large currents are required (>1 mA), which in turn requires wider pass transistor, to get sufficient signal level [65]. In case of a typical current-perpendicular-to-plane (CPP) MTJ structure, shown in Figure 1.16(b), polarization of one magnetic layer is pinned in a fixed direction and resistance of the cell is large hence sense current is reduced in μA range. It allows the use of a minimum size transistor as the isolation device in conjunction with MTJ in an MRAM cell. The MTJ, a spintronics device, is based on quantum mechanical tunneling of spin-polarized electrons through a thin insulating layer.

Function of an MTJ in a memory cell includes storage of state in its parallel and antiparallel orientations of the free magnet with respect to the pinned magnet, and it should also be able to change its state (writing). As in GMR, R_{AP} is larger and R_P is lower in magnitude and a read operation is performed by assessing the MTJ resistance. For write operation, pulsed currents above and below the MTJ provide magnetic fields and orient the free magnet as desired. Materials shown in Figure 1.16(b) have MR in the range of 31%–34% and resistance-area (RA) product of nearly 11 kΩ-μm². However, the behavior of MR and RA of MTJ material depends on AlO_x layer thickness, its uniformity and oxidation time, and uniformity of the MR ratio; and absolute resistance values are critical, as these are compared with that of the reference cell during the read operation. The resistance variation in MTJ cell becomes large with small variation in AlO_x thickness as resistance is exponentially dependent on it.

Figure 1.17(a) shows a schematic of an MRAM cell comprising a transistor and an MTJ bit. The transistor was first on and the MTJ bit was biased to 200 mV for the measurement of switching field, resistance, and MR [66]. Writing a single bit is achieved by applying orthogonal magnetic fields; one along the x-axis or directed along the length of the MTJ called easy axis field and the other along the y-axis or directed along the width of the MTJ, called hard axis field. Figure 1.17(b) shows an ideal switching asteroid curve which shows boundary between switching or not of the orientation of the free magnet. Not only is the practical switching asteroid slightly different from the ideal one, but there are fluctuations among the MTJs of the MRAM, which creates some kind of inconsistency while writing [67].

A 1 Mbit MRAM circuit has been fabricated using a 0.6 μm CMOS technology with less than 50 ns access and cycle times [68]. Because of its excellent performance of fast read/write time, nonvolatility, and unlimited write endurance, MRAM was considered a memory structure of high promise. However MRAM faced many challenges on the technical front. Some of these have been solved satisfactorily, for example, obtaining uniform MTJ

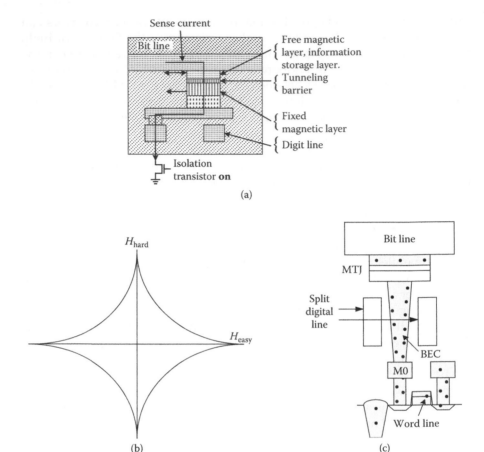

FIGURE 1.17
(a) Diagram of MTJ device in "read" mode with series isolation transistor. ("Recent Developments in Magnetic Tunnel Junction MRAM," S. Tehrani et al., *IEEE Trans. Magnetics*, Vol. 36, pp. 2752–2757, 2000.) (b) Ideal switching asteroid. (Redrawn from "Memories of Tomorrow," W. Reohr et al., *IEEE Circuits and Device Magzine*, pp. 17–27, Sept. 2002.) (c) 8F2 MRAM cell structure with split digital-line. (Modified from "Future Memory Technology including Emerging New Memories," K. Kim and G-H. Koh, Proc. 24th Int. Conf. on Microelectronics (MIEL), Vol. 1, pp. 377–384, 2004.)

resistance and prevention of writing disturbance in the unselected cells [69–71], though MRAM still faced the challenge of scalability with MTJ. Additional metal lines called digit lines for writing and MTJ connection required additional cell area, resulting in typical area of 20–50 F^2 [45]. A modified digit line in split form, as shown in Figure 1.17(c) is able to reduce the cell size to 8 F^2. Another important factor is the requirement of higher switching field as the MTJ size goes down. Large switching field means large current during writing which increases power consumption. If magnetic flux concentrating structure is used in digit lines and bit lines, writing current can be reduced [72]. Hence the degree of scaling considerably depends on the ability of reduction in MRAM's writing current.

1.8.3.1 Spin–Torque–Transfer (STT) MRAM

In conventional MRAMs, writing current was a significant limiting factor, which was effectively overcome through the application of spin-torque transfer magnetizing switching (STS). In the STS scheme, instead of providing magnetic field externally, direct injection of spin polarized electrons was used to reverse the direction of magnetic layer. It was also expected that STS current would decrease with downscaling of technology mode.

A memory cell of the STT MRAM is shown in Figure 1.18, which has a source line instead of a digit line of a conventional MRAM. Operationally, the read process is the same; only the write process is modified in which word line is selected and a positive voltage is applied at bit line or source line of a selected column. A 4 kb STT MRAM was fabricated in 0.18 μm CMOS technology which employed an MTJ with CoFeB ferromagnetic layer and crystallized MgO tunnel barrier layer of 1 nm thickness. With the kind of materials used, an MR ratio of more than 160% at RA product of 20 $\Omega\mu m^2$ was obtained [73]. Fabricated STT MRAM proved the practical feasibility of switching driving force from the spin polarized current other than external induced field and exhibited high speed, low power, and high scalability. Fabrication in smaller technology mode would increase read/write speed and integration density. However, a major challenge in this direction is the process variation affects especially in MTJ. Variations in tunneling oxide thickness and the cross-sectional area of the MTJ cause incorrect read and inability in write process that considerably reduce yield of the MRAMs. Efforts have continuously been made for improving device design and process control [74] though without adequate success, in spite of using advanced fabrication technologies. Alternatively, a circuit level solution is suggested for reducing process variation effects. A detailed study concluded that the

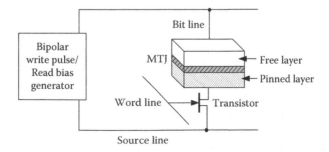

FIGURE 1.18
A schematic STT MRAM memory cell. (Redrawn from "A Novel Nonvolatile Memory with Spin Torque Transfer Magnetization Switching: spin ram," M. Hosomi et al., IEDM Tech. Dig., pp. 459–462, 2005.)

sizing of the n-channel MOSFET (NMOS) used is important [75]. NMOS wider than minimum size reduces write failure probability but increases read failures and incurs area penalty. Reduction in NMOS size affects conversely. A circuit-based scheme was proposed in which one MTJ, two NMOS transistors, one minimum sized for reading, and another NMOS, a bit larger for writing, were used. With an area overhead of ~9%, reduction in read failure probability was 39% and the reduction in write failure probability was 31% [75].

Another way of looking at the read/write failure probability is from a system-performance perspective. Reads occur more frequently than writes and read operations are directly in the critical path too [76]. A design technique called stretched write cycle (SWC) is employed in which architectural modification is made with write operation consuming one more cycle; hence, the requirement of wider NMOS is relaxed, density is improved, and read failure probability is also reduced.

With continued interest in STT-MRAM bit cell design [77–79], low power, less susceptibility to process variation, and reliable memories are highly likely to be available at the high-density level [80].

1.8.4 Phase Change RAM (PRAM)

Compact discs and DVD optical memory discs have extensively used phase change materials. However, a GeSbTe chalcogenide alloy-based memory cell element of an ovonic unified memory (OUM) in 180 nm technology was developed in 2001 [81]. State of the chalcogenide materials changes between being amorphous or crystalline. This change of state is possible with a small amount of heating through the application of a current pulse for a short duration. During amorphous phase, high resistance is offered known as SET state, whereas during crystalline state, low resistance is observed that is known as RESET state, and the difference of resistance in the two states is converted into high and low states of a memory cell. Figure 1.19(a) shows a simplified cross-section of an OUM memory cell element, which is a thin film two-terminal device and can be fabricated using conventional processes. For heating the cell, current pulse of comparatively high value of very short duration (a few ns) is required which melts the chalcogenide material, crystalline nature is totally disturbed and resistance becomes more than 100 kΩ. For conversion to crystalline state, current pulse of comparative small value flows for larger duration (2–50 ns) as shown in Figure 1.19(b), so that enough time is available for crystal growth and resistance becomes less than 1 kΩ.

Because of large resistance difference between two states, sufficient read signal margin becomes available. For the fabricated memory, read/write speed was comparable with contemporary DRAM. Read endurance is unlimited and scaling is mainly dependent on lithography. It is more adaptable for low power because of its low voltage operation [81]. Another OUM 4 Mbit NV memory is realized at 0.18 µm technology node [38]. In a study of its

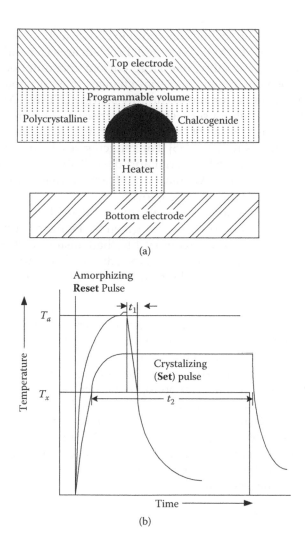

FIGURE 1.19
(a) Simplified cross-section of an ovonic unified memory element. (b) Current pulse width and temperature relationship of the phase change process. ("OUM-A 180 nm Nonvolatile Memory Cell Element Technology for Stand Alone and Embedded Applications," S. Lai et al., IEDM Tech. Dig., pp. 803–806, 2001.)

sequential operations, write/read/write complement/read, cycling at 5 MHz, a reset pulse of 8 ns was applied for ~5 ns falling edge and corresponding cell read resistance was found to be 85 kΩ. A set pulse of 85 ns changes the resistance to 2 kΩ showing wide resistance variation practically, and value of V_{th} at the beginning of set pulse is ~0.6 V. Another distinct advantage of OUM phase change memories is their compatibility with embedded systems because of low temperature processes used after transistor fabrication.

There appears to be no scaling constraint in CMOS technology as far as the fabrication process of PRAM is concerned. Problem area is mainly concerned with its practical functionality, like large current consumption during the reset process when chalcogenic material changes to amorphous state. So it requires large size transistor and eventually a larger cell size. Another problem area is the possibility of set resistance increasing when reset current is decreased. As a consequence of large set resistance, read time is increased and noise immunity is reduced [45]. For reducing reset current, techniques used are modification in GST material and the cell structure, and reduction in cell dimensions. The PRAMs have an advantage that with the cell size reduction, reset current can be scaled down.

Not only to prove the manufacturability of PRAM but also to enhance its performance, a novel structure was given by Pellizzer and others [82], which was CMOS technology-compatible, had faster read/write access time and programming current was also reduced. The important concept used here was that of μTrench, the formation of a trench in which GST is deposited and a thin vertical semimetallic heater as shown in Figure 1.20. A common collector p-n-p BJT access device in vertical cell architecture was realized which results in a compact cell area of 10 F^2 (0.32 μm² at 0.8 μm CMOS technology). Proposed μTrench structure helps in reducing the programming current to 600 μA and endurance of 10^{11} programming cycles. Vertical salicided p-n-p BJT with the storage element comprising of heater and chalcogenide material on top is fabricated at 90 nm technology [83]. Smaller contact area of the μTrench heater reduces the programming current to 400 μA. Using a simple layout with one base contact for every emitter, cell size of 0.096 μm² (equivalent to 12 F^2) is obtained. A similar approach at 45 nm technology, but with one base contact with every four emitters has reduced the cell size to 5.5 F^2 (.015 μm²) [84] while developing 1 Gbit PRAM for the first time.

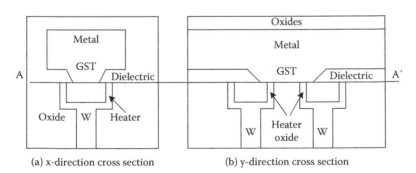

(a) x-direction cross section   (b) y-direction cross section

FIGURE 1.20
Schematic cross-sections of the basic structure of PRAM. ("Novel μTrench Phase-Change Memory Cell for Embedded and Stand-Alone Non-Volatile Memory Applications," F. Pellizzer et al., Symp. VLSI Tech. Dig. Tech. Papers, pp. 18–19, 2004.)

TABLE 1.1

A Comparison of Important Next-Generation Nonvolatile Memories

	FRAM (Ferroelectric)	MRAM (Magnetic)	PCM (Phase Change)
Storage mechanism	Permanent polarization of a ferroelectric material like PZT	Permanent magnetization of ferromagnetic material in a MTJ	Amorphous/polycrystal phases of chalcogenide alloy
Cell size (F^2)	~20	~20	~6
Scalability	Poor	Poor	Good
Endurance	10^{10}	>10^{14}	10^{12}
Application	Embedded, low density	Embedded, low density	Stand-alone, or embedded high density

Source: "Phase Change Memory: Development Progress and System Opportunities," G. Atwood, Interconnect Technology Conference (IITC), International, pp. 1–3, 2010.

With small contact area, μTrench structure is very suitable in reducing the reset current; however, it requires lithography to define the GST dimension and making a small contact for the underlying metal. A cross-spacer phase change memory cell has been developed which provides lithography-independent contact area that is very small (~1000 nm²) resulting in reduced reset current of 230 μA [85]. As already mentioned, reduction of reset current is an important area and further investigations are to continue toward the *wall* heater approach [82]. In addition, finding new chalcogenide material other than GST is also an important research area.

It is desirable to compare important alternative emerging NV memories. Few characteristics have been included in Table 1.1 [86], which may be helpful in selecting a suitable memory for an application.

References

1. J.M. Rabaey, A. Chandrakasan, and B. Nikolic, *Digital Integrated Circuits: A Design Perspective*, Pearson, Prentice Hall, 2003.
2. M. Takada and T. Enomoto, "Review and Prospects of SRAM Technology," *Trans. IEICE*, Vol. E74, pp. 827–839, 1991.
3. S.M. Kang and Y. Leblebici, *CMOS Digital Integrated Circuits—Analysis and Design*, Tata McGraw-Hill, 2003.
4. V.L. Rideout, "One-Device Cells for Dynamic Random Access Memories: A Tutorial," *IEEE Trans. Electr. Dev.*, Vol. ED-26, pp. 839–852, 1979.
5. R.C. Foss, "The Design of MOS Dynamic RAMs," *IEEE ISSCC. Tech. Dig.*, p. 140, 1979.

6. N.-C. Lu and H.H. Chao, "Half-VDD Bit-Line Sensing Scheme in CMOS DRAMs," *IEEE J. Solid State Circuits*, Vol. SC-22, pp. 451–454, 1984.
7. K.U. Stein, A. Sihling, and E. Doering, "Storage Array and Sense/Refresh Circuit for Single Transistor Memory Cell," *IEEE Int. S.S. Circuits Conf.*, pp. 56–57, 1972.
8. Y. Katayama, "Trends in Semiconductor Memories," *IEEE Micro*, pp. 10–17, 1997.
9. B. Prince, *High Performance Memories*, John Wiley and Sons, Chapter 2, 1999.
10. B. Prince, "Application Specific DRAMs Today," *Proc. Int. Workshop on Memory Technology, Design, and Testing* (MTDT), 2003.
11. A.B. Cosoroaba, "Double Data Rate SYNCHRONOUS DRAMs in High Performance Applications," *Proc. IEEE WESCON Conf.*, 1997.
12. Y. Choi et al., "A 16 Mbit Synchronous DRAM with 125 Mbytes/sec Data Rate," *VLSI Circuits Symp. Dig.*, pp. 65–66, 1993.
13. Y. Takai et al., "250 Mbyte/sec Synchronous DRAM Using 3-Stage-Pipelined Architecture," *VLSI Circuits Symp. Dig.*, pp. 59–60, 1993.
14. Elpida Corporation, "DDR II 512 M SDRAM Specification," 2002.
15. H. Fujisawa et al., "1.8-V 800-Mb/s/pin DDR2 and 2.5-V 400-Mb/s/pin DDR1 Compatibly Designed 1 Gb SDRAM with Dual Clock Input Latch Scheme and Hybrid Multi-Oxide Output Buffer," *Symp. VLSI Circuits, Dig. Tech. Papers*, pp. 38–39, 2004.
16. Y.K. Kim et al., "A 1.5 V, 1.6 Gb/s/pin, 1 Gb DDR3 SDRAM with an Address Queuing Scheme and Bang-Bang Jitter Reduced DLL Scheme," *Symp. VLSI Circuits, Dig. Tech. Papers*, pp. 182–183, 2007.
17. T.-Y. Oh et al., "A 7 Gb/s/pin 1 Gbit GDDR5 SDRAM with 2.5 ns Bank to Bank Active Time and Ni Bank Group Restriction," *IEEE. J. Solid State Circuits*, Vol. 46, pp. 107–118, 2010.
18. Ashok K. Sharma, *Semiconductor Memories, Technology, Testing, and Reliability*, IEEE Press, 1997.
19. K. Bernstein and N.J. Rohrer, *SOI Circuit Design Concepts*, Springer Science, 2007.
20. K. Suma et al., "An SOI-DRAM with Wide Operating Voltage Range by CMOS/ SIMO Technology," *IEEE ISSCC Tech. Dig.*, pp. 138–139, 1994.
21. S. Kuge et al., "SOI-DRAM Circuit Technology for Low Power High Speed Multi-Giga Scale Memories," *VLSI Circuits Symp. Dig.*, pp. 103–104, 1995.
22. S. Okhonin et al., "A SOI Capacitor-less 1-T DRAM Concept," *Proc. IEEE. Int. Conf.*, pp. 153–154, 2001.
23. S.K. Moore, "Masters of Memory," *IEEE Spectrum*, pp. 45–49, 2007.
24. H.J. Wann and C. Hu, "A Capacitor-less DRAM cell on SOI Substrate," IEDM Tech. Dig., pp. 635–638, 1993.
25. T. Tanaka, E. Yoshida, and T. Miyashita, "Scalability Study on a Capacitor-less 1T-DRAM: From Single-Gate PD-SOI to Double Fin DRAM," IEDM Tech. Dig., pp. 919–922, 2004.
26. C. Kuo, T.-J. King, and C. Hu, "A Capacitor-less Double-Gate DRAM Cell," *IEEE Electr. Dev, Lett.*, Vol. 23, pp. 345–347, 2002.
27. E. Yoshida, T. Miyashita, and T. Tanaka, "A Study of Highly Scalable DG-Fin DRAM," *IEEE Electr. Dev. Lett.*, Vol. 26, pp. 655–657, 2005.
28. C. Kuo, T.-J. King, and C. Hu, "A Capacitorless Double-Gate DRAM Cell Design for High Density Applications," IEDM Tech. Dig., pp. 843–846, 2002.
29. I. Ban et al., "Floating Body Cell with Independently-Controlled Double Gates for High Density Memory," IEDM Tech. Dig., pp. 1–4, 2006.

30. M.G. Ertosun et al., "A Nanoscale Vertical Double-Gate Single-Transistor Capacitorless DRAM," *IEEE Electr. Dev. Lett.*, Vol. 29, pp. 615–617, 2008.
31. M. Bawedin et al., "A Capacitorless 1T-DRAM on SOI Based on Dynamic Coupling and Double-Gate Operation," *Electron. Dev. Lett.*, Vol. 29, pp. 795–798, 2008.
32. S.-W. Ryu et al., "Unified Random Access Memory (URAM) by Integration of a Nano Crystal Floating Gate for Nonvolatile Memory and a Partially Depleted Floating Body for Capacitorless 1-T DRAM," *Solid State Electronics*, Vol. 53, pp. 389–391, 2009.
33. J-W. Han et al., "Partially Depleted SONOS FinFET for Unified RAM (URAM)-Unified Function for High-Speed 1T DRAM and Nonvolatile Memory," *IEEE Electron. Dev. Lett.*, Vol. 29, pp. 781–783, 2008.
34. J-Y. Lin et al., "Performances of a Capacitorless 1R-DRAM Using Polycrystalline 1T-DRAM Using Polycrystalline Silicon Thin-Film Transistors with Trenched Body," *IEEE Electr. Dev. Lett.*, Vol. 29, pp. 1222–1225, 2008.
35. Y. Yatsuda et al., "Hi-MNOS II Technology for a 64 Kbit Byte-Erasable 5V only EEPROM," *IEEE J. Solid-State Circuits*, Vol. SC-20, p. 144, 1985.
36. J.T. Evans and R. Womack, "An Experimental 512-bit Nonvolatile Memory with Ferroelectric Storage Cell," *IEEE J. Solid State Circuits*, Vol. 23, pp. 1171–1175, 1988.
37. J.M. Daughton, *Thin Film Magnetic RAM Devices*, Honeywell Inc., 1988.
38. M. Gill et al., "Ovonic Unified Memory—A High Performance Nonvolatile Memory Technology for Stand-Alone Memory and Embedded Applications," *ISSCC Tech. Dig.*, pp. 202–204, 2002.
39. R. Pashley and S. Lai, "Flash Memories: The Best of Two Worlds," *IEEE Spectrum*, pp. 30–33, 1989.
40. M. Momodami et al., "An Experimental 4-Mbit CMOS EEPROM with a NAND Structure Cell," *IEEE J. Solid State Circuits*, pp. 1238–1243, 1989.
41. K. Itoh, *VLSI Memory Chip Design*, Springer-Verlog, 2001.
42. S. Pawan et al., "Flash Memory Cell—An Overview," *Proc. IEEE*, Vol. 85, pp. 1248–1271, 1997.
43. R. Bez et al., "Introduction to Flash Memory," *Proc. IEEE*, Vol. 91, pp. 489–502, 2003.
44. H.-H. Tseng et al., "ALD HfO$_2$ Using Heavy Water (D$_2$O) for Improved MOSFET Stability," IEDM Tech. Dig., pp. 83–86, 2003.
45. K. Kim and G-H. Koh, "Future Memory Technology including Emerging New Memories," *Proc. 24th Int. Conf. on Microelectronics* (MIEL), Vol. 1, pp. 377–384, 2004.
46. J.D. Lee et al., "Effects of Floating-Gate Interference on NAND Flash Memory Cell Operation," *Electr. Dev. Letters*, Vol. 23, pp. 264–266, 2002.
47. C.H. Lee et al., "A Novel SONOS Structure of SiO$_2$/SiN/Al$_2$O$_3$ with TaN Metal Gate for Multi-giga Bit Flash Memories," IEDM Tech. Dig., pp. 613–616, 2003.
48. Yan Li et al., "A 16 Gb 3b/cell NAND Flash Memory in 56 nm with 8 MB/s Write Rate," *ISSCC Tech. Dig.*, pp. 506–507, 2008.
49. H. Kurata et al., "Constant-Charge-Injection Programming for 10-MB/s Multilevel AG-AND Flash Memories," *VLSI Circuits Symp. Dig.*, pp. 302–303, 2002.
50. N. Shibata et al., "A 70 nm 16 Gb 16-Level-Cell NAND Flash Memory," *VLSI Circuit Symp. Dig.*, pp. 190–191, 2007.
51. T. Nozaki et al., "A 1 Mbit EEPROM with MONOS Memory Cell for Semiconductor Disc Application," *VLSI Circuit Symp.*, pp. 101–102, 1990.

52. K. Kim, "Future Memory Technology: Challenges and Opportunities," *Int. Symp. VLSI Technology, Systems, and Applications* (VLSI-TSA), pp. 5–9, 2008.
53. C.B. Kim et al., *Memory Cell Capacitor Using Cross Double Patterning Technology for Gigabit Density DRAM*, ISDRS, 2009.
54. Y.C. Shin et al., "A Novel NAND-Type MONOS Memory Using 63 nm Process Technology for Multi-gigabit Flash EEP ROMs," IEDM Tech. Dig., pp. 327–330, 2005.
55. Y. Park et al., "Highly Manufacturable 32 Gb Multi-Level NAND Flash Memory with 0.0098 μm² Cell Size Using TANOS (Si oxide-Al₂O₃-TaN) Cell Technology," IEDM Tech. Dig., pp. 29–32, 2006.
56. D. Kwak et al., "Integration Technology of 30 nm Generation Multilevel NAND Flash for 64 Gb NAND Flash Memory," *VLSI Tech. Dig.*, pp. 12–13, 2007.
57. E-L. Lai et al., "A Multilayer Stackable Thin-Film Transistor (TFT) NAND Type Flash Memory," IEDM Tech. Dig., pp. 37–40, 2006.
58. Y. Fukuzumi et al., "Optical Integration and Characteristics of Vertical Array Devices for Ultra-High Density, Bit-Cost Scalable Flash Memory," IEDM Tech. Dig., pp. 449–452, 2007.
59. W.A. Geideman et al., "Progress in Ferroelectric Memory Technology," *IEEE Trans. Ultrason., Ferr., Freq. Contr.*, Vol. 38, 1991.
60. T. Miyazaki and N. Tezuka, "Giant Magnetic Tunneling Effect in Fe/Al₂O₃/Fe Junction," *J. Magn. Mater.*, Vol. 139, L231, 1995.
61. W.J. Gallagher et al., "Microstructured Magnetic Tunnel Junction," *J. App. Phy.*, Vol. 81, pp. 3741–3746, 1997.
62. M.N. Babich et al., "Giant Magnetoresistance of (100) Fe/(001) Cr Magnetic Superlattices," *Phy. Rev. Lett.*, Vol. 61, pp. 2742–2745, 1988.
63. J.M. Slaughter, E.Y. Chen, and S. Tehrani, "Magnetoresistance of Ion-Beam Deposited Co/Cu/Co and NiFe/Co/Cu/Co/NiFe Spin Valves," *J. App. Phys.*, Vol. 85, pp. 4451–4453, 1999.
64. S. Tehrani et al., "High Density Nonvolatile Magnetoresistance RAM," IEDM Tech. Dig., pp. 193–196, 1996.
65. S. Tehrani et al., "Progress and Outlook for MRAM Technology," *IEEE Trans. Magnetic*, Vol. 35, pp. 2814–2819, 1999.
66. S. Tehrani et al., "Recent Developments in Magnetic Tunnel Junction MRAM," *IEEE Trans. Magnetics*, Vol. 36, pp. 2752–2757, 2000.
67. W. Reohr et al., "Memories of Tomorrow," *IEEE Circuits and Device Magazine*, pp. 17–27, Sept. 2002.
68. S. Tehrani et al., "Magnetoresistive Random Access Memory Using Magnetic Tunnel Junctions," *Proc. IEEE*, Vol. 91, pp. 703–713, 2003.
69. M. Motoyoshi et al., "High Performance MRAM Technology with an Improved Magnetic Tunnel Junction Material," *VLSI Tech. Dig.*, pp. 212–213, 2002.
70. M. Durlam et al., "A 0.18 μm 4 Mb Toggling MRAM," IEDM Tech. Dig., pp. 995–997, 2003.
71. M. Durlam et al., "A Low Power 1 Mbit MRAM Based on 1M 1MTJ Bit Cell Integrated with Copper Interconnects," *VLSI Circuits Dig. Tech. Papers*, pp. 158–161, 2002.
72. M. Durlam et al., "A 1 Mbit MRAM Based on 1 T 1 MTJ Bit Cell Integrated with Copper Interconnects," *J. Solid State Circuits*, Vol. 38, pp. 769–773, 2003.
73. M. Hosomi et al., "A Novel Nonvolatile Memory with Spin Torque Transfer Magnetization Switching: Spin RAM," IEDM Tech. Dig., pp. 459–462, 2005.

74. M. Motoyoshi et al., "A Study for 0.18 μm High-Density MRAM," *VLSI Symp. Dig. Tech. Papers*, pp. 22–23, 2004.
75. J. Lee et al., "Variation-Tolerant Spin-Torque Transfer (STT) MRAM Array for Yield Enhancement," *IEEE Custom Int. Circ. Conf.*, pp. 193–196, 2008.
76. P. Ndai, A. Goel, and K. Roy, "A Scalable Circuit Architecture Co-Design to Improve Memory," *IEEE Trans. VLSI*, Vol. 18, pp. 1209–1219, 2010.
77. C.J. Lin et al., "45 nm Low Power CMOS Logic Compatible Embedded STT MRAM Utilizing a Reverse-Connection 1T/1MTJ Cell," IEDM Tech. Dig., pp. 279–282, 2009.
78. T. Kawahara et al., "2 Mb Spin-Transfer Torque RAM (SPRAM) with Bit-by-Bit Bidirectional Current Write and Parallelizing-Direction Current Read," *Proc. ISSCC*, pp. 480–482, 2007.
79. J. Li et al., "Design Paradigm for Robust Spin-Torque Transfer Magnetic RAM (STT MRAM) from Circuit/Architecture Perspective," *IEEE Trans. VLSI*, pp. 1710–1723, 2008.
80. C. Augustine et al., "Spin-Transfer Torque MRAMs for Low Power Memories: Perspective and Prospective," *IEEE Sensor Journal*, 2011.
81. S. Lai et al., "OUM-A 180 nm Nonvolatile Memory Cell Element Technology for Stand Alone and Embedded Applications," IEDM Tech. Dig., pp. 803–806, 2001.
82. F. Pellizzer et al., "Novel μTrench Phase-Change Memory Cell for Embedded and Stand-Alone Non-Volatile Memory Applications," *Symp. VLSI Tech. Dig. Tech. Papers*, pp. 18–19, 2004.
83. F. Pellizzer et al., "A 90 nm Phase Change Memory Technology for Stand-Alone Non-Volatile Memory Applications," *Symp. VLSI Tech. Dig. Tech. Papers*, 2006.
84. G. Servalli, "A 45 nm Generation Phase Change Memory Technology," IEDM Tech. Dig., pp. 113–116, 2009.
85. W.S. Chen et al., "A Novel Cross-Spacer Phase Change Memory with Ultra-Small Lithography Independent Contact Area," IEDM Tech. Dig., pp. 319–322, 2007.
86. G. Atwood, "Phase Change Memory: Development Progress and System Opportunities," *Interconnect Technology Conference (IITC), International*, pp. 1–3, 2010.

2

DRAM Cell Development

2.1 Introduction

Low cost per bit and high packing density of DRAMs are unmatched by any other semiconductor memory. Within the limitations of fabrication processes, mainly lithography, area of the memory cell layout is very crucial as it has direct impact on the performance and the cost. The initial-stage cell was a simpler entity, of course, in comparison to the complex products of today, in the form of a single access transistor and a storage capacitor laid side by side in two dimensions; it has been called a planar DRAM cell. Overall cell size continued to shrink without any serious obstacles till the 1–4 Mbit range. However, beyond that, shrinking of the cell in two dimensions became impracticable, because planar capacitor alongside access transistor could not hold sufficient charge (or realize enough capacitance) in a small area. Sizing of the transistors was also going to be a challenge with shrinking the cell, but it was a delayed factor at a much lower minimum feature size of around 100 nm. Different alternatives were used while making the cell three-dimensional. One approach was the stacked capacitor cell in which the capacitor was realized on top of the switching transistor. Cells in which the capacitor was fabricated in the substrate came to be known as trench capacitors cells. Varieties of both the stacked capacitor and the trench capacitor cell came into practice.

2.2 Planar DRAM Cell

Single transistor memory cell at the initial stage consisted of a MOSFET switch and a charge storage capacitor as shown in Figure 2.1(a), where total storage consisted of two capacitances: C_{jn} (junction capacitance) and a bias-independent MOS capacitance C_{ox}, which was normally 5 to 10 times larger than C_{jn} [1]. The storage capacitor electrode which connects to the MOSFET source could be either an n-type diffused region or an inversion layer

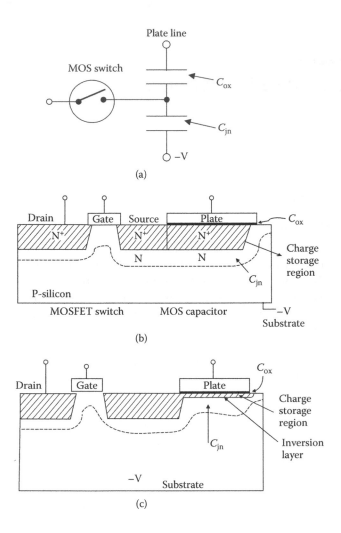

FIGURE 2.1
Planar transistor memory cell. (a) Equivalent circuit, basic structure (b) with diffusion storage, and (c) with inversion storage. (Redrawn from "One-Device Cells for Dynamic Random—Access Memories: A Tutorial," V.L. Rideout, *IEEE Trans. Electr. Dev.* Vol. 26, pp. 839–852, 1979.)

(of electrons for p-substrate in NMOS), as shown in Figure 2.1(b) and (c), respectively. The charge storage capacitor remains (1) in the neutral/stable 0 state while having electrons normally present in n-type doped region as shown in Figure 2.1(b) or in a surface inversion layer under a biased plate electrode as shown in Figure 2.1(c) or (2) in the charged/unstable 1 state by removing electrons through the MOSFET switch. Leakage current through the MOSFET very quickly restores electrons in the 1 state, converting it to the 0 state; hence, removal of these charges or *refreshing* is a must.

During the evolutionary phase of the above-mentioned planar one-transistor DRAM (1T DRAM), several modifications and forms of realizations were found feasible. As the DRAM size increased from 4/16 Kbit to 64 Kbit and then to 1 Mbit per chip, cell structure changed its form due to (1) various MOSFET technologies, and (2) arrangement of the bias and control lines to drain, gate and capacitor plate, and storage capacitor formation itself along with few variations as well in each entity. Details of MOSFETs are discussed in Chapter 4, for advanced MOSFETs, as the conventional two-dimensional MOSFET theory and fabrication process is already well studied. During DRAM development stages two important figures of merit for a DRAM cell were/are given due importance. A number of lithographic squares (i.e., a square whose side equals the minimum feature size) consumed by the cell and the charge transfer ratio, as given in Chapter 1 generally in the range of 5 to 20, form a performance figure of merit [2]. However, other figures of merit like processing complexity (number of lithographic masking operations, etching steps, and high temperature treatments), chip performance (access/cycle time, refresh period, and power dissipation), and cost (chip size, yield, and packaging), must also be considered. Based on the consideration of all figures of merit, enumerated above, it was clear as early as 1979 that polysilicon-to-polysilicon (poly-to-poly or double-poly) storage capacitor [3] shall be a suitable structure for planar cell configuration. It was verified later by many researchers that such a cell was indeed very suitable, however, only up to the 1 to 4 Mbit density level [2].

It is important to know the reasons behind more innovative cell/capacitor structures/organizations involving relatively difficult processing option like a double-poly planar cell. Figures 2.2(a) and (b) show the cross-section

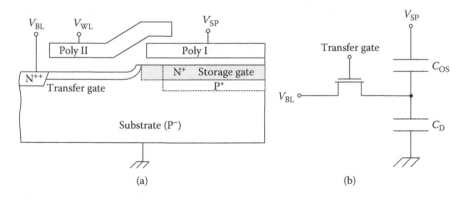

(a)　　　　　　　　　　　　　　(b)

FIGURE 2.2
(a) Cross-section of a double poly one-transistor DRAM cell and modification (within dotted line) to convert as H-C cell and (b) circuit representation. (Redrawn from "Design Parameters of the Hi-C DRAM Cell", Y.A. EL-Mansy and R.A. Burghard, IEEE. J.S.S. Circuits, Vol. SC-17, pp. 951–956, 1982.)

of the double-poly planar cell structure and its equivalent circuit representation. Here, V_{BL}, V_{WL}, and V_{SP} are the voltages applied at the bit line, word line, and the storage plate, respectively. Charge transfer between the bit line and the storage region takes place when the word line is high. The access transistor gate has a higher threshold voltage compared to that of the storage region and its effect is that when word line is high the surface potential below the transfer gate is lower than both the electrostatic potentials of the bit line and the storage area for the same applied voltage to these regions [4].

It is observed that the amount of charge Q in the storage region corresponding to low-state, Q (low), for a full-charge well, is a function of C_{os}, the oxide capacitance per unit area for the storage region; N_B, bulk doping concentration; P_f, the surface potential corresponding to full charge well; and V_{SP}. The charge Q remaining in the storage area corresponding to an empty-cell, Q (high) is a function of C_{os}, N_B, V_{SP}, and P_e, the surface potential in the transfer gate region at gate voltage V_{WL}. So, the maximization of charge signal $\Delta Q = Q(\text{low}) - Q(\text{high})$ for a given storage area (or increasing the storage charge capacity) is done through, (1) increase of C_{os}, (2) reduction in threshold voltage in the storage gate area, and P_f, and (3) increase of P_e. Increase in the depletion capacitance C_D through increase of bulk doping concentration N_B is another option. All these options effectively require reduction of oxide thickness both at the storage and the transfer gate region, and reduced doping concentration below the storage gate. Obviously, reduction in SiO_2 thickness could be attempted up to a certain minimum level. Even if breakdown does not occur, carriers can pass through very thin oxide layer. Increase in permittivity through use of another dielectric was not given a consideration at that stage, which later became a strong point. The other option was to increase C_D, which required an increase in the doping concentration below the storage gate—a requirement contradictory to the earlier options, as it reduces storage gate threshold voltage. A modified Hi-C cell was introduced to satisfy both the requirement of increasing the substrate doping in the storage area for high depletion capacitance while maintaining low threshold voltage by Sodini and Kamins [5] and by Tasch et al. [6]. A boron layer is implanted in the storage area to increase substrate doping (and hence increases C_D); then the resulting increase in threshold voltage is offset by implanting a layer of n-type impurities at the surface shown as modification in Figure 2.2. The Hi-C cell has also been shown to operate with grounded plate ($V_{sp} = 0$), which is desirable for high-density memory chips.

The packing density and capacity of 1T DRAM cells continued to increase in conformity with the device scaling theory of Dennard [7]. It was predicted as early as 1975 that planar MOS capacitor 1T cell with a two-dimensional structure would encounter following problems:

1. Scaling down limitation is serious because MOS capacitor area could not be scaled down to have minimum signal charge ΔQ in the range of 200 fC corresponding to the storage capacitance of 65 fF at V_{DD}

of 3.3 V. With scaled-down insulator thickness of 10 nm, estimated needed capacitor area was approximately 20 μm^2, which was about twice the cell area for a 4 Mbit DRAM chip of 75 mm^2, for which cell area has to be below 10 um^2 (assuming ~60% of the chip area earmarked for the core).

2. Another problem of a serious nature was in the form of signal reduction due to the minority carriers generated in the silicon substrate. This occurs because the charges in the inversion layer and the depletion layer of the storage capacitor are affected by the minority carriers, which are thermally or optically generated, or generated by the irradiation of cosmic rays like alpha particles.

3. Fabrication of access transistor, whether by the side of storage capacitor, on top of it, or below it, was also going to be a serious problem, with increasing DRAM density and shrinkage of minimum feature size. To keep transistor performance as per requirements, DRAM cell was also going to be three-dimensional, thereby changing its shape drastically.

As a result of the number of fabrication and operational difficulties mentioned, further advancement in DRAM density was not possible beyond 1 4 Mbit using the planar DRAM cell. Higher density 1T DRAMs were realized through innovation in cell design, mainly through various forms of three-dimensional capacitors and cells, such as trench and stacked structure, at least up to 1 Gbit density level. Be it trench or stacked capacitor, capacitance had to be enhanced without increasing projected area (in fact area was to be decreased) with increasing DRAM density. Schemes for the capacitance enhancement are discussed in Chapter 5, whereas development of a few early stage three-dimensional cells is discussed in the following sections. First-generation trench capacitor cells laid the foundation of the fabrication of capacitors below the surface of the wafer. Various kinds of the first-generation trench capacitors have been discussed in Section 2.3.1. To save the trench capacitors from likely intercell punch-through and for the minimization of the effects of cosmic ray irradiation inverted trench capacitor cells were realized. These cells, which kept the storage node inside a trench, are discussed in Sections 2.3.2 and 2.6. In almost all such cells access transistor was fabricated by the side of the trench capacitor which consumed chip area. A modified cell structure in which chip area is saved by fabricating access transistor above the trench capacitor is given in Section 2.4. Section 2.5 describes a DRAM cell in which both the vertical transistor and storage capacitor are realized in a trench. In late 1980, fabrication of vertical transistor was not mature and few cells were fabricated with planar transistors and capacitors buried under them. However, a popular choice for higher density DRAM became available in the form of stacked capacitor cells in which storage capacitor was stacked above the access transistor. Stacked capacitor cell basics and some variations are discussed in Section 2.8. Though their properties, characteristics, and advantages like using metal-insulator-metal (MIM) form capacitor and/or capacitor-over-bit line (COB) structure are discussed in Chapter 6.

2.3 Three-Dimensional Capacitor DRAM Cell

Exceeding a 4 Mbit conventional planar access transistor and a planar capacitor in a two-dimensional structure was highly unpractical so cell structures had to use advanced processing techniques to meet the design criteria. It became necessary to move toward a third dimension for the capacitor as well as for the cell structure. As mentioned before, main categorization was in terms of trench or stacked form. First, trench capacitor cells of the first generation, that have storage plates of the capacitor outside the insulated trench are described.

2.3.1 First-Generation Trench Capacitor Cell

As the name suggests, storage of charge is done in a trench that is below the surface level and inside the substrate in trench capacitor cell, thereby reducing the projected area of the capacitor. Although this kind of structure was attempted in vertical MOS (VMOS) DRAM [8,9] earlier, it did not succeed due to the sharp edges at the bottom, which resulted in the thinning of the gate oxide at corners, making it unmanufacturable. The main reason for this problem in VMOS structure was the use of orientation based etching. Whereas reactive ion etching (RIE) was used in the realization of trench capacitor with breakdown characteristics similar to the planar capacitor. Attempts were made to get better performance with a combination of trench etch control, edge rounding technique and the use of the dielectrics which were a combination of thermal and deposited layers [10]. However, the first reports on manufacturable trench capacitor appeared in 1982–83 [11,12] in the form of corrugated capacitor cells, which are discussed in the following section. A few more first-generation trench capacitor cells are then discussed.

2.3.1.1 Corrugated Capacitor Cell

Sunami [11] gave the concept and structure of a corrugated capacitor cell (CCC) having a capacitor formed in a dry etched moat into silicon substrate but a more physically realizable trench cell form was given later which is shown in Figure 2.3 [10]. As the trench capacitor stores charge along the

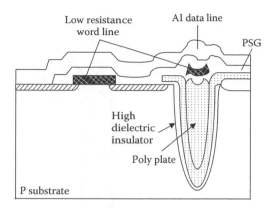

FIGURE 2.3
Trench capacitor cell (CCC-Hitachi). ("Trench and Compact Structures For DRAM's",
P. Chaterji et al., Proc. IEDM, pp. 128–131, 1986.)

outside walls as well as at the bottom of the trench, the storage area is greatly
enhanced with even reduced planar projected surface for the capacitor. The
main advantage of the trench idea was that by simply increasing the depth
of the trench, capacitance could be increased. Obviously, the fabrication of
a huge number of uniformly etched trenches was a challenge without having dielectric breakdown. However, the objective was successfully achieved
[13] and 1 Mbit to 4 Mbit DRAM chips using trench capacitor were realized.
Modified and improved versions of trench capacitor cells have also been successfully manufactured, in spite of the fact that some design concerns were
expressed in the original trench capacitor structure of ref. [11]. A major concern was that as the charge was stored around the trench surface in the substrate silicon, the larger storage areas became highly susceptible to minority
carrier leakage and prone to high soft error rate (SER). High-leakage currents could also arise between adjacent trenches and the active device of the
next cell due to the punch-through or surface conduction. Furuyama et al.
suggested a solution of the problem in an experimental 4 Mbit DRAM [14].
To make the cell less susceptible to soft errors due to the irradiation of the
α-particles, an n-channel depletion type trench cell is fabricated in a p-well
on a p-substrate similar to the one in Figure 2.3. In addition, p-well impurity concentration was increased that narrowed down the depletion region
and suppressed leakage current between adjacent cells. With increased doping concentration, leakage is reduced but it was likely to cause avalanche
breakdown. Trench spacing was limited by the diffusion depths of doping
profiles and depletion regions. Another option used was to make a deeper
trench, however, these were difficult to fabricate, hence more problematic.
The problem of leakage in a deep trench was solved to some extent in a 4
Mbit DRAM by T. Sumi et al. [15] that is shown in a cross-sectional view of

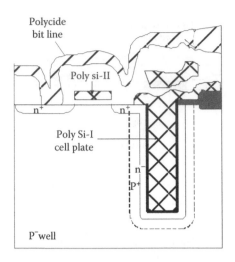

FIGURE 2.4

Cross-sectional view of a deep trench cell. (Adapted T. Sumi et al., "A 60 ns 4Mb DRAM in a 300 mil DIP," ISSCC Dig. Tech. Papers, pp. 282–283, 1987.)

the cell (Figure 2.4). In deep trench capacitors, the bottom of the electrode is more susceptible to leakage on account of lower p-well concentration. To overcome it, arsenic was doped at the sides and bottom of the trench to form N-type regions, preventing the decrease in memory cell capacitance. In addition, boron was doped to form a Hi-C structure. A memory cell of 8.0 μm^2 (1.7 μm × 4.6 μm) with a 40 fF as capacitance and a 4 μm trench depth was fabricated which could suppress leakage current between the adjacent cells even at $V_{cc} = 7$ V.

2.3.1.2 Folded Capacitor Cell

To maintain required capacitance for higher-density DRAM, the trench needs to be made deeper in CCC, giving rise to the leakage problem via punch-through between neighboring trenches, in addition to the practical problem of etching a deeper trench and suffering from high SER. To minimize the problems cell structures where the trench also serves part of the isolation function have been proposed. One such practical realization for this purpose is a folded capacitor cell (FCC). Instead of using the four sidewalls and bottom of the trench capacitor for charge storage, it was proposed to make the storage capacitor on the vertical wall of a long trench [16]. Increase in the surface area is achieved by folding the capacitor of the conventional trench cell over the isolation edge. The FCC could increase the storage capacitance considerably, therefore, it was then used for realizing larger-density DRAMs. Neighboring capacitors were effectively shielded from each other by high-threshold voltage transistors formed at the bottom; hence, capacitors could be placed closer to each other.

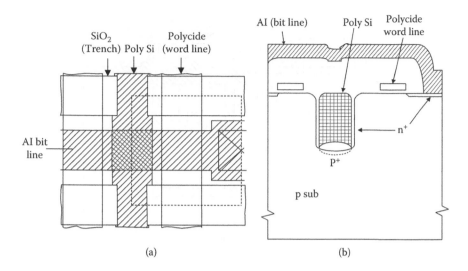

(a)   (b)

FIGURE 2.5

(a) Top view and (b) cross-sectional view of a vertical cell. (Adapted from T. Furuyama and J. Frey, "A Vertical Capacitor Cell for ULSI DRAMs," Symp. on VLSI Tech. Dig. Tech. Papers, pp. 16–17, 1984.)

2.3.1.3 Vertical Capacitor Cell

Another trench capacitor with a design similar to that of FCC is a vertical capacitor (VC) cell, which has been realized by Furuyama and Frey [17]. As shown in Figure 2.5, storage MOS capacitor is located vertically only on one sidewall of the trench. Another trench filled with SiO_2 and orthogonal to the first provides field isolation between cells. In addition, the bottom part of the trench also isolates capacitors of the neighboring cell, similar to FCC. The VC cell has been shown to have advantage in comparison to a conventional trench cell, in terms of capacitor area and cell area as a function of minimum feature size F. The cell is shown to work satisfactorily for the realization of 16 Mbit DRAM on a 100–120 mm^2 chip with a 0.5 µm process, a 2 µm deep trench, and a 6.0 nm thick oxide or equivalent insulator for the capacitor.

2.3.1.4 Isolation-Merged Vertical Capacitor and Buried Isolation Capacitor Cells

Similar to the FCC cell, merging of storage node and isolation is done in isolation-merged vertical capacitor (IVEC) cells [18,19]. As shown in Figure 2.6, the storage node is a poly layer inside the trench instead of being housed in bulk silicon, improving on soft error rate. Both the storage and the plate electrodes are in a grid-type trench surrounding the access transistor. The storage capacitance is large, but parasitic channel sidewall leakage can

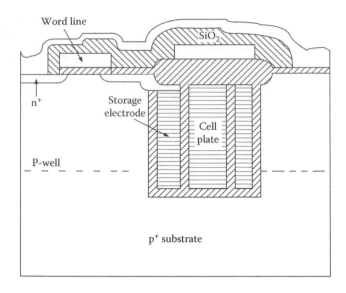

FIGURE 2.6
Cross- section of an isolation-merged vertical capacitor cell. (Redrawn from "Advanced cell structures for dynamic RAMs", N.C.C. Lu, *IEEE Circuits and Devices Magazine*, pp. 27–36, Jan. 1989.)

increase due to a gated-diode structure across the ends of the device channel [18]. Cell process is complex and provision of layer of high dielectric insulator may be difficult between poly layers. First 16 Mbit DRAM using IVEC was realized occupying 4.9 μm^2 area in 0.7 μm technology involving 20 masking steps [19]. Overcoming some of the limitations of IVEC, like avoiding the possible sidewall leakage or sacrificing some performance like having a little less capacitance, similar cells have been realized, such as the surrounding Hi-C capacitance cell in a 4 Mbit chip [20].

The basic concept of merging the capacitor and isolation region into a buried capacitor formed in a trench is applied in another cell by Nakamura et al. [21]. In the buried isolation capacitor (BIC), cell sidewall of the trench and the polysilicon buried within the trench form two plates of the capacitor. The structure is free from punch-through between adjacent cells; thus, a memory cell of nearly 7 μm^2 was realized in 0.8 μm technology, which could make the production of 4 Mbit chip practical. Storage capacitor in each cell is formed on a part of the trench sidewall, which surrounds every two-bit cell periphery, reducing considerably its horizontal surface area. A B+ implantation is done into the bottom of the trench to provide isolation between adjacent cells and silicon dioxide is filled into the trench to complete the fabrication.

In all the implementations where merging of isolation and capacitor is done in a trench, like FCC, IVEC, BIC, and so on, some walls of the trench must act as the cell capacitor while other walls of the trench must act as insulator.

Charging (discharging) of the capacitor is done from one edge of the trench as sidewalls are otherwise used; hence, all parasitic paths from the other edges of the trench must be plugged. Different methods were used to achieve this objective such as selective oxidation of the bottom of the trench. Filling different walls of the trench with polysilicon and oxide was the technique employed in the first generation of these cells. The second generation of these cells has employed ion implantation at an angle to selectively dope different faces of the trenches depending on applications [22,23]. Use of these techniques provides some flexibility in reducing/increasing the trench depth.

2.3.1.5 FASIC Cell

An advanced 4-MBit density DRAM has been realized occupying 72.3 mm² chip area, using 0.8 µm design rule, in which different faces of the trenches have been selectively doped forming a folded bit line adaptive side-wall isolator capacitance (FASIC) cell [22,23]. FASIC cell achieves effectively larger sidewall area in a grid-type trench structure as shown in Figure 2.7 and uses several advanced processing steps such as oblique implantation and deep isolation with groove refill. Using oblique implantation, shallow n⁺ junction at the sidewall is formed uniformly and the junction depth can be controlled by ion beam incident angle. Shallower junction depth depresses the collection of excess electron by 40%. The isolation characteristics of cell are

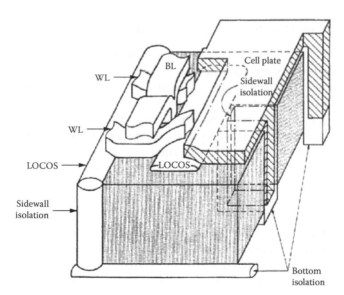

FIGURE 2.7
Perspective view of the FASIC cell. (From "A 90 ns 4Mb DRAM in a 300 mil DIP", K. Mashiko, IEEE Intern. Solid-State Circuits Conf., pp. 312–315, 1987.)

excellent with parasitic leakage current being nearly of the same order as that of a planar memory cell. Formation of cell plate is easier than conventional isolation-merged cell as it is formed on a planar area rather than in the bottom of the trench. However, the cell needs higher number of masks and some signal charge is stored in the bulk.

2.3.2 Second-Generation Inverted-Type Trench Cells

In spite of few improvements in design and advancements in structures as discussed in the last section, first generation trench cells were not able to realize DRAMs beyond 4 Mbit density levels, without using deeper trench cells, though some attempts have been made at 16 Mbit level [19,24]. As a major difference, trenches were developed with the storage electrode placed inside the trench instead of outside. The main purpose was to shield the storage electrodes from each other to reduce punch-through and soft error problems. These were mentioned as inverted-type trench cells.

These problems also could be solved by placing the cell array within a CMOS well, but it limits considerably the depth of the trench as well as the shallowness of the well. A similar problem is encountered if the trench cells are implemented in an epitaxial CMOS technology with heavily doped substrate for reducing substrate noise and probability of latch-up. A few examples are now discussed using inverted type trench cells.

2.3.2.1 Substrate Plate Trench-Capacitor Cell

The substrate plate trench-capacitor (SPT) cell [25–27] employs polysilicon in a trench as the storage node and the bulk silicon surrounding the trench acts as the plate electrode of the capacitor. Figure 2.8 shows the cross section of the SPT cell, where the access transistor is a PMOS transistor within an n-well in a p-type epi on p+ substrate. The p+ source region of the access transistor and the p+ polysilicon electrode inside the trench are connected through a conducting strip. The substrate is connected to ground and the n-well is biased at or above V_{DD}. A composite insulator of 10 nm equivalent thickness of SiO_2 is formed between the polysilicon filled in the trench and the bulk silicon substrate. A major component of the capacitor thus formed is primarily due to four trench sidewalls in the p+ substrate and the trench bottom, whereas the rest of the realized capacitance is formed in the n-well region as shown in Figure 2.8.

In spite of larger capacitance area (and value) in comparison to planar capacitor cells, soft error rate (SER) is much less, again due to the location of the storage node inside the trench, which is insulated by the composite dielectric layer.

In the conventional trench cell with diffused or inversion-layer storage nodes, it is required that capacitor resides completely within the lightly doped region of the array bulk, which means suffering from doping and

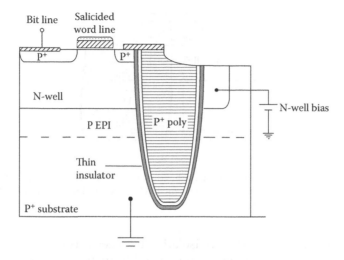

FIGURE 2.8
Schematic of the SPT cell cross section. (Redrawn from "A Substrate-Plate Trench-Capacitor (SPT) Memory Cell for Dynamic RAM's", N.C.C. Lu et al., IEEE. J.S.S Circuits, Vol. SC-21, pp. 627–633, Oct. 1986.)

geometry limitations. In SPT cell there is no such restriction, thus allowing use of CMOS epitaxial thickness and depth of the well according to optimum device design considerations. Therefore, array of PMOS in an n-well CMOS technology results in several advantages, including greatly reducible minimum distance between adjacent cells; greater cell packing density can be achieved. In addition, configuration of the array using PMOS devices also has a metallurgical junction between the substrate and the array that provides a barrier to minority carriers generated in the substrate. Thus the SPT cell is inherently a low leakage, low SER structure. However, the cell cannot easily employ a half-V_{DD} plate, as this approach loses some storage signal [26], and the insulator thickness could not be scaled down in proportion to the field stress when effective $t_{ox,eq}$ was less than 10 nm.

For the SPT cell, an important design effort is to minimize the access device sub-threshold leakage in the well without compromising performance. This has been achieved in a 65 ns 4 Mbit DRAM with a cell area of 10.5 μm^2 with 0.8 μm design rules [27]. In another realization using the SPT cell with its storage capacitor merged into the transistor isolation area, this area could be shrunk to 4 μm^2 at 0.5 μm design rule and 16 Mbit chips were made available.

The SPT cell contains a vertical parasitic p^+ (access transistor drain) n (n-well) p^+ (substrate) FET along the trench wall, gated by the node polysilicon as shown in Figure 2.9, which is not present in the other trench cells. As gate and source are connected together, this parasitic device should remain in the limit of weak inversion, making it safe. In addition, making depletion region narrower is also attempted by using moderately doped retrograde n-well implant, making it safer. Although parasitic, the FET has its

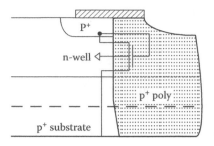

FIGURE 2.9
Parasitic FET device along the SPT cell trench sidewalls. (Redrawn from "A Substrate-Plate Trench-Capacitor (SPT) Memory Cell for Dynamic RAM's", N.C.C. Lu et al., IEEE. J.S.S Circuits, Vol. SC-21, pp. 627–633, Oct. 1986.)

usefulness as it provides effective isolation between access transistor source region and the substrate, which can be used to further reduce the cell size by butting the source region to the trench wall. The leakage is thus smaller, because the n-well can either reject or sink carriers to reduce the disturbance to the storage node [28].

If the SPT cell is not located within a well like first generation and other inverted trench cells, some extra design efforts would be needed. It is normal and essential in DRAMs to apply an array back-gate bias to reduce minority-carrier injection into the array [29]; however, if this back-gate bias is generated on chip, its application would generate considerable noise [30]. If the substrate is used both as the array bulk and the capacitor plate, this noise would be coupled directly to the storage node, and noise margin would be considerably reduced [26,30], whereas in the SPT cell located within a well the p⁺ substrate is connected to the low-noise ground and the n-well serves as the bulk array to which the generated back gate bias is applied.

In some advanced cell designs of this type, both the plate and storage electrode are fabricated inside the trench opening, which allows both electrodes to be completely oxide isolated. This provides the cell with punch-through protection even if the cells are placed very close to each other and reduces the soft-error rates relative to other inverted trench cells. A few of these are described below in brief.

2.3.2.2 Dielectrically Encapsulated Trench Cell

Capacitor structure of a dielectrically encapsulated trench (DIET) cell realized for a 16 Mbit DRAM is shown in Figure 2.10. The storage capacitor is formed with a poly-to-poly structure in the trench, and an insulator layer surrounds it to provide isolation between cells. As a result, it claims to have potential for achieving extremely small distance between cells with the advancement of fabrication process. The storage electrode formed in a trench cell is surrounded by a cell plate and storage electrode adjoins only a small

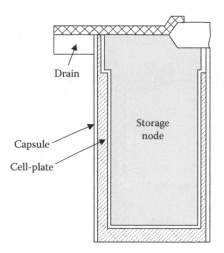

FIGURE 2.10
Schematic cross section of a DIET cell capacitor. (Redrawn from "Dielectrically Encapsulated Trench Capacitor Cell", M. Taguchi et al., *IEDM*, pp. 136–138, 1986.)

depletion layer around the source/drain region of MOSFET. Such a structure leads to immunity from punch-through and drastically lowers the soft error rate compared to that for conventional trench capacitor cell. Doped impurities within the cell plate are prevented from diffusing out during the heat cycle in fabrication because the dielectric layer covers the cell plate. Without this, a depletion layer shall be present in series with the storage capacitor and its value shall be reduced. The cell plate bias voltage is fed from the substrate through a hole opened at the bottom of the dielectric capsule. This structure allows the cell to be made small, but there is a drawback that it is difficult to work with half-V_{DD} biasing of the cell plate. Realization of a 16 Mbit DIET cell required a capacitor size of 0.87 μm x 1.6 μm with a depth of 4.0 μm that provided a capacitance of 28 fF with 25 nm thick oxide [31].

2.3.2.3 Half-V_{cc} Sheath-Plate Capacitor Cell

The half-V_{cc} sheath-plate capacitor (HSPC) cell employs an inner storage node and half-V_{cc} plate operation through the use of buried plate-wiring which controlled capacitor plate voltage. The inner storage node has the advantage of low inter-trench current leakage and good immunity against injection of minority carriers due to cosmic irradiation. The half-V_{cc} plate operation allows the capacitance to be doubled and a negative substrate voltage to be used. Due to its highly self-aligned cell structure, it occupies only a small cell area of 4.2 μm² (1.3 μm × 3.2 μm) for a 16 Mbit DRAM, in 0.6 μm technology, in which Si_3N_4-SiO_2 composite film of 5.5 nm equivalent thickness was used to get 51 fF capacitance. The sheath plate is formed inside SiO_2 sheath and is connected to the buried plate-wiring at the bottom of the groove. The word

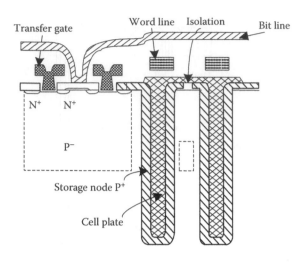

FIGURE 2.11
Schematic drawing of a DSP cell. (Redrawn from "Double Stacked Capacitor with self-aligned poly source/drain transistor (DSP) cell for megabit DRAM", K. Tsukamoto et al., Proc. IEDM, pp. 328–331, 1987.)

line is isolated from the storage node by the SiO_2 layer, which is fabricated by using a self-aligned oxidation technique [32].

2.3.2.4 Double Stacked Capacitor with Self-Aligned Poly Drain/Source Transistor Cell

A schematic diagram of a double stacked capacitor with self-aligned poly drain/source transistor (DSP) cell is shown in Figure 2.11. The storage capacitance, which is a combination of two capacitors, (1) between first polysilicon and substrate, and (2) between first and second polysilicon, which are stacked in a trench, results in a larger capacitance. The access transistor has a self-aligned poly source/drain, where the source/drain region is formed by a dopant diffusion into the substrate from the first polysilicon. Using unique self-aligned structure, 0.7 µm technology results in a cell of 5.92 µm². With a trench depth of 4 µm, a storage capacitance over 50 fF could be obtained [33].

2.4 Access Transistor Stacked above Trench Capacitor Cell

In trench cell designs, access transistor is placed beside the trench and occupies a significant area of the cell, similar to planar access transistor cell where the capacitor was occupying a large part of the cell area. A more efficient use of cell area is achieved in a key development with transistor being stacked above the trench capacitor to form a vertical cell. It was first tried in VMOS DRAM cell [9] but could not succeed in manufacturability due to technological/processing problems in realizing sharply edged v-grooves with good oxide integrity. However, advances in process technology made it possible to realize three-dimensional buried trench cell for 64 Mbit and higher densities. Examples of such cell are described now.

2.4.1 Trench Transistor with Self-Aligned Contact Cell

The trench transistor with self-aligned contact (TSAC) cell has been developed for 4 Mbit and beyond density level, using following technologies: (1) a shallow trench forming the channel of the access transistor, as formation of such a transistor channel is for the improvement of short and narrow channel effects, and (2) bit line contact is formed with self-alignment to the word line; hence, the name TSAC. In the structure, marginless contact design capability enables substantial reduction of the memory cell size. In fact, TSAC revived efforts toward the formation of trench transistor cell which were terminated after the experience of VMOS discussed earlier [34].

2.5 Trench Transistor Cell

The concept of storing signal charge on the poly was heavily attempted between 1984 and 1986. One of such configurations is the trench transistor cell (TTC) given by W.F. Richardson and others and soon used by A.H. Shah and others [35,36]. Both implemented TTCs were in 4 Mbit chips with cell areas of 10.6 μm^2 in 0.8 μm technology and 9 μm^2 in 1.0 μm technology, respectively. The TTC realizes a vertical transistor and capacitor in a deep trench. Earlier most of the cell designs focused on decreasing the planar area of the capacitor and/or optimizing the configuration of the capacitor to the access device. This occurred because well-researched and practiced planar technology gave excellent results when used in the fabrication processing of the devices and components. However, it became obvious to DRAM researchers that further scaling down of the cell area was not possible and change from planar to vertical device was unavoidable [37].

A schematic diagram of a trench transistor cell is shown in Figure 2.12. An 8 μm deep trench (1.3 μm x 1.5 μm) is etched in a 4.5 μm p-type epitaxial layer on a p^+ substrate. A vertical access transistor is made in the top 2 $\mu m s$ whereas the storage capacitor is below the transistor in the trench. The inner plate of the capacitor is an n^+ polysilicon plug, and the outer plate is the p^+ substrate. The source of the transistor is connected to the n^+ polysilicon plug

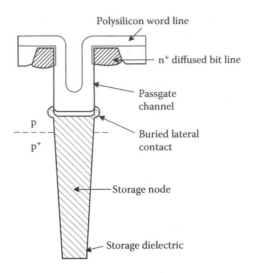

FIGURE 2.12

Trench transistor cell schematic cross-section. (Redrawn from "Advanced cell structures for dynamic RAMs", N.C.C. Lu, IEEE Circuits and Devices Magazine, pp. 27–36, Jan. 1989.)

of the capacitor by a buried lateral contact which is made by oxide under-cut etch and refilled with polysilicon [35]. The drain, gate, and source of the trench pass transistor are made of diffused-buried n^+ bit line, n^+ polysilicon word line, and a buried lateral contact, respectively. The gate oxide thickness and channel length are around 25 nm and 1.5 µm, respectively, while the transistor width is determined by the perimeter of the trench.

The early DRAMs employing TTC used the open bit line architecture for the array, however the signal-to-noise ratio deteriorated with higher level of integration. The double ended adaptive folded (DEAF) bit line architecture has been used to obtain a 4 Mbit DRAM, which is in the form of a truly cross-point cell array [36]. The DEAF architecture has all the advantages of the conventional folded bit line scheme, and at the same time realizes a very dense cross-point memory cell array [38].

Characteristics of the access transistor in the TTC cell as well as the properties of the charge storage element have been studied in detail. Banerjee and others [39,40] reported the characteristics of the vertical trench transistor like that of reference [35] which was characteristically different from a planar transistor. Simulation of the trench transistor shows good agreement (within 10%) in both the linear and the saturation regions, but the threshold of the transistor is controlled by deep boron implant, having a nonuniform Gaussian profile. It indicates good sub-threshold slope, which allows low sub-threshold leakage, and statistical data from a large number of transistors indicate good uniformity of device characteristics and parameters.

In the earlier versions of TTCs, there were certain limitations. The trench sidewalls got oval cross section after patterning and reactive ion etching

(RET), which resulted in different crystallographic orientation and hence gave different oxide thicknesses. The cell stored charge on the poly inside the trench, which was not located within a well and, therefore, could suffer substrate-plate bump effect if a substrate bias generator was used [40]. It was also observed that though thickness of thermal oxide dielectric was nonuniform around the oval trench perimeter, the oxide quality and leakage were comparable to planar oxide, and the capacitance was sensitive to the trench edge angle and depth [41]. Both the TSAC and TTC cells suffer high mask count due to using vertical devices in the array but having planar devices in peripheral circuits [36].

It is well known that thermally grown oxide becomes thinner both at top and bottom corners of trenches. It was extremely difficult to control precisely the growth of thin thermal oxide film without impurity redistribution using conventional furnace at high temperature. However, by using rapid thermal oxidation (RTO) process, highly reliable thin gate oxide can be grown reproducibly to realize the high performance megabit class DRAMs [42].

2.6 Buried Storage Electrode Cell

Buried storage electrode (BSE) cell is another example of an inverted trench cell in which storage electrode is inside a trench [43,44]. Basic structure of the BSE cell is shown in Figure 2.13(a), where a buried polysilicon electrode stores the signal charge. The storage node polysilicon refilled into the capacitor trench is connected to the access transistor electrode. The heavily doped substrate of a p/p^{++} epi wafer serves as the capacitor plate [43]. Most of the portion of the capacitor trench is deeply penetrated into the p^{++} substrate. Because of the use of heavily doped substrate, depletion layer at the capacitor surface is very thin, and hence inversion layer would not form. The cell is inherently free of punch-through between neighboring trenches and immune to soft errors. Due to its better immunity to soft errors, smaller storage capacitor is allowable. In a test cell in 0.8 µm design rules, an 8.8 µm² cell realized 35 fF capacitance. With 0.8 µm trench isolation between capacitor trenches show negligible interference between adjacent cells, which is a major advantage at 4-MBit level in comparison to the first generation trench cells [13]. BSE cells and internal constant voltage (V_{BLS}) convertor combined with half bit line voltage precharge, and high signal-to-noise ratio sensing, a 4 Mbit DRAM have been obtained [44]. Figure 2.13(b) shows a top layout view of BSE cells with folded aluminum bit line structure. The cell size is 2.3 µm × 4.6 µm with a 50 fF storage capacitance using 5 µm deep trenches. In 4 Mbit DRAM design using BSE cells the bit line driving high voltage must be limited to prevent the characteristic degradation of 1 µm gate cell transistors on account of hot carrier effects and to suppress voltage bounce of the cell

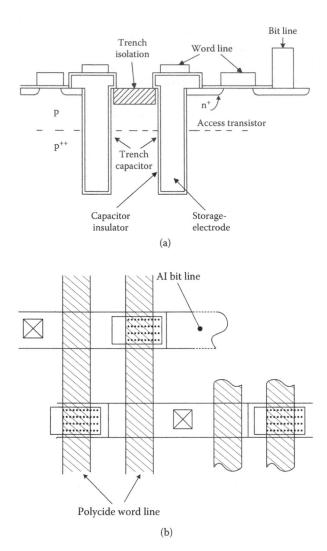

FIGURE 2.13
(a) Schematic cross-section of a BSE cell. (b) Layout of a BSE cell. (Both figures redrawn from"Buried Storage Electrode (BSE) Cell for Megabit DRAMs", M. Sakamoto. et al., Proc. IEDM, pp. 710–713, 1985.)

capacitor electrode substrate. A suitable internal voltage converter could easily solve the problem.

Because of the work-function difference between the p^{++} substrate and the n^+ poly used in BSE cell capacitor, a higher voltage is dropped across its insulator compared to p^+ poly in the SPT cell. The problem of voltage drop across the insulator is enhanced when a boosted word line is used or a substrate bias is used for the BSE cell. The constraint limits the scalability of the insulator thickness; otherwise the dielectric might break down. In addition, since along the trench sidewall in the cell there is a gated diode structure, leakage can become a serious concern, especially when the storage node voltage is high [45].

2.7 Buried Capacitor or Stacked Transistor Cell

Vertical access transistor has been used at 4–16 Mbit density DRAM level, like trench transistor cell [35], to reduce the cell area. However, the vertical transistor fabrication processes could not match the advantage of self-aligned planar technology, like better device uniformity, shallow junctions, shallow threshold implant adjustment, lightly doped drains, etc., which result in high performance. Alternatively a few cells have been designed in which capacitor is buried under the access device or the access transistor is stacked over the capacitor. Examples of stacked cell included a silicon-on-insulator (SOI) substrate-plate trench-capacitor (SS-SPT) cell [46] and a stacked switching transistor in SOI (SSS) cell [47]. The cells could be realized in a small area, as well as with very low SER and leakage on account of having SiO_2 as the substrate for the planar transistor; however, its quality depended on the technology advancements in SOI technology. Obtaining SOI films of fine quality for submicron devices was difficult in the 1980s. In addition, a floating-substrate transistor has different characteristics in comparison to the bulk planar transistor.

Development of a self-aligned epitaxy-over-trench (SEOT) technology in a submicron CMOS process has realized a three-dimensional buried trench (BT) cell having a potential for high-density DRAMs. The technology allowed a bulk horizontal access transistor formed on top of the trench transistor with self-aligned vertical via connection to the buried capacitor. Hence, the BT cell showed a high performance transistor and small size, with possibility of realizing DRAM of 64 Mbit densities and beyond [48].

Figure 2.14 shows the schematic view of the BT cell in which the trench storage capacitor is accessed through the horizontal transistor, and it is completely isolated from the heavily doped substrate, which serves as grounded plate. The bulk horizontal transistor is fabricated in an epitaxial layer over the BT capacitor. The self-aligned vertical via connection between access transistor and top of the buried capacitor is known as *neck*. A single-crystal p-type silicon layer is grown over the p^{++} substrate till it completely surrounds and reaches the top level of the trench capacitor. A self-aligned window is formed on the top of the trench and the oxide on top of the trench is etched; neck is formed on top of the exposed polysilicon in the trench, and then a second epi film is grown. A flat surface results after a certain

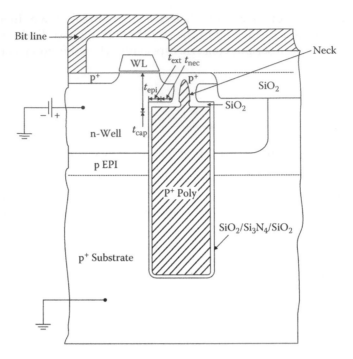

FIGURE 2.14
Schematic of a BT cell, showing possible parasitic backside leakage path as well. (Redrawn from "A Buried-Trench DRAM Cell Using a Self-Aligned Epitaxy over Trench Technology", N.C.C. Lu et al., IEDM Tech. Dig., pp. 588–590, 1988.)

minimum amount of epitaxial growth, followed by isolation, and a MOS transistor is fabricated.

Operation of the cell is similar to that of a conventional SPT cell [49]. However, the BT cell suffers from a minimal backside leakage path between the storage node and the bit line node, i.e., between the source and the drain of the access transistor gated by the node polysilicon inside the trench as shown in Figure 2.14. The possible leakage depends on the n-well doping level and its bias, oxide capacitor thickness t_{cap}, epitaxial film thickness t_{nec} and the trench extension from the source edge t_{ext}. Simulated results have shown that the backside leakage component can be made negligible [48].

2.8 Stacked Capacitor Cells

In planar MOS capacitor cells, signal charges were considerably reduced in the inversion and depletion layers due to the influence of the minority carriers generated in silicon substrate. Koyanagi tried to eliminate the inversion

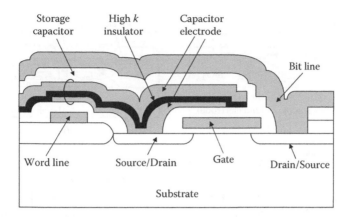

FIGURE 2.15
Basic structure of stacked capacitor cell (STC). ("The Stacked Capacitor DRAM Cell and Three Dimensional Memory", M. Koyanagi et al., IEEE SSCS News, pp. 37–41, Winter 2008.)

capacitance and the depletion capacitance by employing a different passive capacitance known as metal insulator metal structure as a storage capacitor in a three-dimensional (3-D) cell in 1978 and named it stacked capacitor cell (STC). Figure 2.15 shows the basic structure of a stacked capacitor cell where the storage capacitor is three dimensionally stacked on a switching transistor [50]. Self-aligned contact is used to connect the bottom electrode of the capacitor to the source/drain region, and to connect the bit line to source/drain of the switching transistor. By using the stacking, not only was the area occupied by the capacitor reduced, but the use of high dielectric material as capacitor insulator also became practically feasible. First Si_3N_4 was used instead of SiO_2 as capacitor insulator and polycrystalline silicon for the two electrodes to increase the storage capacitance and then Ta_2O_5 film, having dielectric constant five to six times that of SiO_2 was employed [51]. A discussion on MIS/MIM capacitors is included in Chapter 5.

Memory cells with three-dimensional capacitors such as stacked capacitor cells [52] were realized for obtaining sufficient storage capacitance at still higher memory density levels. Starting from the basic idea of the STC cell, further progress was made; cylindrical capacitor [53] and double stacked capacitor [54] are typical examples, and shall be studied later. Another significant advantage of stacked capacitor was that the signal charges stored in the MIS/MIM capacitor were not affected by the carriers generated in the substrate. This means that the structure is considerably less affected by soft errors.

The trendsetter stacked capacitor [52] had some limitations as well. It required the connection of bit-line to the access transistor's drain through the window in the plate electrode. If alignment tolerance limit between the plate and the bit line contact hole is taken into consideration, the storage node pattern design results in the decrease of storage capacitance. The limitation

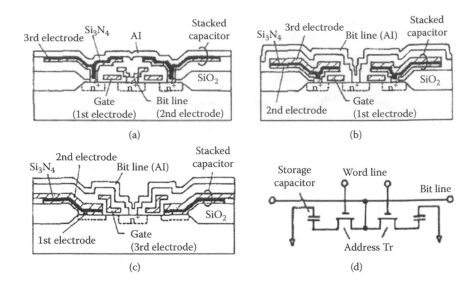

FIGURE 2.16

Cross-sectional configuration of three types of STC cells. (a) Top capacitor type, (b) intermediate capacitor type, (c) bottom capacitor type, and (d) its equivalent circuit. ("A 5-V Only 16-kbit Stacked-Capacitor MOS RAM", M. Koyangi et al., IEEE. Journal of Solid State Circuits, Vol. SC-15, No. 4, 1980.)

was overcome by forming stacked capacitor structure by a number of studies in the 1980s at the DRAM level of 4 Kbit to 64 Kbit. Koyanagi and others [55] gave three similar stacked capacitor cell structures with slight difference in the position of the stacked capacitor as shown in Figure 2.16. In Figure 2.16(a) the storage capacitor had a poly-Si,-Si_3N_4-Al layer stacked on the access transistor, bit line and field oxide; in Figure 2.16(b), the storage capacitor as a poly-Si-Si_3N_4-poly-Si structure is stacked on the access transistor and field oxide, and in Figure 2.16(c) capacitor is stacked on the field oxide only. Stacked capacitor of Figure 2.16(c) was used for the realization of 16 Kbit memory because of better compatibility with the conventional fabrication process. Another stacked capacitor cell was realized by forming storage capacitor on bit line using a diagonal active stacked-capacitor with a high-packed storage node (DASH) for use in a 16 Mbit DRAM [56]. In this scheme active area is set diagonal to both the word line and bit line. Figure 2.17 shows some process steps of the DASH. After word line formation, the polycide bit line is formed directly on the word line. When SiO_2 sidewall is formed, the storage node contact area is automatically opened. No precise patterning is needed; the bit lines are covered by the storage node and the plate electrode. Storage capacitance of 35 fF/bit was obtained in a cell size of 3.4 μm^2 using 0.5 μm design rule with 5 nm thin composite dielectric film of SiO_2 and Si_3N_4. Similar attempt was made by using double layered storage node structure with slightly modified fabrication procedure of exchanging the order to open

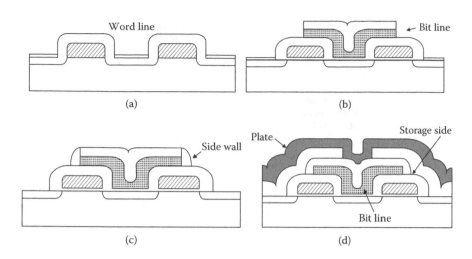

FIGURE 2.17
Process flow of the DASH. (a) Transfer gate formation, (b) bit line formation, (c) bit line isolation by sidewall SiO_2, and (d) capacitor formation by storage node and plate. ("A New Stacked Capacitor DRAM Cell Characterized by a Storage Capacitor on a Bit-Line Structure", S. Kimura et al., Proc. IEDM, pp. 596–599, 1988.)

contact window and to deposit poly-Si of the lower electrode to get sufficient cell capacitance in 16 Mbit DRAMs [57].

Realizing *capacitor over bit lines* (COB) changed the memory cell structure and the fabrication steps considerably. These capacitor-over-bit line (COB) cells were found to have immunity to inter-bit-line noise as the bit lines were shielded by the storage electrode. Structure has shown potential for realizing gigabit DRAM by introducing high dielectric constant capacitor film [58]. Earlier COB structures [59,60] required complex active area pattern to form the capacitor contact due to which DRAM in gigabit density could not be produced commercially. However, in a straight-line-trench isolation and trench-gate transistor (SLIT) simple straight line layout of the active area is formed [62]. In addition, isolation width and the gate length are also reduced to less than design rule using the sidewall-aided fine pattern etching (SAFE) technique.

With Pt/Ta as capacitor electrode, $(Ba_xSr_{1-x})TiO_3$ film was used for the first time to realize a 256 Mbit DRAM in COB structure [58]. A 100 nm thick dielectric film was equivalent to 8 Å of SiO_2. Combination of COB and high dielectric constant material in MIM capacitors then progressed fast. Stacked capacitor cell was employed in a 1 Mbit DRAM production for the first time by Fujitsu, with a cell area of 26.46 μm^2 with having 55 fF capacitance [62]. Increase in the cell capacitance from 35 fF to 55 fF was possible by utilizing the curvature and sidewall of second layer poly-Si as shown in Figure 2.18. First layer polycide forms the word line and the second layer poly-Si, which is extended over its own word line and the next word line forms the storage

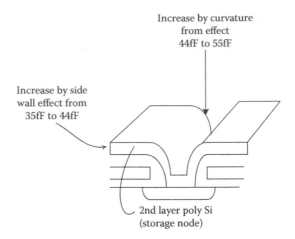

Increase by curvature
from effect
44fF to 55fF

Increase by side
wall effect from
35fF to 44fF

2nd layer poly Si
(storage node)

FIGURE 2.18
Increase of cell capacitance using three dimensions, a projected area of 10.0 um² realizing
a 35 fF capacitance. (Redrawn from "A 1Mb DRAM with 3-Dimensional Stacked Capacitor
Cells", Y. Takemae et al., IEEE. Int. Solid State Circuit Conference, pp. 250–251, 1985.)

node. Cell capacitor is formed between the second and third layer poly-Si,
which forms the cell plate.

Hitachi produced a 4 Mbit, 1-Mbit × 4 DRAM using a stacked capacitor
cell in 1987 using a twisted driveline sense amplifier (TDSA) scheme and the
multiphase drive circuit (MPD) on a 0.8 μm double-well CMOS technology
[63]. In the TDSA scheme two drive lines of the CMOS amplifiers cross each
other in the middle of the array, which reduces the total capacitance associated with each drive line to half of its value. The MPD circuit for PMOS
drives reduces peak charging current of bit lines and maintains at nearly
constant value. Combination of TDSA and MPD helped in obtaining the
access time of 65 ns.

Further scaling of stacked capacitor cell (STC) from 4 Mbit level to 64 Mbit
was hardly able to satisfy both the cell area and storage capacitance requirement because dielectric films with less than 3 nm SiO_2-equivalent thickness
was required at 64 Mbit DRAM. To overcome this limitation one of the methods applied was to use poly-Si film with rugged surface of electrode; that
is, morphology of the poly-Si storage electrode decides increase in effective
surface area/capacitance realized. A second alternative was to use more than
one layer of storage electrode and/or by using both sides of the electrode,
surrounded by thin dielectric and top capacitor plate; effective capacitance
was to be increased without increasing the projected area. Polysilicon storage electrodes, which are called membranes or fins, can expand horizontally,
vertically, or in both directions. It was found that 0.1 μm poly-Si film deposited at 570°C had 2.5 times larger rugged surface area than the film deposited
at conventional temperature of 620°C, giving larger capacitance [64].

2.8.1 Horizontal and Vertical Fin Structured Capacitor Cells

Examples of the horizontal fin structures include double stacked [54, 60], and spread stacked [65]. The storage node completely overlaps the storage contact in the double stacked horizontal fin structure; however, it also needed two additional masking steps. Since 16 Mbit DRAMs were not feasible with single fin stacked capacitors, a curved storage electrode was used by Takemae and others [62]. It is observed that sidewall storage effect on capacitance realized is more significant, but thick storage polysilicon makes further processing difficult. To overcome this problem a new cell architecture in which bit lines are formed before storage electrode is realized [60]. A comparison of the cell architecture with a conventional cell for comparison is shown in Figure 2.19. It is observed that large capacitor in this fin type capacitor cell is obtained

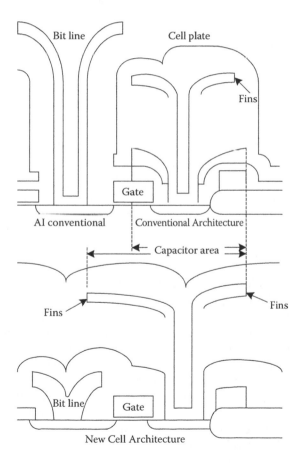

FIGURE 2.19

Comparison of conventional and new cell architecture in which bit lines are formed before storage electrode. (Redrawn from "3-Dimensional Stacked Capacitor Cell For 16M And 64M DRAMs", T. Ema et al., IEDM Tech. Dig., pp. 592–595, 1988.)

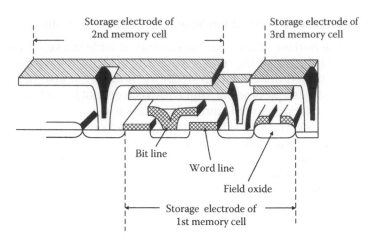

FIGURE 2.20
Schematic cross-section of the spread stacked capacitor cell. (Redrawn from "A Spread Stacked Capacitor (SSC) cell for 64 Mb DRAMs" S. Inone et al., in Proc., IEDM, p. 31, 1989.)

TABLE 2.1

Comparison of Storage Capacitance Realized in Conventional STC Cell and SSC Cell

Cell Size (μm²)		1.0	1.5	2.0	2.5	3.0
	Oxide Thickness (nm)	Realized Capacitance (fF)				
Conventional cell	10	5	6.5	8	9.5	12
	4	9	13	17	21	24
SSC cell	10	13	17	21	25	27
	4	24	32	40	51	60

as bit line contact hole is not needed. In addition, bit line coupling can be reduced because the cell plate acts as a shield. Another significant example of horizontal fin structure is spread stacked capacitor (SSC) cell with its schematic cross section shown in Figure 2.20 [65]. As shown in Figure 2.20, storage electrode of a memory cell is expanded to the neighboring memory cell area, so each storage electrode occupies two memory cell areas, enhancing the storage capacitance to 1.8 times that of a conventional cell. Table 2.1 shows the advantage of the storage capacitance of the SSC cell over the conventional STC cell. For 64 Mbit DRAMs, cell size must be less than 1.3 μm² with minimum feature size of 0.3 μm and a 4 nm SiO_2-equivalent dielectric film. The SSC cell can achieve satisfactory value of 27 fF whereas conventional STC cell capacitance is less than 15 fF.

Figure 2.21 shows the schematic cross-section view of a dual cell plate (DCP) cell [66] used in a 64 Mbit DRAM. To increase the storage capacitance, dual cell plates completely surround the whole surface of the storage poly-Si.

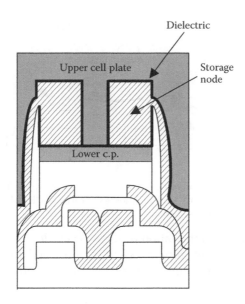

FIGURE 2.21
Schematic cross-sectional view of a DCP CELL. (Redrawn from "A Spread Stacked Capacitor (SSC) cell for 64 Mb DRAMs" S. Inone et al., in Proc., IEDM, p. 31, 1989.)

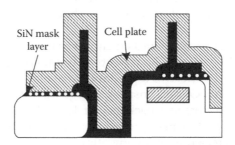

FIGURE 2.22
Simplified view of a cylindrical cell. (Adapted from "Novel Stacked Cell for 64 Mb DRAM", W. Wakamiyia et al., Symp. VLSI Techn., Dig. Tech. Papers, p. 31, 1989.)

Using 0.5 μm lithography, a fabricated cell needs 2.88 μm^2 chip area and realizes 50 fF capacitance with 5 nm thin dielectric and 600 nm thick storage poly-Si. The DCP cell has a potential to realize more than 25 fF in 1.3 μm^2 cell area when 0.3 μm design rule is used. The DCP cell may be considered as one of the variations of the fin-type stacked cell.

Vertical fin structure cells at initial phase of development are now discussed briefly. In 1989, Wakamiya and others [53] gave the cylindrical storage node cell. Figure 2.22 shows cylindrical storage node cell in simplified form used by Mitsubishi [67]. About 0.5 μm high cylinder realizes 25 fF capacitance using 1.5 μm^2 cell area, with a condition that the shielded bit line electrode is combined with the cylindrical capacitor. The cell is realized in fully

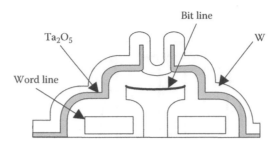

FIGURE 2.23
Cylindrical storage node using polymide reverse pattern and poly-Si CVD. (Adapted from "A 1.28 µm2 Bit-Line Shielded Memory Cell Technology for 64 Mb DRAMs", Y. Kawamoto et al., VLSI Symp. Tech. Digest, pp. 13–14, 1990.)

self-aligned manner and fabrication process is simple. Another advantage of the cell is that difference of steps between memory cell area and peripherals circuits are reduced, improving interconnections. Another bit line shielded stacked capacitor cell with storage capacitor formed over bit line cylindrical storage node structure is realized by Kawamoto and others in 1990 [59]. The cylindrical storage node is formed using polymide sacrificial layer and poly-Si using CVD on it. In addition, planarized bit line wiring reduces the bit line resistance making high-speed operation possible. The planarized bit line along with cylindrical storage node is shown in cross-section form in Figure 2.23.

A self-aligned crown-shaped stacked capacitor cell has been developed by T. Kaga and others [68], using 0.3 µm electron-beam lithography having a cell area of 1.28 µm². A large effective capacitance area of 3.7 µm² and a Ta_2O_5 film equivalent to a 2.8 nm SiO_2 film ensures a large capacitance charge of 33 fC, which is equivalent to storage capacitance of 44 fF at 1.5 V (Vcc/2) operation. Layout of the CROWN cell is an improved version of the DASH cell [69]. Larger plane area could be created without any alignment problems because storage capacitors are formed on the openings among the data lines and the word lined after these lines are formed. Figure 2.24 shows the cross section of the CROWN cell.

A double cylindrical storage spread-vertical capacitor (SVC) cell has been given for 64 Mbit and 256 Mbit DRAM [70,71]. As shown in Figure 2.25, an exploded view of SVC illustrates that storage capacitor spreads to the neighboring cell area. Each storage electrode A′ and B′ of the two adjacent memory cells (A, B) is connected through a poly-Si via to the memory cells A and B, respectively. The storage electrodes A′ and B′ form the inner wall loop and the outer wall loop, respectively. Structure of SVC provides a simple fabrication and 24 fF capacitance for 256 Mbit DRAM with only 0.4 µm capacitor height. Since storage capacitance is proportional to the peripheral length of the foot prints of the capacitor including two adjacent memory cells, it becomes three times that of the conventional stacked capacitor (STC).

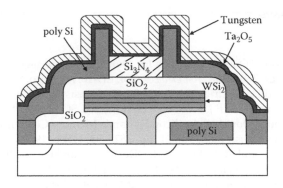

FIGURE 2.24
Schematic view of a CROWN cell. (Redrawn from "Crown-Shaped Stacked-Capacitor Cell for 1.5V Operation 64-Mb DRAMs", T. Kaga et al., IEEE Trans. Electron. Devices, Vol. 38, No. 2, pp. 255–260, 1991.)

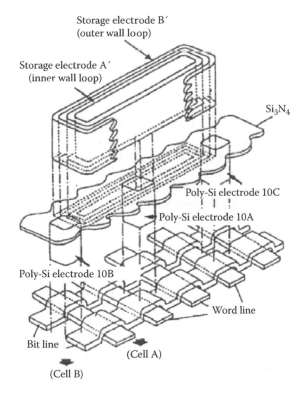

FIGURE 2.25
An exploded view of the spread vertical capacitor cell. ("Spreaded Vertical Capacitor Cell (SVC) for beyond 64 MBit DRAMs", N. Matsuo et al., IEDM, pp. 473–476, 1991.)

References

1. V. Leo Rideout, "One-Device Cells for Dynamic Random-Access Memories: A Tutorial," *IEEE Trans. Electron Devices*, Vol. ED-26, no. 6, pp. 839–852, June 1979.
2. K.U. Stein and H. Friedrick, "A 1-mil² Single Transistor Memory Cell in Silicon-Gate Technology," *IEEE J. Solid State Circuits*, Vol. SC-8, pp. 319–323, Oct. 1973.
3. V. Leo Rideout, "Double Polysilicon Dynamic Memory Cell with Polysilicon Bit Line," *IBM Tech. Disc. Bull.*, Vol. 21, pp. 3828–3831, Feb. 1979.
4. Y.A. El-Mansy and R.A. Burghard, "Design Parameters of the Hi-C DRAM Cell," *IEEE. J.S.S. Circuits*, Vol. SC-17, pp. 951–956, 1982.
5. C.G. Sodini and T.I. Kamins, "Enhanced Capacitor for One Transistor Memory Cell," *IEEE Trans. Electron Devi.*, Vol. ED-23, p. 1187, 1976.
6. A. Tasch et al., "The Hi-C RAM Cell Concept," *Trans. Electron Devi.*, Vol. ED-25, p. 33, 1978.
7. R.H. Dennard et al., "Design of Ion Implanted MOSFETs with Very Small Physical Dimensions," *IEEE J. Solid State Circuits*, Vol. SC-9, pp. 256–268, Oct. 1974.
8. F.B. Jenne, U.S. Patent 4,003,036, 1977.
9. J.J. Barnes, S.N. Shabde, and F.B. Jenne, "The Buried Source VMOS Dynamic RAM Device," *Electron Devices Meeting*, (10.1109/IEDM.1977.189135) Vol. 23, pp. 272–275, Dec. 5–7, 1977.
10. P. Chaterji et al., "Trench and Compact Structures for DRAM's," *Proc. IEDM*, pp. 128–131, 1986.
11. H. Sunami et al., "A Corrugated Capacitor Cell (CCC) for Megabit Dynamic MOS Memories," *IEEE IEDM Technical Digest*, pp. 806–808, Dec. 1982.
12. T. Mano et al., "Submicron VLSI Memory Circuits," *IEEE ISSCC Tech. Digest*, pp. 234–235, Feb. 1983.
13. H. Sunami et al., "Scaling Consideration and Dielectric Breakdown Improvement of Corrugated Capacitor Cell (CCC) for Future DRAM," *IEDM Dig. Tech. Papers*, pp. 232–235, 1984.
14. T. Furuyama et al., "An Experimental 4Mb CMOS DRAM," *ISSCC Dig. Tech. Papers*, pp. 272–273, 1986; also *IEEE J. Solid State Circuits*, Vol. 21, pp. 605–611, Oct. 1986.
15. T. Sumi et al., "A 60 ns 4Mb DRAM in a 300 mil DIP," *ISSCC Dig. Tech. Papers*, pp. 282–283, 1987.
16. M. Wada, K. Heida, and S. Watanabe, "A Folded Capacitor Cell (FCC) for Future Megabit DRAMs," *Proc. of IEDM*, pp. 244–247, 1984.
17. T. Furuyama and J. Frey, "A Vertical Capacitor Cell for ULSI DRAMs," *Symp. on VLSI Tech. Dig. Tech. Papers*, pp. 16–17, 1984.
18. S. Nakajima et al., "An Isolation Merged Vertical Capacitor Cell for Large Capacity DRAMs," *IEDM Dig. Tech. Papers*, pp. 240–243, 1984.
19. T. Mano et al., "Circuit Technologies for 16 Mb DRAMs," *ISSCC Dig. Tech. Papers*, pp. 22–23, 1987.
20. H. Kotani et al., "A 4 Mb DRAM Design Including 16-Bit Concurrent ECC," *Symp. on VLSI Circuits Dig. Tech. Papers*, pp. 87–88, 1987.
21. K. Nakamura et al., "Buried Isolation Capacitor (BIC) Cell for Megabit MOS Dynamic RAM," *IEEE IEDM Tech. Digest*, pp. 236–239, Dec. 1984.

22. M. Nagamoto et al., "A High Density 4M DRAM Process Using Folded Bitline Adaptive Sidewall Isolated Capacitor (FASIC) Cell," IEDM Tech. Dig., pp. 144–147, 1986.
23. K. Mashiko, "A 90 ns 4Mb DRAM in a 300 mil DIP," *IEEE International Solid-State Circuits Conference*, pp. 312–315, 1987.
24. P. Bakeman et al., "A High Performance 16-Mb DRAM Technology," *Symp. on VLSI Technology*, pp. 11–12, 1990.
25. N.C.C. Lu et al., "A Substrate-Plate Trench-Capacitor (SPT) Memory Cell for Dynamic RAM's," *IEEE J.S.S Circuits*, Vol. SC-21, pp. 627–633, Oct. 1986.
26. N.C.C. Lu, H.H. Chao, and W. Hwang, "Plate Noise Analysis of an On-Chip Generated Half-V_{DD} Biased Plate PMOS Cell in CMOS DRAMs," *IEEE J. Solid-state Circuits*, Vol. SC-20, no. 6, pp. 1272–1276, 1985.
27. R. Parent et al., "A 4 Mb DRAM with Double-Buffer Static Column Architecture," *ISSCC Dig. Tech. Papers*, pp. 14–15, 1987.
28. W. Noble et al., "Parasitic Leakage in DRAM Trench Storage Capacitor Vertical Gated Diode," *IEDM Dig. Tech. Papers*, pp. 340–343, 1987.
29. H. Masuda et al., "A 5V Only 64K Dynamic RAM Based on High S/N Design," *IEEE J. Solid State Circuits*, Vol. 15, pp. 846–854, Oct. 1980.
30. M. Takada et al., "A 4 Mb DRAM with Half Internal Voltage Bitline Precharge," *ISSCC Dig. Tech. Papers*, pp. 270–271, 1986.
31. M. Taguchi et al., "Dielectrically Encapsulated Trench Capacitor Cell," *IEDM*, pp. 136–138, 1986.
32. T. Kaga et al. "A 4.2 μm^2 Half-V_{CC} Sheath Plate Capacitor DRAM Cell with Self-Aligned Buried Plate-Wiring," *IEDM*, pp. 332–335, 1987.
33. K. Tsukamoto et al., "Double Stacked Capacitor with Self-Aligned Poly Source/ Drain Transistor (DSP) Cell for Megabit DRAM," *Proc. IEDM*, pp. 328–331, 1987.
34. M. Yanagisawa, K. Nakamurs, and M. Kikuchi, "Trench Transistor Cell with Self-Aligned Contact (TSAC) For Megabit MOS DRAM," *IEDM Dig. Tech. Papers*, pp. 132–135, 1986.
35. W.F. Richardson et al., "A Trench Transistor Cross Point DRAM Cell," *Proc. IEDM*, pp. 714–717, 1985.
36. A.H. Shah et al., "A 4 Mbit DRAM with Trench Transistor Cell," *IEEE J. Solid-State Circuits*, Vol. 20, pp. 618–626, Oct. 1986.
37. N.C.C. Lu, "Advanced Cell Structures for Dynamic RAMs," *IEEE Circuits and Devices Magazine*, pp. 27–36, Jan. 1989.
38. A.H. Shah et al., "A 4 Mb DRAM with Cross-Point Trench Transistor Cell," *IEEE Int. Solid-State Circuits Conf.*, pp. 267–268, 1986.
39. S.K. Banerjee et al., "Characterization of Trench Transistors for 3-D Memories," *VLSI Tech. Symp.* pp. 79–80, May, 1986.
40. S. Banerjee, et al., "Leakage Mechanism in the Trench Transistor DRAM cell," *Trans. Electr. Dev.*, Vol. ED-35, pp. 108–116, 1988.
41. K. Rao et al., "Trench Capacitor Design Issues in VLSI DRAM Cells," *IEDM Dig. Tech. Papers*, pp. 140–143, 1986.
42. K. Yoneda et al., "Trench Capacitor Using RTO for Megabit DRAMs," *VLSI Techn. Dig. Tech. Papers*, pp. 95–96, 1987.
43. M. Sakamoto et al., "Buried Storage Electrode (BSE) Cell for Megabit DRAMs," *Proc. IEDM*, pp. 710–713, 1985.
44. M. Takada et al., "A 4 Mb DRAM with Half Internal Voltage Bitline Precharge," *IEEE Int. S.S.C. Conf.*, pp. 270–271, 1986.

45. S. Banerjee et al., "A Band-to-Band Tunneling Effect in the Trench Transistor Cell," *Symp. VLSI Tech., Dig. Tech. Papers*, pp. 97–98, 1987.
46. N.C.C. Lu, "Three Dimensional Dynamic RAM Cell," U.S. Patent 620,667, 1984.
47. M. Ohkura et al., "A Three Dimensional DRAM Cell of Stacked Switching Transistor in SOI (SSS)," *IEDM Dig. Tech. Papers*, pp. 718–721, 1985.
48. N.C.C. Lu et al., "A Buried-Trench DRAM Cell Using a Self-Aligned Epitaxy Over Trench Technology," IEDM Tech. Dig., pp. 588–590, 1988.
49. N.C.C. Lu et al., "The SPT Cell—A New Substrate-Plate Trench Cell for DRAMs," *IEDM Dig. Tech. Papers*, pp. 771–772, 1985.
50. M. Koyanagi et al., "The Stacked Capacitor DRAM Cell and Three Dimensional Memory," *IEEE SSCS News*, pp. 37–41, Winter 2008.
51. M. Koyanagi et al., "Novel High Density, Stacked Capacitor MOS RAM," *Extended Abs. 10th Conf. Solid State Devices*, 1978.
52. M. Koyanagi et al., "Novel High Density, Stacked Capacitor MOS RAM," IEDM Tech. Dig., pp. 348–351, Dec. 1978.
53. W. Wakamiyia et al., "Novel Stacked Cell for 64 Mb DRAM," *Symp. VLSI Techno., Dig. Tech. Papers*, p. 31, 1989.
54. T. Kisu et al., "Novel Storage Capacitance Enlargement Structure Using a Double Stacked Storage Node in STC DRAM Cell," *SSDM Ext. Abs.*, p. 581, 1988.
55. M. Koyangi et al., "A 5-V Only 16-kbit Stacked-Capacitor MOS RAM," *IEEE Journal of Solid State Circuits*, Vol. SC-15, pp. 661–666, 1980.
56. S. Kimura et al., "A New Stacked Capacitor DRAM Cell Characterized by a Storage Capacitor on a Bit-Line Structure," *Proc. IEDM*, pp. 596–599, 1988.
57. H. Watanable, K. Kurosawa, and S. Sawada, "Stacked Capacitor Cell for High-Density Dynamic RAMs," *Proc. IEDM*, pp. 600–603, 1988.
58. K. Koyama et al., "A Stacked Capacitor with (Ba_xSr_{1-x}) TiO_3 for 256M DRAM," *IEDM. Tech. Dig.*, pp. 823, 1991.
59. Y. Kawamoto et al., "A 1.28 μm^2 Bit-Line Shielded Memory Cell Technology for 64 Mb DRAMs," *VLSI Symp. Tech. Digest*, pp. 13–14, 1990.
60. T. Ema et al., "3-Dimensional Stacked Capacitor Cell for 16M and 64M DRAMs," IEDM Tech. Dig., pp. 592–595, 1988.
61. M. Sakao et al., "A Straight-Line-Trench Isolation and Trench-Gate Transistor (SLIT) Cell for Giga-bit DRAMs," *IEEE Symp. VLSI Tech.*, 1993.
62. Y. Takemae et al., "A 1Mb DRAM with 3-Dimensional Stacked Capacitor Cells," *IEEE Int. Solid State Circuit Conference*, pp. 250–251, 1985.
63. K. Kimura et al., "A 65-ns 4 Mbit CMOS DRAM with a Twisted Driveline Sense Amplifier," *IEEE J. of Solid State Circuits*, Vol. SC-22, no. 5, Oct. 1987.
64. M. Yoshimura et al., "Rugged Surface Poly-Si Electrode and Low Temperature Deposited Si_3N_4 for 64 MBit and Beyond STC DRAM Cell," *Proc. IEDM*, pp. 659–662, 1990.
65. S. Inone et al., "A Spread Stacked Capacitor (SSC) Cell for 64 Mb DRAMs," *Proc. IEDM*, p. 31, 1989.
66. H. Arima et al., "A Novel Stacked Capacitor Cell with Dual Cell Plate for 64 Mb DRAMs," *Proc. IEDM*, pp. 651–652, 1990.
67. T. Matsukawa and T. Nakano, "Stacked Capacitor Cells for 64 Mb DRAM," *VLSI Tech. Systems and Appl.*, pp. 264–267, 1991.
68. T. Kaga et al., "Crown-Shaped Stacked-Capacitor Cell for 1.5V Operation 64-Mb DRAMs," *IEEE Trans. Electron. Devices*, Vol. 38, no. 2, pp. 255–260, 1991.

69. S. Kimura et al., "A Diagonal Active-Area Stacked Capacitor DRAM Cell with Storage Capacitor on Bitline," *IEEE Trans. Electron Devices*, Vol. 37, pp. 737–743, 1990.
70. N. Matsuo et al., "Spreaded-Vertical-Capacitor Cell (SVC) for beyond 64 MBit DRAMs," *IEDM*, pp. 473–476, 1991.
71. N. Matsuo et al., "Spread-Vertical-Capacitor Cell (SVC) for High-Density DRAM's," *IEEE Trans. Electron Devices*, Vol. 40, no. 4, pp. 750–753, 1993.

3

DRAM Technologies

3.1 Introduction

DRAM is a volatile memory since information stored is destroyed once the power supply to it stops even for a short duration. Information is also lost because of charge leakage from the storage capacitor unless it is periodically charged or refreshed. In spite of this complexity of operation and comparatively larger data access time with respect to SRAM, DRAMs have been produced in largest quantities because of their low cost per bit [1]. There are quite a few reasons for the low cost of DRAM memory. Prominent among the reasons is the changeover from 6/4 transistor SRAM to 3 transistor DRAM and eventually to single-transistor–single-capacitor DRAM. Consumed chip area per bit further continued to reduce on account of reduction in design rules. Obviously smaller cell size fabrication could become possible only with the advancements in technology at each generation. However, circuit innovations and advancements connected with DRAMs were no less important. At the earliest stages, it was the deployment of sense amplifiers that not only improved performance of the DRAM but also ensured its functionality. A brief mention of the same is included in Section 3.2 and in some detail in Chapter 8. Section 3.2.1 includes the importance of conversion to mixed NMOS/CMOS from the pure NMOS DRAMs, which also facilitated half-V_{DD} precharge of bit-line technique. It resulted in great boost to the DRAM deployment with reduced power consumption. Initially planar DRAM cell was simple in realization with a planar capacitor and the access transistor placed side by side. It continued up to nearly 4 Mbit DRAM density level but beyond that technology had to change to three-dimensional realization of capacitor with adoption of Si_3N_4 and Ta_2O_5 dielectrics. At the same time sensing techniques were improved for the detection of small signals during read operation. Some other features, like use of redundancy and folded bit line architecture, were also important developments at this stage. Main technological developments for 16 Mbit to 1 Gbit DRAM range have been discussed in Section 3.4. One major concern of obtaining ~25 fF capacitance in as small chip area as possible is also discussed. Though its advanced technology is described in Chapter 5, Sections 3.4.1 and 3.4.2 discuss new capacitances in metal insulator-metal (MIM) form especially in CROWN shape and those using higher-dielectric material like Ta_2O_5. Capacitor-over-bit-line (COB) type of DRAM cell, a special form of a cell that was only briefly mentioned with other kinds of DRAM cells in Chapter 2, is now taken up in some detail in Section 3.5.

3.2 DRAM Technology—Early Stage Development

In 1972, Intel manufactured a 1 Kbit DRAM chip commercially. This chip had 32 rows and columns each, and every intersection had a silicon-gate p-channel type cell having three MOS transistors. Refreshing was done every 2 ms and the chip worked on supply voltage of −17 V with +3 V biasing of the n-type substrate and the p-type source [2]. Soon an n-channel 4 Kbit DRAM was manufactured at 10 μm feature size, using a three-transistor cell with single-polysilicon, single aluminum metallization scheme.

Changeover from three transistors to one transistor cell was an extremely important development as it considerably improved the cell density. One transistor cell became almost universal for fabricating DRAMs up to the highest density level. Though 4 Kbit DRAM cell would become industry standard in 1973 [3], it was Dennard's patent of 1968 [4] which gave first memory cell using only one transistor. One of the major problems of detecting small cell signal was soon solved by the use of a differential amplifier in the middle of the bit line called a sense amplifier, which structurally divided the bit line in two halves [5]; it automatically restored/refreshed the cell as well. The first commercial product using one transistor DRAM cell with gated flip-flop sense amplifier was at 4 Kbit level produced by Texas Instruments using power supply of 12 V with −3 V substrate bias. A very important development reported in 1973 and which was converted to reality in 1975 was address multiplexing [6], again at the 4 Kbit level. A row address strobe ($\overline{RAS}$) first selects all the cells in a row, and activates sense amplifiers. Then column address strobe ($\overline{CAS}$) was applied to select a desired cell and the sensing signal is detected by the sense amplifier for reading or data is made available through I/O bus for writing. As RAS strobe selected all cells in a row, $\overline{RAS}$ only refresh became standard due to multiplexing. A *page* of data term was also coined for the cells of a row and later used for rapid *page mode* reading of memory.

A 16 Kbit DRAM created by Proebsting became the industry standard [7], and it was the 16 Kbit generation which used 12 V, +5 V, and −5 V supplies before the next generation of 64 Kbit DRAMs, which used only 5 V supply—an important development. Beyond 16 Kbit, development was mainly through scaling down cell dimensions. However, decrease in the size of the access transistor, and storage capacitor, and reduction in the power supply to 5 V, necessitated the development of better sensitivity sense amplifiers. A number of modifications/improvements were made on the basic gated flip-flop sense amplifier including use of *dummy cell* as a reference. One important improvement in the sense amplifier design came in the form of charge-transfer cross-coupled circuit, which could sense smaller signals and was also useful in minimizing the problem of threshold mismatch of driver transistors of the sense amplifier [8]. Another important development was the word line boost by at least a threshold voltage above the new standard 5 V supply to compensate the back gate effect of the source to substrate potential.

As mentioned, the sense amplifier was placed in the middle, which divided the bit line in equal halves in the open bit-line structure. In this structure noise generated on two sides of the sense amplifier was unbalanced. There were many more reasons for the noise generation, like imbalance of the sense amplifier, transistor threshold difference due to the process variations, noise generated in nearby peripheral circuits, and also a considerable noise due to capacitive coupling between the bit line and word line. At 4 to 64 Kbit DRAM level, *folded bit-line* architecture was used in which the bit line itself was folded back, forming a pair, and the sense amplifier was placed at the end. It eliminated the common mode noise generated on the column decoders. Folded bit line arrangement also had the advantage of simpler bit line shorting for precharge and provided simpler column access.

3.2.1 Mixed NMOS/CMOS DRAM

The 256 Kbit DRAM generation was the stage when early CMOS DRAM was manufactured, but NMOS DRAMs were equally important. Folded bit line did show improvements in noise reduction, but open bit line was also used in good amount. In the mixed NMOS/CMOS technology the access transistor and storage capacitor were normally fabricated in the NMOS technology and the peripheral circuits in CMOS technology. Apart from the well-known advantage of lower power consumption, CMOS design improved refresh characteristics in comparison to NMOS design mainly because of providing restriction on thermally generated charges from substrate defects close to the memory cells. Access time in CMOS design could also be reduced by using *static column decoding*, which was better suited than NMOS design.

In spite of significant advantages of CMOS technology, some teething trouble stalled the full development of CMOS processing till 1 Mbit generation. From the functionality point of view, latch-up could occur because of noises and parasitics, leading to malfunctioning or even breakdown. However, a major problem was the rising cost in terms of increase in the number of masks and processing steps. Biasing of substrate was found to be useful in minimizing latch-up by reducing minority carrier injection. At the same time prohibitively high power consumption per bit level for high density DRAMs forced minimal use of NMOS technology. When the memory capacity increased from 1 Mbit to 64 Mbit die size increased from 5×10 mm^2

to 12.7×20.4 mm^2. The word line and bit line lengths increased in higher density DRAMs which increased their resistance and capacitance and hence resulted in (1) increased delay associated with the charging and discharging of these lines, and (2) the sensing signal in the bit line reduced in direct proportion as it depended on the ratio of storage cell capacitance to bit line capacitance. It necessitated development of better sensing methods and better sense amplifier [9]. With DRAM density reaching 1 Mb, planar storage capacitors were employed and trench and stacked capacitor were becoming necessities for reducing the die size and retaining sufficient charge on the storage capacitor with even smaller size. Boosting the word line was a standard approach at +5 V supply voltage level. However, in this approach some of the nodes needed observation where greater voltage level was likely to pose problems. Although this was not a major issue with feature size used at 1 Mbit level, it came into focus at latter stage with further reduced channel length. A pumped supply level set by feedback control was employed for the word line driver as a better solution in place of double bootstrap voltages [10]; the approach was successful at 4 Mbit level as well. However, big boost to CMOS came in the form of half-V_{DD} bit line sensing with n-well CMOS DRAMs in 1984 [11] giving a number of advantages. More details were given in Chapter 2.

Another important feature of CMOS DRAM was the introduction of redundancy at 256 Kbit level, that is, use of additional rows and columns available for replacing defective ones for improving the memory yield through error correction circuits. The system came to be universally adopted with modifications and improvements and shall be discussed a little more in Chapter 8.

3.3 Two-Dimensional DRAM Cell

From 16 Kbit to 1 Mbit density level mostly a double-poly planar cell was used. However, for higher-density DRAMs, area occupied by this two-dimensional capacitor crossed a limit where the simple structure was not affordable and three-dimensional capacitors like trench [12] and stacked capacitors [13] were used. A two- or three-dimensional sensing signal can be increased by reducing noise, bit line capacitance and storage capacitor dielectric thickness, and by increasing capacitor area, its dielectric strength and storage signal. In the trench capacitors, which were mostly used at the 4 Mbit level and beyond, capacitor area was increased by increasing the trench depth and the capacitor dielectric formed on the trench wall was made as thin as practically feasible. Thermally grown silicon dioxide and CVD silicon nitride composite dielectric films were used for capacitance enhancement and prevention of leakage due to oxide breakdown. The main problem with the trench capacitor is the difficulty in fabricating a deeper trench for further capacitance enhancement. Stacked capacitor was also used at the 4 Mbit level, but it was preferable over trench capacitor at higher density level because of its easier fabrication and being less prone to noise and soft errors. Composite SiO_2/Si_3N_4 films were better than only SiO_2, but beyond 16 Mbit, even the composite layer was not able to provide enough capacitance and practical limitations forced a search for other dielectrics. At 64 Mbit CROWN-type memory cell was produced using Ta_2O_5 film and at 256 Mbit density Ta_2O_5 as well as $Ba_{0.5}Sr_{0.5}TiO_3$ [14] films were successfully used by the manufacturers.

Some of the features which helped consolidating the manufacture of 1 Mbit CMOS DRAM were better sense amplifiers, use of address transition detection and redundancy, and suppression of noise through the folded bit line architecture, etc. At further increased DRAM density, storage capacitance was realized using trench or stacked capacitors but the DRAMs started to become slow on account of increased parasitic resistance and capacitance of almost unchanged number of cells connected to ever-thinning word and bit lines. Increased charging–discharging currents created reliability problems due to larger voltage drop in the word and bit lines. Apart from the established mentioned features of 1 Mbit density level new design techniques and changes in architecture were required for 4 Mbit and higher density level.

With continued reduction in feature size, smaller channel length of transistor made them susceptible to hot electrons at the then standard 5 V power supply. One of the solutions was to reduce voltage levels for the DRAM cell array using on-chip voltage converters. One such example at the early stage is the 16 Mbit DRAM given by Hitachi which used 3.3 V internal voltage for 0.6 µm MOS channel transistors [15], while peripheral circuits used +5 V supply with transistor channel length at 0.9 µm. However, reduced voltage required further improvement in bit line sensing and more reduction in noise, especially the bit-line coupling noise.

3.4 16 Mbit–256 Mbit, 1 Gbit DRAM Development

For realizing 64 Mbit DRAM, cell area must be less than 1.5 µm². Because of the smallness of the cell area and intercell distance, trench capacitors have a tendency to suffer from leakage of charge, and lithographic limitation did not allow the fabrication of stacked capacitors like fin structure or cylindrical structure unless some relatively complex processing steps were taken. A comparatively planar topography is used in dual cell plate (DCP) stacked capacitor cells to overcome the lithographic limitations [16]. Planar area of the storage node is maximized by using data line shielded structure [17]. The storage node comprises two cubical parts of thick polysilicon and one cylindrical part of thin polysilicon, and the whole top and sidewalls of storage node are enclosed by top cell plates and the bottom side is covered by the lower cell plate to maximize the storage capacity as shown in Figure 2.21. For a cell having an area of 2.88 μm^2, realized capacitance was about 50 fF with 5 nm thick dielectric and 600 nm thick storage polysilicon. In a 64 Mbit DRAM, for cell area to be less than 1.3 μm^2, 0.3 μm technology was to be used; calculated capacitance became 25 fF with 5 nm (oxide-nitride-oxide) ONO dielectrics and 0.5 μm thick storage polysilicon. It is important to note that at 64 Mbit level or beyond, reliability of thin gate oxide became critical as the operating word line voltage was not scaled down with every DRAM generation, resulting in stress field across the gate oxide. Hence, time-dependent dielectric breakdown (TDDB), which is affected by the thinness of the gate oxide along with operating temperature and supply voltage, needed to be in check. It was observed that an ONO TDDB characteristic of the DCP cell is almost the same as that of a conventional stacked capacitor cell. Because of the inherent variations in oxide thickness during fabrication process and the operating temperature, degradations affect TDDB as well. Toshiba has given word line architecture for a 64 Mbit DRAM which could compensate for such degradations [18].

The preceding paragraph showed example of a 64 Mbit DRAM which emphasized achieving sufficient storage capacitance in a small projected chip area. At a working voltage of 3.3 V, and the kind of noise generation in early 1990s structures, 45 fF capacitor was essential for a 64 Mbit level DRAM so as to get satisfactory signal for sense amplifier and good soft-error immunity. Hence, a three-dimensional capacitor became essential which required enhanced photolithographic techniques, in addition to other improvements in fabrication processes. Getting higher value of cell capacitance also suggested the use of higher dielectric constant insulators like tantalum pentoxide (Ta_2O_5). It was not only the storage capacitor, but fabrication of reliable access transistor at around 0.35 μm technology with thin gate oxide and shallow junction depth at low thermal budget, also became a challenge.

3.4.1 Crown-Shaped Stacked Capacitor Cell Technology

An experimental 64 Mbit DRAM was fabricated in 0.3 μm technology, which was operated at a low voltage of 1.5 V and used a newly developed crown-shaped stacked capacitor [19]. In the CROWN cell, shown in Figure 2.24, a crown-shaped storage electrode combined with a smaller self-aligned cell structure produced a cell capacitance of 40 fF in a chip area of 1.28 μm^2. A CROWN cell with a single wall crown-shaped electrode using a 3 nm equivalent CVD-Ta_2O_5 film with tungsten plate realized more than 40 fF capacitance, which was extendible to 70 fF for a double-wall CROWN structure. In the newly developed fabrication process for the CROWN cell, a planar WSi_2 surface is obtained using etch-back method on polycrystalline surface, as shown in Figure 3.1, along with a nonplanarized structure. Planarization of

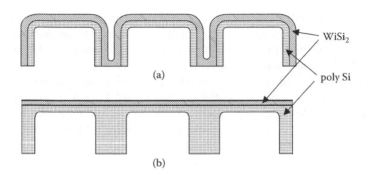

FIGURE 3.1
Structural comparison between (a) conventional nonplanarized data line, and (b) developed planarized data line. (Redrawn from "Crown-Shaped Stacked-Capacitor Cell for 1.5 V Operation 64-Mb DRAMs," T. Kaga et al., IEEE Electr. Dev. Vol. 38, pp. 255–260, 1991.)

WSi_2 allows formation of low resistance (15Ω Sq.) data line on the word lines. The data lines are realized using anisotropic dry etching.

The crown-shaped storage node and the selective polycrystalline silicon contact formation is done through self-aligned process which requires highly selective anisotropic dry etching for the thick Si_3N_4 film. Hence, a CH_2F_2 plasma etching [20] was used which has high selective etching rate of ~40 between Si_3N_4 and SiO_2 film. Despite the drawback of severe topography and requirement of large depth of focus from lithographic tools, CROWN structure using Ta_2O_5 dielectric has been shown to be the most promising candidate at 64/256 Mbit DRAM density level [21]. In the referred study, the conclusion is based on a comparison among a variety of 16 Mb/64 Mb DRAM structures on the basis of a figure of merit, "the ratio of the capacitor area to the cell area."

An MIM crown-shaped capacitor has also been used to fabricate a 1 Gbit DRAM at 0.16 µm design rule, using 0.29 µm^2 chip area [22]. Substituted tungsten (SW) electrode technology was developed in which electrode was formed substituting tungsten in place of polysilicon as shown in Figure 3.2. Shape and thickness of the electrode can easily be controlled in SW, which

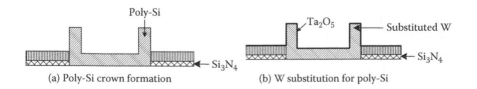

(a) Poly-Si crown formation   (b) W substitution for poly-Si

FIGURE 3.2
Substituted W electrode for Ta_2O_5 CROWN capacitor. (Redrawn from "A .29 µm^2 MIM-CROWN Cell and Process Technologies for 1 Gigabit DRAMs," T. Kaga et al., IEDM Tech. Dig., pp. 927–929, 1994.)

was not possible with polysilicon at smaller design rules. Another requirement at 0.16 μm technology was that of fine isolation. A single-Si_3N_4-spacer (SSS) offset local oxidation (OSELO), an improved version of OSELO II [23], was developed that reduced fabrication steps and provided better isolation than conventional structures. In STC-type cells, lithographic resolution becomes a problem because of the memory cell height. Even the use of recessed array (RA) technology [24] was not able to decrease the height difference sufficiently. Therefore, plate-wiring-merge (PWM) was introduced to decrease the height difference to about 0.5 μm, which was within the limits of the depth of focus in KrF-excimer phase-shift lithography.

A 64 Mbit DRAM using 1.28 $μm^2$ crown-shaped stacked capacitor has been fabricated to work at low voltage of 1.5 V, while maintaining speed performance [25]. Better speed performance was obtained by using a complementary current-sensing scheme and a word line driver circuit, which allowed the data line voltage to swing to its full value. CROWN cell was found to be suitable at low voltage as data line interference noise is reduced due to the shielding of data lines by either the plate or storage layer.

Noise minimization/elimination has always been a priority item in DRAMs. Folded-bit line architecture was an important step in this direction, but it has limitations in typical cross-point cell memory arrays. A divided/shared bit line (DSB) scheme was developed at 64 Mbit levels for signal sensing, which eliminated common-mode noise [26]. In DSB scheme folded bit-line scheme is used in a cross-point cell arrangement, which not only eliminates common mode noise but also provides higher density. Inter-bit coupling noise was minimized through the use of a twisted bit line technique.

3.4.2 Tantalum Pentoxide (Ta_2O_5)–Based Storage Capacitor

DRAM density range of 64 Mbit to 1 Gbit saw a marked interest in the investigation and usage of high-density dielectrics especially tantalum pentoxide (Ta_2O_5) in place of SiO_2 or ONO layers for storage capacitor [19,22,25,27–34]. At 64 Mbit level CROWN cell used a 3-nm equivalent CVD-Ta_2O_5 film to obtain 40 fF in a cell area of 1.28 $μm^2$ [19,25]. However, reliability in Ta_2O_5 CVD technology was enhanced by Kamiyama and Saeki [31] by introducing some finer processing steps. Rapid thermal nitridation (RTN) of native SiO_2 on the stacked poly-Si was done at 900°C for 60 sec in NH_3 atmosphere, which helped in reducing the equivalent oxide thickness. After low-pressure CVD (LPCVD) Ta_2O_5 film deposition rapid thermal annealing (RTA) was done at 700°C–900°C to make the film dense. In addition, TiN plate electrode was used on the Ta_2O_5 film to reduce leakage current through it.

Application of Ta_2O_5 film in storage capacitor was further extended to 1 Gbit DRAMs and almost all reports appeared in 1994 [22,27,32–34]. Shibahara and others [27] proposed a capacitor-over-bit-line (COB) cell for 1 Gbit DRAM occupying 0.375 $μm^2$ chip area. A hemispherical grain silicon (HSG) cylindrical capacitor employing Ta_2O_5 film realized a capacitance of

28.5 fF. A 10 nm Ta_2O_5 film deposition was done by LPCVD with $Ta(OC_2H_5)_5$ and O_2 at a low temperature of 450°C [33]. The film was annealed using rapid thermal oxidation; SiO_2 equivalent thickness was 2.5 nm. Low-temperature integrated (LTI) process was able to reduce leakage current by nearly four orders of magnitude to that with high temperature process and resulted in a 10 times improvement in data retention time. For 1 Gbit DRAM and beyond, a Ta_2O_5 capacitor was realized with a TiN/polysilicon top electrode with the plasma enhanced CVD-WN [34]; though it required a not-so-simple three-dimensional capacitor structure for which better step coverage was needed.

Other high dielectric constant materials have also been used, though in a small way, for 256 Mb–1 Gb DRAM level. A 100 nm thin film of $(Ba_xSr_{1-x})TiO_3$, which was equivalent to 8 Å of SiO_2, was used probably for the first time for a practical 256 Mbit DRAM [14]. A low-temperature ECR MOCVD $SrTiO_3$ film was used on the side walls of the storage electrode for 1 Gbit DRAM [35]. A cell capacitance of 25 fF with a leakage current density of 8×10^{-7} A/Cm^2 was obtained using RuO_2/TiN storage node. Barium strontium titanate (BST) was again used, at 1 Gbit DRAM level in 1995 with an SiO_2 equivalent thickness of 0.4 nm [36]. Using corrugated cylinder structure a capacitance of more than 40 fF was realized for 256 Mbit (extendible to 1 Gbit) DRAM with leakage current density of 1×10^{-8} A/cm^2.

Transition from 64 Mbit affected large voltage bounce in chip power lines due to narrow but lengthy metal lines on a bigger size chip. In addition, wide-bit I/Os were also needed for high data transfer rates in high-capacity DRAMs. A 256 Mbit DRAM with 16 I/Os, 30 ns access time with operating current of 30 mA at 60-ns cycle time was developed to meet specifications while using a multidivided array structure [37]. DRAM cell array consisted of eight blocks of 32 Mbit cell array and each block was further divided into 256 and 128 Kbit subarrays. Using a sub-word decoder-driver, size of the activation array became 256 Kbit. Inclusion of a time-sharing refreshing scheme also helped in reducing the power-line voltage bounce.

Another important contribution at 256 Mbit level was the development of a NAND structured cell (NAND DRAM) which used four memory cells connected in series. Figure 3.3 shows a comparison between a NAND structured cell and a conventional cell [38]. Cell area was 0.962 μm^2 at 0.4 μm CMOS technology, which was 63% compared to a conventional cell. Reduction in cell area was possible due to the reduction in isolation area and bit line contact area. A time division multiplex sense amplifier (TMS) architecture, in which sense amplifier was shared by four bit lines was also helpful in reducing the chip size to 464 mm^2, which was 68% compared to other 256 Mbit DRAM cells.

Sub-threshold current (I_{sub}) reduction has always remained in focus and shall be discussed in Chapter 7. At 256 Mbit DRAM level, a self reverse biasing scheme for word driver and decoders was used, which suppressed I_{sub} to nearly 3% of the conventional methods [39].

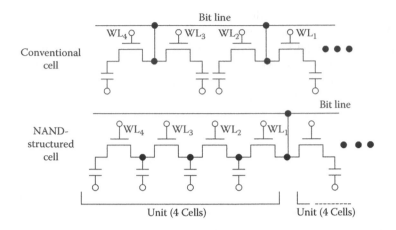

FIGURE 3.3
Comparison of the NAND-structured cell with a conventional cell. (Redrawn from "An Experimental DRAM with a NAND-Structured Cell," T. Hasegawa et al., *IEEE. J. Solid State Circuits*, Vol. 28, pp. 1099–1104, 1993.)

Up to 16 Mbit DRAM generation conventional LOCOS isolation was used but beyond that bird's beak scaling faced serious problems. For the cell pitch less than 0.4 μm modified LOCOS isolation techniques were employed [40]. Polysilicon spacered LOCOS (PSL) is a simple isolation scheme based on conventional LOCOS, in which drawn isolation size of 0.2 μm could be achieved [41]. PSL isolation had low bird's beak encroachment and good vertical profile. Another isolation technology at 64–256 Mbit range is self-aligned LOCOS trench (SALOT) [42], which has the isolation structure of a poly-buffered LOCOS field oxide and a self-aligned trench at the center of a field region. However, for gigabit density DRAM, requirement of the scalability and the planarity could be achieved using shallow trench isolation (STI) with dual slope trench etching [43] followed by chemical mechanical polishing (CMP).

By 1998, 256 Mbit DRAMs were in mass production while 1 Gbit and even 4 Gbit experimental DRAMs were presented in conferences [43]. A highly manufacturable 1 Gbit SDRAM was fabricated, probably for the first time in 0.18 μm technology. The SDRAM used retrograded twin-well, shallow trench isolation; Ta_2O_5 based storage capacitor with HSG, a $TiSi_2$ gate, W plug and W wiring process followed by relaxed double metallization and CMP process for global as well as local topography [44]. It is observed that most of the technologies employed were in line with the expected technology trends for DRAM process modules shown in Table 3.1 in a study by Nitayama and others [45]. These expected processes were thought to be based on the expected DRAM market needs of low cost, high speed, and so on, and preferably in embedded (logic) DRAM form. It was desired to use minimum chip area at the least possible temperature process and with metal word and bit lines.

TABLE 3.1

Some Technology Trends for DRAM Process Module in a Gigabit Era

Module	Concerns	Possible Actions
Lithography	• Narrow process margin • KrF To ArF • Optical proximity effect • Thin resist process	• Fully planarized process/mask critical dimension (CD) control • Optical proximity correction
Capacitor (STC)	• Step coverage/patterning • SN contact resistance, plasma damage	• CVD BST/CVD Ru • BEOL process optimization
STI	• Trench fill • Stress reduction	• HDP • High temp./thin oxidation
Gate, S/D	• Low resistance for gate • Low resistance for S/D	• Poly/metal gate (ex. W/WN/Poly) • Co silicide
Contact/MOL dielectric	• Low temp. process • Void-less • Self-aligned contact	• W dual damascene • HDP-PSG, TEOS-03 PSG, SAUSG • Selective etch to SiN cap
Interlayer dielectric	• Low k • Water absorption	• FSG, k = 3.3–3.7 • Siloxane SOG, k = 2.6–2.8
Metallization	• Low resistance interconnect • Patterning/gap fill	• Cu wiring • Damascene

3.5 Capacitor over Bit Line (COB) DRAM Cell

A simple idea of forming capacitor over bit line changed the memory cell structure and fabrication steps of the three-dimensional capacitor like the stacked capacitors and trench capacitors to a great extent. The COB technique enabled the memory cell to maintain sufficient storage capacitance with decreasing feature size. Up to around 4 Mbit DRAM density level, bit line was formed above the storage electrode. To get higher value of capacitance, storage electrode necessarily had to be taller in the three-dimensional case, as the capacitance was proportional to the total capacitor surface area including top/bottom and sidewalls. However, it made the formation of bit line contact and maximization of the storage electrode pattern difficult as shown in Figure 2.15 [46]. A unique structure for a COB cell was given by Kimura in 1988, wherein contact of the storage electrode to the active regions was formed through adjacent bit lines by a self-aligned process as shown in Figure 2.17 [17]. Inter-bit-line noise was almost eliminated because the bit lines were shielded by the storage electrode and holes needed in the plate for making bit line contact with the active region were no longer required.

A number of DRAM cells were realized based on COB technique. Some of these were mentioned in Section 2.8 and a few more shall be discussed now along with the issues involved in scaling down of COB stack DRAM cells.

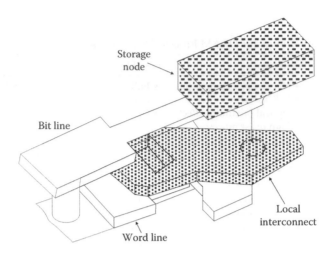

FIGURE 3.4
Perspective of the capacitor-over-bit-line (COB) cell with the hemispherical-grain (HSG) poly-Si storage node. (Redrawn from "A Capacitor-Over-bitline (COB) Cell with a Hemispherical-Grain Storage Node for 64 Mb DRAM," M. Sakao et al., IEDM Tech. Dig., pp. 655–658, 1990.)

Feasibility of a COB cell for 64 Mbit DRAM was verified by Sakao and others in 1990 [47]. The given cell occupied chip area of 1.8 μm^2 at 0.4 μm design rules, obtaining storage capacitance of 30 fF with storage node height of 0.5 μm. Use of hemispherical grains on the storage node allowed using reduced node height (to nearly half of that without hemispherical grain), for obtaining comparable value of storage capacitance. Perspective view for a COB cell with HSG poly-Si storage node is given in Figure 3.4. To make the COB cell structure, an additional local interconnect layer is laid for forming a connection between capacitor with the active area [47]. This additional layer formation allowed the use of a rectangular area against diagonal active area. Sakao and others gave another COB cell structure with straight-line-trench isolation and trench-gate transistor (SLIT) having potential for giga-bit DRAMs. As mentioned in Chapter 2, the SLIT cell overcame the limitation of complex active area pattern in the formation of capacitor contact in some previously reported COB structures [48]. Figure 3.5 shows a three-dimensional cross-section of the SLIT cell in which two types of trenches, the isolation trenches and the gate trenches, are formed orthogonally, defining the rectangular active area. Straight-lined isolation trenches isolate adjacent cells in one direction and the trench gate serving as field plate provides isolation in the orthogonal direction. Layout of the SLIT cell having a size of 6 F^2 is shown in Figure 3.6(a). In this layout there is no alignment margin space, and it is produced using a simple pattern shrinking technique SAFE (side-wall aided fine pattern etching). Process sequence of SAFE technique is given in [48], which allows the trench to reduce to less than half of the design rule and layout of SLIT cell after the processing is shown in Figure 3.6(b). Storage node

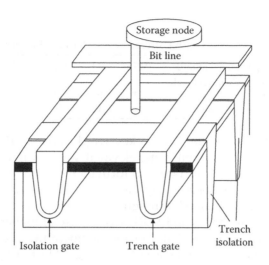

FIGURE 3.5
Simplified schematic view of the SLIT cell. (Redrawn from "A Straight-Line-Trench Isolation and Trench-Gate Transistor (SLIT) Cell for Giga-bit DRAMs," M. Sakao et al., IEDM Tech. Dig., pp. 19–20, 1992.)

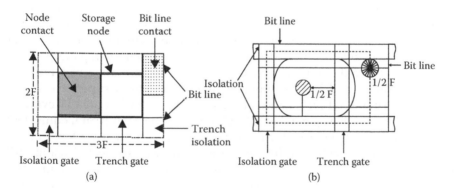

FIGURE 3.6
(a) Layout of the (2F*3F) SLIT cell; (b) plan view after processing. (Redrawn from "A Straight-Line-Trench Isolation and Trench-Gate Transistor (SLIT) Cell for Giga-bit DRAMs," M. Sakao et al., IEDM Tech. Dig., pp. 19–20, 1992.)

electrode is also made of poly-Si including scaled capacitor contact using the SAFE technique. An experimental cell array with 0.96 μm^2 cell size was realized in 0.4 μm design rule.

A simple 4 Gbit DRAM is realized using high aspect-ratio pillar capacitor for COB type STC cell. The storage node and the peripheral vias are fabricated simultaneously with high aspect-ratio pillars, which allows using ON dielectric or Ta_2O_5; and surface is planar for metal wiring [49]. A pillar height of 1.5 μm provides a storage capacitance of 17 fF/cell for ONO film and 22 fF

TABLE 3.2

Realized Capacitance with Different Cylinder Heights [50]

Projected capacitor area (μm^2)		0.075	0.1	0.15	0.2	0.3
Capacitance	Cylinder height					
per cell (fF)	1.5 μm	22	29	41	48	59
	1.2 μm	14	19	26	32	41

using Ta_2O_5. Self-aligning was also a key word in the fabrication as it was used for patterning of the cell plate, elevated source/drain by damascene and Ti salicidation of their surface, and formation of various metal contacts.

Maximization of poly-Si-based capacitor electrode surface area and hence cell capacitance was obtained through innovations in structures having fin, cylindrical, pillar shapes and HSG surface, use of higher dielectric constant materials like Ta_2O_5, and use of low temperature processing having low resistance, and fully metal-based technology. Most of these ideas were combined in a cylindrical full metal capacitor (FMC) electrode with integrated metal contact-hole plug for gigabit DRAMs [50]. In the proposed FMC technology-based cell conventional CVD-W/TiN/Ti was used for metal-based capacitor contact-hole plug and, for the multi-surface bottom electrode two different W cylinder formation processes were evaluated; an exterior process which employs an SiO_2-core, PVD+CVD-W body and W-RIE, whereas the interior process employed an SiO_2 frame, PVD+CVD-W body and W-CMP with a protective plug. Its results show that interior cylinder process is better for gigabit densities as it provides improved lithographic/RIE pattern stability, the PVD-W adhesion layer for CVD-W enables formation of a uniform W-only cylinder and W-CMP implementation results in superior cylinder shape. The single cell capacitance of the projected capacitor area W cylinder could go over 40 fF as per Table 3.2 [50], while using 1.6 nm SiO_2 equivalent thick layer of Ta_2O_5 deposited with oxidation treatment such as O_2 plasma.

A COB memory cell with 0.1 μm design rule was fabricated using SRP (SAC spacer removal after plug implantation) a novel SAC process and DMO (dual molded oxide) capacitor process. The SRP process improves data retention time characteristics and minimizes short channel effect in cell transistor. Dual spacer and downstream surface cleaning process was used to cure the surface defects and minimize junction leakage current [51]. In the DMO process two oxide layers (BPSG+PE–TEOS) are deposited on etch stopping layer of Si_3N_4, then the two oxide layers are etched anisotropically. The bottom area of the storage node is enlarged during the precleaning process by about 20% due to different wet etching rates of the two oxide layers. Finally a 400 Å thick poly-Si was deposited to form one cylindrical storage node capacitor as shown in Figure 3.7. Mechanical stability of the COB stack cell becomes important for below sub-quarter micron technology and mechanical stability of the storage node decreases rapidly with scaling down. It results in putting a limit on the stack height of the memory cell capacitor according to the

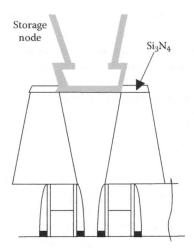

FIGURE 3.7
DMO OCS node capacitor. (Modified from "Highly Extendible Memory Cell Architecture for Reliable Data Retention Time for 0.10 μm Technology Node and beyond," J. Lee et al., Proc. ESSDRC, pp. 571–574, 2002.)

TABLE 3.3

Permissible Stack Height of Cylinder-Type COB Stack
Cell at DRAM Technologies with Mechanical Stability

Design rule (μm)	0.07	0.09	0.11	0.13	0.15
Maximum stack height (μm)	0.5	0.75	1.05	1.3	1.6

estimate given in Table 3.3; otherwise, storage node will be shortened [52]. It is this reason that the DMO process, which has better mechanical stability, is employed. The data retention characteristics of DMO capacitor are considerably better than that of a conventional one, and the improvement factor is at least three.

COB stack cells studied so far had some key features like planar structure of memory cell transistor with uniformly doped channel and three-dimensional structure of memory cell capacitor. In such a configuration independent design and optimization of memory cell transistor and storage node are not possible and needed new ideas to make it practical beyond 100-nm technology. Some other closely involved required technologies are advanced lithographic techniques and novel MIS/MIM capacitors. Another important consideration is to choose between open or folded bit line cell architecture. It has been observed that in spite of smaller capacitor and chip area ($4F^2/6F^2$) used in open bit line architecture, the folded bit line architecture was preferable on account of better noise immunity, though its cell used up 8 F^2 of chip area. Additionally in the COB stack cell lower capacitor area was achievable. It was therefore, correctly assumed that a COB stack DRAM cell, combined with folded bit line architecture would be a good choice for further DRAM density advancements, while using below 100-nm technology, however with a rider that suitable solutions were available for foreseeable difficulties [53]. Issues connected with the process technologies like isolation dimension scaling, memory cell transistor parasitics, reduction in cell capacitance, and non-reduction of bit line capacitance with scaling down have been identified. If problems are attended to satisfactorily, scaling-down effects on degrading DRAM cells such as increased parasitic resistance and node junction leakage current, variation of threshold voltage, and so on, could be overcome.

A common practice for cell-to-cell isolation is the implantation of channel stop using either LOCOS or STI. But this practice degrades data retention line and the problem becomes severe in sub 100-nm technology node. In addition, serious cross talk occurs among memory cells and severe inverse narrow width effect, etc. takes place. To overcome such a host of problems, a metal shield embedded STI (MSE-STI) [54] for isolation technology beyond 100 nm node is suggested. The MSE-STI is able to eliminate the cell-to-cell cross talk and with a provision of negative bias to the embedded metal, channel stop implantation would be avoided, so the degradation of data retention time should be halted.

A number of new issues crop up when the dimensions of the cell transistor become very small. Constraints of the lithographic and other fabrication processes produce variation in physical dimensions and other features like doping profile, which results in large fluctuations in transistor parameter like threshold voltage [55]. The amount of the variation in the threshold voltage tends to be inversely proportional to the minimum feature size. It is important to control dimensional and other variations, so that variations in threshold voltage of the cell transistor are minimized. Cell transistor is also degraded due to short channel effect and narrow width effect at smaller dimensions. Another important issue for the cell transistor is leakage current of the storage node junction. Beyond the 100 nm technology node, the junction leakage current is primarily decided by perimeter-type junction leakage current [56]. The most effective way to reduce the junction leakage current is to lower the substrate doping density underneath storage node junctions. However, it is contradictory to the requirement of increasing the channel doping density for keeping threshold voltage of nearly 1 V. Moreover, current drive capability of the cell is also reduced with unscalable threshold voltage and reduced channel mobility if its doping is increased. One alternative to overcome this problem is to use local channel implantation technology for selective doping of the channel of the memory transistor leading to spike-doped channel technology; only a small portion of channel has high doping concentration and the other areas have low concentration. For example, in 70 nm memory cell transistor, the substrate doping density underneath storage node junction can be reduced from 1×10^{18} cm^{-3} with a conventional channel to 1×10^{17} cm^{-3} in spike channel [53]. The spike-domed channel has the added advantage of reducing the short channel effect and increased robustness in terms of fluctuations of transistor parameter due to variations in its dimensions, width, and length. Since the spike-doped region located at the center of the channel suppresses the electric field penetration from the drain to the source, it also reduces the drain-induced barrier lowering (DIBL) effect.

Another issue for sub-100 nm COB stack DRAM cell is that of realizing a minimum capacitance of 25 fF/cell, a problem common to other types of cells. Therefore the solutions are also on the same lines which are applicable in other cases, that is, use of thinner/different dielectric and use of innovative MIS or MIM capacitors. In addition, increase of stack height of cylinder-type COB stack cell sometimes results in mechanical stability problems, leading to inter-storage node shorting or twin bit failure. Thus the minimum height of the stack is limited as was mentioned in Table 3.3. Mechanical stability of the storage node becomes a serious issue. It will be discussed in Chapter 6, especially for more COB stack DRAM cells. A suggested combination for COB stack capacitor technology for DRAM technology is given in Table III of ref. [53]. It is observed that though HSG continued to be used, the storage node of box-shape was replaced by a cylinder-shaped one at 0.13 μm technology which itself was replaced by merged structure of cylinder-box shape. Changeover from MIS capacitor to MIM was nearly complete around 100 nm stage but equivalent oxide thickness remained nearly constant up to 0.11 μm (37 Å to 30 Å) which decreased considerably later to 15 Å at 90 nm and ~5 Å at 70 nm, respectively, with the use of much higher dielectric constant material and metal electrodes.

One of the challenging issues for future technology is to keep parasitic bit line capacitance to a low value as well as maintaining a low resistance. It is observed that up to around 0.15 μm technology bit line parasitic capacitance follows set rules for its evaluation, but beyond that it tends not to be reduced in spite of dimension scaling and remains nonscalable beyond 100 nm. This nonscalable nature of bit line capacitance is due to increased inter-bit line capacitance and capacitance between bit lines and other conducting layers. This problem of nonscalable bit line capacitance can be overcome with a hierarchical bit line scheme, a combination of global and segmented bit lines. The global bit lines do not have parasitic capacitance between bit line and other conducting lines like word lines. Since the total parasitic capacitance is the sum of global bit line and segmented bit line capacitance, overall capacitance gets reduced. DRAM technologies employed by 2002 were discussed here briefly. Few representative reports were discussed without detailed descriptions. It is expected that readers will get a fair idea of the sequence of technology developments. Advanced DRAM technologies will be discussed in Chapter 6. But before that, the advancements in the storage capacitor formation and the access transistor structures are addressed in Chapters 5 and 4, respectively, for better understanding of the technologies discussed in Chapter 6.

References

1. C. Mazure et al., "Technology Challenges and Solution for 1 Gbit and Beyond," *Integrated Ferroelectronics*, Vol. 21, pp. 15–25, 1998.
2. W.M. Regitz and J. Karp, "A 3-Transistor Cell 1024 bit 500ns MOS RAM," *Dig. of Tech. Papers*, ISSCC, 1970.
3. L. Boonstra et al., "A 4096-b One Transistor per Bit Random Access Memory with Internal Timing and Low Dissipation," *IEEE J. Solid State Circuits*, Vol. SC-8, p. 305, 1973.
4. R.H. Dennard, "Field-Effect Transistor Memory," U.S. Patent 3,387,286, June, 1968.
5. K.U. Stein et al., "Storage Array and Sense/Refresh Circuit for Single-Transistor Memory Cells," *Dig. Tech. Papers*, ISSCC, pp. 56–57, 1972.
6. R. Proebsting and R. Green, "A TTL Compatible 4096-Bit n-Channel RAM," *Dig. Tech. Papers*, ISSCC, pp. 28–29, 1973.
7. R.R. Schroeder and R.J. Proebsting, "A 16k × 1 Bit Dynamic RAM," *Dig. Tech. Papers*, pp. 12–13, 1977.
8. L.G. Heller and D.D. Spampinato, "Cross-Coupled Charge-Transfer Sense Amplifier," U.S. Patent 4,039,861, Aug. 1977, also L. Heller, "Cross-Coupled Charge Transfer Sense Amplifier," *IEEE ISSCC Tech. Dig.*, p. 20, 1979.
9. H. Miyamato et al., "A Fast 256k × 4 CMOS DRAM with a Distributed Sense and Unique Restore Circuit," *IEEE J. Solid-State Circuits*, Vol. SC-22, 1987.
10. P. Gillingham et al., "High-Reliability Circuit Design for Megabit DRAM," *IEEE JSS Circuits*, Vol. 26, no. 8, pp. 1171–1175, 1991.
11. N.C. Lu and H.H. Chao, "Half-V_{DD} Bit-Line Sensing Scheme in CMOS DRAMs," *IEEE J. Solid State Circuits*, Vol. SC-19, p. 451, 1984.
12. H. Sunami et al., "A Corrugated Capacitor Cell (CCC) for Megabit Dynamic Memories," *Proc. IEDM*, pp. 806–808, 1982.
13. W. Wakamiya et al., "Novel Stacked Capacitor Cell for 64 Mb DRAM," *Symp. VLSI Technol. Dig. Tech. Papers*, pp. 31–32, 1989.
14. K. Koyama et al., "A Stacked Capacitor with (Ba_xSr_{1-x}) TiO$_3$ for 256 M DRAM," IEDM Tech. Dig., pp. 823–826, 1991.
15. M. Aoki et al., "A 60 ns 16-Mbit CMOS DRAM with a Transposed Data Line," *IEEE J. Solid State Circuits*, Vol. 23, pp. 1113–1111, 1988.
16. H. Arima et al., "A Novel Stacked Capacitor Cell with Dual Cell Plate for 64 Mb DRAMs," *IEDM*, pp. 651–654, 1990.
17. S. Kimura et.al, "A New Stacked Capacitor DRAM Cell Characterized by a Storage Capacitor on a Bitline Structure," *IEDM*, pp. 596–599, 1988.
18. T. Takashima, "Word-Line Architecture for Constant Reliability 64 Mb DRAM," *Proc. Symp. VLSI Circuits*, p. 57, 1991.
19. T. Kaga et al., "Crown-Shaped Stacked-Capacitor Cell for 1.5 V Operation 64-Mb DRAMs," *IEEE Electr. Dev.*, Vol. 38, pp. 255–260, 1991.

20. T. Kure et al., "VLSI Device Fabrication Using a Unique, Highly-Selective Si_3N_4 Dry Etching," IEDM Tech. Dig., pp. 757–759, 1983.

21. H. Shichijo, "DRAM Technology Trend and Prospect," *Proc. VLSI-TSA*, pp. 349–353, 1991.

22. T. Kaga et al., "A.29 µm² MIM-CROWN Cell and Process Technologies for 1 Gigabit DRAMs," IEDM Tech. Dig., pp. 927–929, 1994.

23. T. Kaga et al., "Advanced OSELO Isolation with Shallow for High-Submicrometer ULSIs," *Trans. IEEE Electr. Dev.*, Vol. 35, pp. 893–898, 1988.

24. K. Sagara et al., "A o.72 µm² Recessed STC (RSTC) Technology for 256 Mbit DRAM Using Quarter-Micron Phase-Shift Lithography," *Symp. VLSI Techn. Dig.*, pp. 10–11, 1992.

25. Y. Nakagome et al., "An Experimental 1.5 V 64-Mb DRAM," *IEEE J. Solid State Circuits*, Vol. 26, pp. 465–472, 1991.

26. H. Hidaka et al., "A Divided/Shared Bit-Line Sensing for VLSI DRAMs," *IEEE J. Solid State Circuits*, Vol. 26, pp. 473–478, 1991.

27. K-Shibahara et al., "1 GDRAM Cell with Diagonal Bit-Line (DBL) Configuration and Edge Operation MOS (EOS) FET," IEDM Tech. Dig., pp. 639–642, 1994.

28. M. Saito, T. Mori, and H. Tamura, "Electrical properties of thin Ta_2O_5 films grown by chemical vapour deposition," IEDM Tech. Dig., pp. 680–683, 1986.

29. Y. Numasawa et al., "Ta_2O_5 Plasma CVD Technology for DRAM Stacked Capacitors," IEDM Tech. Dig., pp. 43–46, 1989.

30. H. Shrinki and M. Nakata, "UV-O_3 and Dry O_3. Two-Step-Annealed Chemical Vapour Deposited Ta_2O_5 Films for Storage Dielectrics of 64 Mb DRAM," *IEEE Trans. Electr. Dev.*, Vol. 38, pp. 455–462, 1991.

31. S. Kamiyama and T. Saeki, "Highly Reliable 25 nm Ta_2O_5 Capacitor Process Technology for 256 Mbit DRAM," IEDM Tech. Dig., pp. 827–830, 1991.

32. H.K. Kang et al., "Highly Manufacturable Process Technology for Reliable 256 Mbit and 1 Gbit DRAMs," IEDM Tech. Dig., pp. 635–638, 1994.

33. Y. Takaishi, M. Sakao, and S. Kamiyama, "Low-Temperature Integrated Process below 500°C for Thin Ta_2O_5 Capacitor for Giga-Bit DRAMs," IEDM Tech. Dig., pp. 839–842, 1994.

34. K.W. Kwon et al., "Ta_2O_5 Capacitors for 1 Gbit DRAM and Beyond," IEDM Tech. Dig., pp. 835–838, 1994.

35. P-Y. Lesaicherre et al., "A Gbit-Scale DRAM Stacked Capacitor Technology with ECR MOCVD $SrTiO_3$ and RIE Patterned RuO_2/TiN Storage Nodes," IEDM Tech. Dig., pp. 831–834, 1994.

36. C.V. Teng, "DRAM Technology Trend," *Proc. VLSI Techn. Systems and Appl.*, pp. 295–299, 1995.

37. T. Sugibayashi et al., "A 30-ns Mb DRAM with a Multidivided Array Structure," *IEEE J. Solid State Circuit*, Vol. 28, pp. 1092–1098, 1993.

38. T. Hasegawa et al., "An Experimental DRAM with a NAND-Structured Cell," *IEEE J. Solid State Circuits*, Vol. 28, pp. 1099–1104, 1993.

39. G. Kitsukawa et al., "256-Mb DRAM Circuit Technologies for File Applications," *IEEE J. Solid State Circuits*, Vol. 28, pp. 1105–1113, 1993.

40. C.G. Hwang, S.I. Lee, and M.Y. Lee, "The State of the Art and Future Trends in DRAMs," *Proc. ESSDRC*, pp. 37–44, 1995.

41. D.H. Ahn et al., "A Highly Practical Modified LOCOS Isolation Technology for the 256 Mbit DRAM," IEDM Tech. Dig., pp. 679–682, 1994.

42. T. Park et al., "Self-Aligned LOCOS/Trench (SALOT) Combination Isolation Technology Planarized by Chemical Mechanical Polishing," IEDM Tech. Dig., pp. 675–678, 1994.
43. K.P. Lee et al., "A Process Technology for 1 Giga-Bit DRAM," IEDM Tech. Dig., pp. 907–910, 1995.
44. T. Murutami et al., "A 4-Level Storage 4 Gb DRAM," ISSCC Tech. Dig., pp. 74–75, 1997.
45. A. Nitayama, "Future Directions for DRAM Memory Cell Technology," IEDM Tech. Dig., pp. 355–358, 1998.
46. M. Koyanagi et al., "Novel High Density, Stacked Capacitor MOS RAM," IEDM, pp. 348–351, 1978.
47. M. Sakao et al., "A Capacitor-Over-Bitline (COB) Cell with a Hemispherical-Grain Storage Node for 64 Mb DRAM," IEDM Tech. Dig., pp. 655–658, 1990.
48. M. Sakao et al., "A Straight-Line-Trench Isolation and Trench-Gate Transistor (SLIT) Cell for Giga-bit DRAMs," IEDM Tech. Dig., pp. 19–20, 1993.
49. S. Nakamura et al., "A Simple 4 G-bit DRAM Technology Utilizing High-Aspect-Ratio Pillars for Cell-Capacitors and Peripheral-Vias Simultaneously Fabricated," IEDM Tech. Dig., pp. 29–32, 1997.
50. J.M. Drynan et al., "Cylindrical Full Metal Capacitor Technology for High Speed Gigabit DRAMs," Symp. VLSI Techn. Dig. Tech. Papers, pp. 151–152, 1997.
51. J. Lee et al., "Highly Extendible Memory Cell Architecture for Reliable Data Retention Time for 0.10 μm Technology Node and Beyond," Proc. ESSDRC, pp. 571–574, 2002.
52. Y. Park and K. Kim, "COB Stack DRAM Cell Technology beyond 100 nm Technology Node," IEDM Tech. Dig., pp. 391–394, 2001.
53. K-Kim and M-Y. Jeong, "The COB Stack DRAM Cell at Technology Node below 100 nm-Scaling Issues and Directives," IEEE Trans. Semiconductor Manufacturing, Vol. 15, pp. 137–143, 2002.
54. J.H. Sim, J.K. Lee, and K. Kim, "High-Performance Cell Transistor Design Using Metallic Shield Embedded Trench Isolation (MES-STI) for Gigabit Generation DRAM's," IEEE Trans. Electr. Dev., Vol. 46, pp. 1212–1217, 1999.
55. T. Mizuno et al., "Performance Fluctuations of 0.1 μm MOSFETs—Limitations of 0.1 μm ULSI's," Proc. Symp. VLSI Techn. Dig. of Tech. Papers, pp. 13–14, 1994.
56. J.Y. Lee, D.W. Ha, and K. Kim, "Novel Cell Transistor Using Retracted Si_3N_4-Linear STI for the Improvement of Data Retention Time in Gigabit Density DRAM and Beyond," IEEE Trans. Electr. Dev., Vol. 48, pp. 1152–1158, 2001.

4

Advanced DRAM Cell Transistors

4.1 Introduction

As DRAM technology crosses the sub-100 nm node, planar transistors are not able to maintain level of on-current. Also, transistors do not satisfy the overall leakage current limit, mainly due to high electric field on account of shrinking of channel length, while array voltage was not changing with every lithographic generation. A common remedy used in transistors was to design with increased channel doping concentration to increase threshold voltage, so as to reduce sub threshold current a substantial component of the leakage current. However, increased doping level causes high junction leakage current at the storage node and also reduces the current drive capability of the transistor. Due to increased leakage current a significant problem appears in the form of reduced data retention time. To face the challenge of achieving low level of leakage current, while not increasing doping density and maintaining sufficient drive, various cell transistor structures and processes have been investigated. Table 4.1 shows a trend in this direction in the use of DRAM cell transistors [1]. In addition to these changes in structures, other techniques are also used, such as having elevated source/drain structure with selective epitaxial growth (SEG) process [2]. Characterization of these developed transistors in CMOS is a separate subject. This chapter offers only a bird's-eye view of some technological details of few such transistors mainly used in DRAM circuits beyond the conventional planar MOSFETs. The first attempt in this direction was in the form of recessing the channel, mainly to increase the channel length without increasing projected chip area; this is discussed in Section 4.2. However, FinFET structure soon caught up and remained in use in quite a few generations of DRAMs. Variations in the FinFET, in the form of body-tied/bulk FinFETs, multichannel, and saddle FinFETs, came into use and tried to overcome the technological and performance limitations; these are discussed in Sections 4.4 to 4.8. A different kind of structure was made available in the form of surrounding gate transistors, mainly to improve current drivability without compromising other characteristics. Stacked surrounding gate transistors helped in further increasing DRAM density even below 4 F^2 cell area as shown in Section 4.10.1. Deep-trench DRAM cells in

TABLE 4.1

Trend in the Use of DRAM Cell Transistors

	Planar Tr	RCAT/S-RCAT	FinFET	Vertical Tr
Technology	>100 nm	90–50 nm	50–30 nm	<30 nm

Section 4.11 in the form of BEST and VERIBEST cells and the vertical transistor cells further improved DRAM density and performance. Recently some attempts also have been made to combine recessing the channel with FinFETs and studies are ongoing to overcome technological limitations so as to deploy these devices in sub-20 nm DRAM cells.

4.2 Recess-Channel-Array Transistor

The simplest remedy of the problem of increased junction leakage current due to the increased electric field is to increase effective channel length (L_{eff}) of cell transistor. A few attempts have been made in this direction, and one of the significant ones is the realization of a double-gate vertical access transistor, which was basically suitable for trench capacitor structure [3]. Here gate length was decided independent of the technology design rule. However, to utilize the fabrication process simplicity of planar cell array transistor, a vertical transistor, recess-channel-array transistor (RCAT) structure was given in stack capacitor cells [4]. The basic concept of the RCAT is to increase the L_{eff} by recessing the channel from silicon surface. So, channel doping density is allowed to be reduced, thereby reducing source/drain resistance of memory cell transistor, and enhancing carrier mobility, in addition to the main advantage of reduced leakage.

Because of the different voltage levels and characteristics requirements in the core and the peripherals, the recess channel transistor was used in cell array regions, whereas the conventional planar transistors were used elsewhere at peripherals, requiring suitable changes in fabrication steps. Recess channel is formed with a thin oxide layer and thick poly layer deposited on the substrate as a mask layer. Poly-Si hard mask is a cost-effective solution for the integration of RCAT with planar-type transistors because during recessed channel etching process poly-Si hard mask is removed from the regions on which conventional planar transistors are fabricated. One of the key processes used in RCAT fabrication is chemical dry etching (CDE), which is used for bottom rounding and for removing Si sidewall and some damaged Si during the dry etching.

This innovative and simple transistor RCAT structure added a few more advantages. For example, for drain-induced barrier lowering (DIBL) and source/drain breakdown voltage (BV_{DS}), the RCAT shows improvement of more than 300 mV and 3 V, respectively at recessed channel depth of 150 nm. The most important improvement is the reduction of array junction leakage by more than one order, which resulted in great improvement of data retention time. On the flip side, parasitic capacitance of word line increased by nearly 58% due to the lengthening of the channel. Increased mobility also offsets some decrease in the on-current (I_{on}) due to the increased L_{eff}.

Application of RCAT continued to show excellent results in terms of data retention time even down to 70 nm node. However, further scaling down created certain problems, and RCAT could be successfully used only when these problems were minimized. The RCAT has to match the requirements of the combination of low power and high speed of the next-generation DRAM. For speeding up the DRAM, a suggestion was to reduce word line resistance and increase the drive of the support transistor. Out of the technological options, "dual work function gate for the RCAT structure to convert n-type doped poly-silicon into a p-type doped polysilicon by counter doping implantation on the PMOS region using only one photolithography" has been used [5]. In addition, use of tungsten gates in place of WSi gates reduced word line resistance nearly one-third. From the low-power device consideration at reduced array voltage level, threshold of RCAT was to be reduced; but it adversely affected sub-threshold leakage. Under this consideration negative word line (NWL) scheme was most appropriate, but its use affected GIDL and the data retention time. Due considerations have shown that GIDL and I_{on} can be improved by having a "combination of high doping cell contact landing PAD and low post heat treatment" [5].

Another major problem is that the characteristic of transistor with recessed channel is heavily determined by the shape of the bottom curvature. Due to the rise in the potential barrier at the bottom curvature, threshold voltage is increased locally and effective channel length is decreased. Figure 4.1 shows schematic diagram of the recessed channel and its numerical analysis gives the expression of threshold voltage by the following relation:

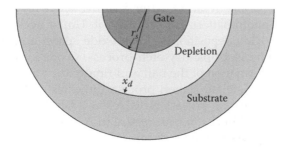

FIGURE 4.1
Schematic diagram of a recessed channel of the RCAT. (Modified from "S-RCAT (Sphere-Shaped-Recess-Channel-Array-Transistor) Technology for 70 nm DRAM Feature Size and beyond," J.Y. Kim et al., Symp., VLSI Techn. Dig. Tech. Papers, pp. 34–35, 2005.)

TABLE 4.2

Simulated Variations of V_{th} and DIBL with Design Rule for RCAT

Design rule (nm)	110	100	90	80	70	60	50
V_{th} (V)	1.05	1.07	1.1	1.15	1.20	1.24	1.40
DIBL (mV)	40	50	60	100	120	150	200

$$V_{th} = V_{FB} + 2\varphi_B + \frac{Q_B(\varphi_s)}{C_{ox}}, \varphi_s = 2\varphi_B + V_{SB}$$

where

$$C_{ox} = \frac{\varepsilon_{ox}\varepsilon_o}{r_s\left(\ln(r_s) - \ln(r_s - t_{ox})\right)}$$

Q_B and φ_s are functions of r_s, x_d, and N_A with r_s and x_d being the gate curvature radius and depletion region radius, respectively. N_A is the doping concentration of the substrate, and C_{ox} is the gate oxide capacitance density, with other terms having their usual meanings. Here Q_B and φ_s increase while C_{ox} decreases with reduction in r_s as the technology is scaled down. Its effect is that threshold and DIBL both increase when the minimum feature size decreases the way as shown in Table 4.2 and the effective channel length L_{eff} gets decreased [6]. One option to increase the L_{eff} is to increase the recess depth further but in practice it does not help and increasing the recess depth does not solve the problem. Another attractive alternative was developed in the form of sphere-shaped RCAT, which overcomes the curvature problem and maintains effective channel length [6].

4.2.1 Sphere-Shaped RCAT

Sphere-shaped RCAT (S-RCAT) which is an improved form of RCAT features a spherical ball part on the bottom of the recessed channel. Figure 4.2(a) shows schematic diagram of the S-RCAT structure, prominently showing the shape of the bottom curvature. After forming active regions, hard mask is deposited and photo pattern is formed on it. Upper neck part is formed followed by thin oxide deposition to form oxide spacer and then spherical ball part is made by isotropic dry etch process. Oxide is removed by wet cleaning after the formation of the ball and finally the gate stack is formed. Observation of complete fabrication process indicates quite a few simplifications over the fabrication process of RCAT. A 2 GB density DRAM using SRCAT was developed in 70 nm feature size in 2005 [6].

S-RCAT structure has a number of improvements over that of RCAT. The mean value of V_{th} is decreased by 200 mV and the V_{th} distribution is improved to 250 mV from 350 mV mean value. With the same channel doping density S-RCAT hence needs just 0.9 V through improvement in DIBL and subthreshold swing (SS), in comparison to the requirement of 1.2 V in RCAT to

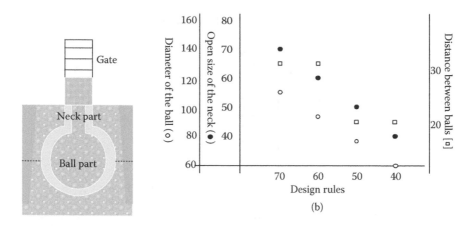

FIGURE 4.2

(a) Basic S-RCAT structure. (b) Scalability of the S-RCAT (all dimensions are in nm). (Modified from "S-RCAT (Sphere-Shaped-Recess-Channel-Array-Transistor) Technology for 70 nm DRAM Feature Size and Beyond," J.Y. Kim et al., Symp., VLSI Techn. Dig. Tech. Papers, pp. 34–35, 2005.)

keep data retention for an arbitrary data pattern. DIBL is improved from 80 mV to 40 mV and SS is improved from 125 mV/dec to 105 mV/dec. Body effect is also improved from 600 mV/Vb to 470 mV/Vb and it helps lower the word-line boosting voltage. Junction leakage current also decreased to half as compared to RCAT. In RCAT leakage current is minimized by lowering channel-doping level, but in S-RCAT junction area is also reduced, resulting in further leakage current reduction. Figure 4.2(b) shows the S-RCAT dimensions which are reduceable beyond sub-50 nm. Though other transistors have been employed in sub-50 nm level, interest in S-RCAT is continuing. In a different kind of behavior of S-RCAT, negative DIBL was observed at 90°C unlike at 60°C. Reasons for such behavior have been studied also [7] suggesting continued use of S-RCAT beyond 50 nm.

4.3 Vertical Depleted Lean-Channel Transistor Structure

Much before the arrival of RCAT/SRCAT, limitations of the planar transistor even other than mentioned in Section 4.1 were becoming obvious. Device sizes and isolation lengths were becoming nearly equal to depletion layer widths. Short channel effect (SCE) and parasitic MOS effects were becoming serious problems. Sub-threshold swing was also to be controlled for maintaining I_{on}/I_{off} ratio within safe limits. Consequently, three-dimensional device structures which increased the effective length of the channel came into existence in the late 1980s. There were several novel vertical structures

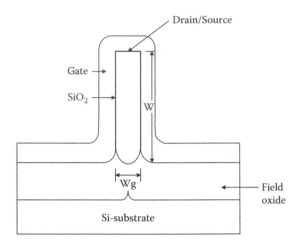

FIGURE 4.3

Schematic cross- section of a DELTA structured transistor. (Redrawn from "Impact of Vertical SOI "DELTA" Structure on Planer Device Technology," D. Hisamoto, T. Kaga and E. Takeda, IEEE Trans. Electr. Devices, Vol. 38, No. 6, June, 1991.)

using complex fabrication procedures in terms of the technology of that time that became standard later on like cylindrical and surrounding gate transistor [8,9]. In another line of thinking the channel was intentionally depleted using thin-film SOI technology. The technique had several advantages like having better immunity to SCE, high transconductance, reduction in the substrate floating effects of vertical transistor structure. In addition, it also provided different device characteristics through the interaction between front and back channels like higher trans-conductance and minimization in the kink effect which was the result of substrate floating effect. At that point in time formation of ultrathin SOI film was a difficult process; therefore, to overcome this limitation a fully depleted lean-channel transistor (DELTA) was proposed [10], which is schematically shown in Figure 4.3. On a 25-nm thermal oxide pad, 200 nm of CVD nitride is deposited and Si island is formed through the use of anisotropic reactive ion etching. Next, 10 nm of thermal oxide is grown and 100-nm CVD nitride layer is deposited, nitride spacers are formed, and oxidizing the substrate at 1100°C forms an SOI structure. The gate structure, which steps over the Si island, is formed and finally, using gate as an arsenic ion mask, the source and drain are implanted. Effective channel width becomes larger because current flows on both, vertical as well as on top of the horizontal surface and it results in better conductance. It is observed that DELTA structure has strong immunity to short channel effects and has superior sub-threshold characteristics. In spite of its vertical structure, DELTA has high compatibility with conventional MOSFET fabrication process flow. Soon the structure became the starting point of FinFETs, which led to generations of practical/manufacturable DRAMs.

4.3.1 Double Gate MOSFET

Double gate MOSFET structure has widely been studied as it offers distinct advantages for scaling to very short gate lengths [11,12], though the complexity of the fabrication process was a major obstacle in adopting double gate CMOS (DG CMOS) architecture. Additionally, the use of ultrathin body devices suffered from parasitic resistance and threshold voltage controllability [13]. However, applying new ideas, especially through the use of DELTA devices now commonly referred to as the FinFETs, significant advances in DGCMOS usage have been obtained.

In double gate (DG) FETs, a second gate opposite to the control gate is added, and it is the basis of better control over SCEs. Figure 4.4 shows a simplified representative view of double gate FET along with a conventional MOS. As the channel length of the FET is reduced, drain depletion region becomes comparable to it and the control gate even loses its control to shut off the channel current. One method to overcome this problem is to make gate oxide thinner; however, the gate oxide could be made thinner up to a certain stage and no more, because after that power drain due to gate leakage becomes comparable to the power used for switching of the transistor. Another technique is to have thinner depletion depth below the channel, so that channel gets shielded from the drain but further decrease of the depletion region degrades gate influence on the channel and makes the transistor slower.

The second gate being closer to the channel than the control gate has greater effect, and it is more effective in screening the electric field generated by the drain potential. It results in reduced short channel effect and in particular the drain-induced barrier lowering (DIBL) and improves sub-threshold swing.

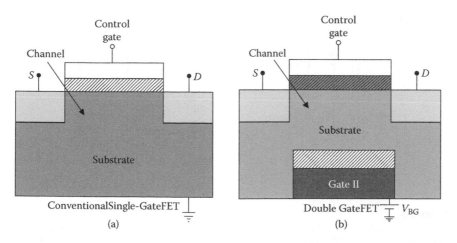

FIGURE 4.4
Simplified view of (a) conventional MOSFET and (b) a conventional double gate MOSFET.

Therefore, at a stage where conventional planar CMOS scaling becomes prohibitive due to excess leakage current, DGCMOS can go much beyond that. In single gate, increased body doping could reduce DIBL; however, at a certain doping level sub-threshold swing would also increase. In turn it needs higher threshold voltage to keep sub-threshold current within desired limits. A decrease in body doping acts in opposite way, hence a compromise is necessary in conventional MOSFET, to decide the level of body doping.

A large number of DGFET structures have been reported. These may be classified into three categories. (1) The planar DGFET shown in Figure 4.5(a) is a simple extension of single-gate FET, with a second gate buried in the bulk. (2) In the vertical DGFET, which is shown in Figure 4.5(b), the body of the FET has been rotated to a vertical orientation on the wafer with the source/drain being on the top and bottom, and the gate on either side. (3) In nonplanar FinFET, the body has been rotated on its edge into a vertical orientation with source and drain region placed horizontally about the body, as shown in Figure 4.5(c). While fabricating the DG-FET and transforming it

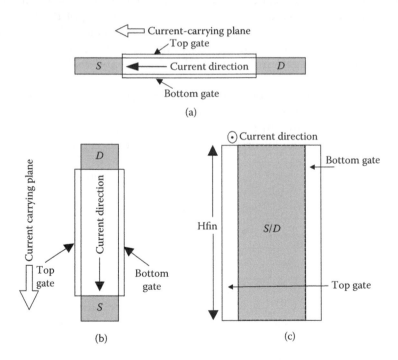

FIGURE 4.5
Simplified view of (a) planar FinFET, (b) vertical DGFET, (c) non planar DGFET. (Adapted from "High Performance Double Gate Device Technology Challenges and Opportunities," M. Ieong et al., Proc. Int. Symp. Quality Electron Device, 2002.)

into a manufacturable device, quite a few problems were encountered, which are given below:

1. Both the gates are not mirror images.
2. Self-alignment of the source/drain regions to both the gates is difficult.
3. Both the gates are difficult to align.
4. An area-efficient means of connecting the two gates with a low resistance path is difficult.

Type (a) or (b) DGFETs (Figure 4.5) are able to solve some of the mentioned problems but are not able to address all the problems whereas type (c) can. Vertical fin-type DG-FETs (b) have the advantage of having access to both gates and both sides of source and drain, from the front of the wafer but suffer from worse channel thickness uniformity than planar films.

4.4 FinFET—A Self-Aligned DG-MOSFET

To suppress short channel effect, a self-aligned double-gate MOSFET, FinFET based on DELTA [10] structure was proposed by Hisamoto and others [14]. The quasiplanar nature of FinFET makes its fabrication easier using the conventional planar MOSFET process technologies. A typical layout is shown in Figure 4.6 with schematic cross-sectional structure shown in the last part of Figure 4.7 [14]. The following are the main features of the FinFET:

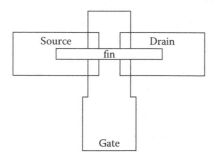

FIGURE 4.6
Typical layout of a FinFET. (Redrawn from "FinFET—A Self-Aligned Double-Gate MOSFET Scalable to 20 nm," D. Hisamoto, et al., IEEE Trans. Electron Devices, Vol. 47, No. 12, pp. 2320–2325, Dec. 2000.)

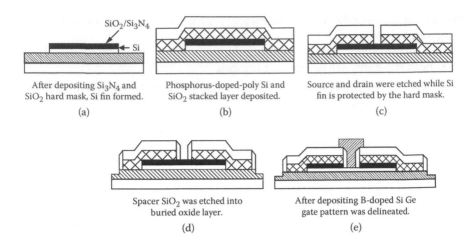

(a) After depositing Si_3N_4 and SiO_2 hard mask, Si fin formed.

(b) Phosphorus-doped-poly Si and SiO_2 stacked layer deposited.

(c) Source and drain were etched while Si fin is protected by the hard mask.

(d) Spacer SiO_2 was etched into buried oxide layer.

(e) After depositing B-doped Si Ge gate pattern was delineated.

FIGURE 4.7
FinFET fabrication process flow. ("FinFET—A Self-Aligned Double-Gate MOSFET Scalable to 20 nm," D. Hisamoto, et al., IEEE Trans. Electron Devices, Vol. 47, No. 12, pp. 2320–2325, Dec. 2000.)

1. The DG-MOSFET is built around the ultra thin silicon fin (~10 nm), which is the most important component for suppression of short channel effects.

2. Raised source/drain configuration [15], which was used in *delta* configuration, and the poly-Si film which wraps around the fin reduced the S/D resistance.

3. Self-aligned gates, which are aligned to the S/D, provide better performance.

4. As no high temperature steps are required after gate deposition, low-temperature gate process with high-k dielectrics can be used.

5. Boron doped $Si_{0.4}Ge_{0.6}$ as a gate material provides proper threshold voltage for the ultrathin body MOSFET.

Figure 4.7 shows the process flow for the fabrication of FinFET. A 50 nm silicon film on a SOI wafer with 400 nm thick buried oxide layer can be used as the starting material and a CVD Si_3N_4/SiO_2 layer is deposited on it to make a hard mask that protects the fin later on. Nearly 20 nm wide Si fin was patterned using electron beam (EB) lithography. With SiO_2 hard mask present, etching exposes Si only at the sides of the Si fins. *In situ* p-doped, 100 nm amorphous Si is deposited for S/D pads at 480°C and a 300 nm SiO_2 film is then deposited. The process of realizing fins and S/D areas is very similar to that used to define trench isolation in CMOS. Again using EB lithography, the S/D pads are marked and a gap is formed between them.

Spacers are formed around the S/D pads through CVD SiO_2 and then spacers on the sides of the Si fin are removed while Si fin remains protected. At this stage fins whose width was 20 nm initially reduce to 15 nm, and are located in a gap between two spaces of light oxide. The gap is nearly 50 nm, and it determines the gate length which is now 20 nm. A 2.5 nm film of gate oxide is grown through oxidation, which simultaneously helps diffusing phosphorus from the S/D pads into the Si fin. Finally *in situ* boron-doped SiGe (60% Ge) was deposited as the gate material at 470°C and probing windows are etched; then no high-temperature process is needed, providing compatibility with the use of high-k dielectric.

FinFET processing using SOI wafer employs standard planar manufacturing process modules. Spacer lithography, sidewall image transfer (SIT), is used for etching those ultrathin fins, which always generates an even number of fins. If an odd number of fins are to be fabricated, an extra step is needed to *trim* a fin. It is to be noted that the transistor derives current as many times as the number of fins realized.

If a planar design is to be converted for processing in the FinFET technology, the FinFET height H_{Fin} together with the fin pitch defines the FinFET device width within the given silicon width of the planar device to get a better or same device strength.

Variation in device width and threshold voltage and self-heating must also be considered while designing with FinFETs. Having multiple thresholds and multiple gate oxide thickness was supposed to provide full advantage of the FinFETs. Since FinFETs have reduced channel and gate leakage currents, a power reduction factor of two is possible when a planar design is converted to a FinFET technology. Power can also be saved when both gates are controlled separately. The second gate controls the threshold voltage of the device, allowing fast switching as well as reduced leakage current when circuits are idle.

Typical output characteristics of a 30-nm gate length FinFET show no kink effect due to floating body nature due to its ultra thinness [14]. Measured sub-threshold leakage current has been well suppressed for low channel impurity concentration of $2 \times 10^{16}/cm^3$. For a 30-nm gate length device, channel current is measured with respect to the D/S voltage, with Si fin width as a parameter. It is observed that larger Si fin widths show a stronger DIBL effect and Si fin widths just smaller than the gate length may be sufficiently small to suppress DIBL [15]. Sub-threshold swing is a little large in the range of 70–100 V/dec when fin width increases from 15 nm to 55 nm. Its transconductance peaks at 30 nm Si fin width. Another distinct feature of adapting SiGe with 60% Ge as gate material is gate work function control, which determines the threshold voltage; other metals like TiN and Ta were not found suitable [16].

Suppression of SCE has been demonstrated by using sub-50 nm NMOS [14] and PMOS FinFET [17], but their fabrication process is complex and results in large overlap capacitance between the gate and S/D regions. A sub 20-nm gate-length CMOS FinFET given by Choi and others [18] overcame the problem of large S/D-gate overlap capacitance by using fin formation by spacer

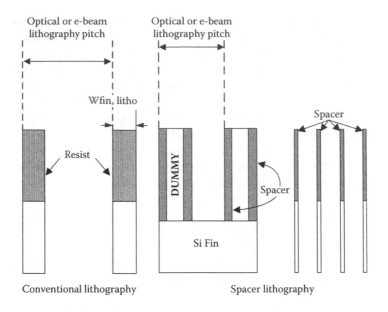

FIGURE 4.8
Comparison of fin density between conventional lithography and spacer lithography technology. (Redrawn from "Sub 20-nm CMOS FinFET Technologies," Y.K. Choi, et al., IEDM, pp. 421–424, 2001.)

lithography and raised S/D by selective Ge deposition. Initially, a hard mask of 50 nm silicon nitride is deposited on the 4 nm thin pad oxide and then 200 nm sacrificial layer of LPCVD $Si_{0.4}Ge_{0.6}$ is deposited on it and for spacer FinFETs. After that, phosphosilicate glass (PSG) is deposited and etched and $Si_{0.4}Ge_{0.6}$ layer is also removed. The PSG spacers are used as a hard mask to define the narrow Si fins as the PSG thickness determines the fin width. For a given lithography pitch, the fin density is doubled, resulting in principle, doubling the drive current as shown in Figure 4.8. Figure 4.9 shows the spacer FinFET structure, where the Si fins are defined by PSG spacers. Threshold voltage, V_{th}, is found to be less sensitive to body doping type and concentration compared to bulk CMOS and more strongly controlled by gate work function. Selective Ge deposition and raised S/D improve transistor drive current.

FinFETs are considered promising candidates for sub-50 nm range in CMOS scaling because of their small threshold swing and DIBL and good SCE immunity resulting from excellent gate controllability with thin Si substrate. Most of the basic issues connected with the FinFET fabrication and performance have been attended to, such as realization of the narrow and uniform fins, lowering of the series resistance from the extension and contact regions, threshold voltage control and its variation, and so on [19,20]. However, placement of the fins at a fine pitch to make efficient use of layout area was studied by H. Shang and others in 2006 [21], wherein e-beam lithography was used

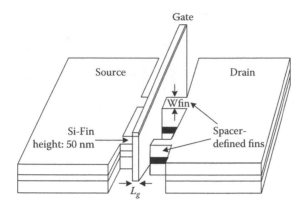

FIGURE 4.9
Schematic diagram of spacer defined fins in a FinFET. ("Sub 20-nm CMOS Fin FET Technologies," Y.K. Choi, et al., IEDM, pp. 421–424, 2001.)

for 20 nm wide multiple fins and 120 nm pitch at 32 nm node technology. In addition, spacers are formed only along the gate side, which allowed raised S/D fabrication using selective epitaxial on the sides of the fins. For the formation of multiple fins, S/D landing pads were used for connecting fins in parallel, which was not conducive for future technologies. In this scheme, S/D pads were eliminated, though it required strapping of parallel fins by local interconnects and trade-off takes place between parasitic resistance and capacitance. Raised S/D helps in reducing series resistance and presence of a hard mask on top of the fin minimizes parasitic capacitance.

4.5 Body Tied MOSFETs/Bulk FinFETs

The main attraction of FinFETs was their scalability and the process compatibility with the conventional planar MOSFETs in spite of the fact the FinFETs are basically SOI MOSFETs [21–23]. However, high wafer cost, high defect density, and heat transfer problems led to overcoming these problems. Moreover, SOI FinFETs become three terminal devices because of the floating body, which restricts the circuit operation window. Hence four terminal FinFETs were evolved in which fin body was connected to the Si substrate. These four-terminal FinFETs were built on bulk Si wafer, and were called body-tied FinFETs, though at early stage these were called Omega MOSFETs because the cross section of the body resembled Greek letter Ω [23–25]. For distinction's sake, the body-tied FinFETs are called bulk FinFETs. The bulk FinFETs keep the advantage of SOI FinFETs of compatibility with the conventional planar MOSFETs fabricated on bulk Si wafers, as well as the property

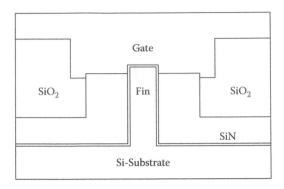

FIGURE 4.10
Front view of an Omega MOSFET. (Adapted from "Fabrication of Body-Tied FinFET (Omega MOSFETs) Using Bulk Si Wafers", T. Park et al., Symp. VLSI Tech. Dig., Tech. Papers, pp. 135–136, 2003.)

of scalability. In addition the bulk FinFETs have much less heat dissipation problem [26] than SOI FinFETs and problems like wafer cost, defect density, floating-body effect are also minimized/eliminated [27]. Another important advantage of bulk FinFETs is much less back-bias effect than conventional planar transistors have.

First body-tied FinFET (Omega MOSFET) or bulk FinFET on Si wafer instead of SOI wafer was fabricated in 2003 and its simplified front view is shown in Figure 4.10 [24]. A few important fabrication steps are as follows. On p-type (100) wafer, trenches were etched up to a depth of nearly 300 nm. Thermal oxidation was carried out which was completely etched away and only thin fins remained standing vertically in which the channel and the S/D were formed. A thermal oxide layer and a SiN layer (50 nm) were deposited and trenches were then filled with high-density plasma (HDP) CVD SiO_2. The SiN layer was additionally recessed (depth ~100 nm) and fin sidewalls were opened. Gate electrodes were patterned and after p^+ ion implantation for LDD, 45 nm thick SiN spacer was formed. The rest of the process steps were the same as for the conventional DRAM realization. It is observed that Omega MOSFET has much lower value and smaller deviation in sub-threshold swing, and very small body-bias dependence. DIBL of 24 mV/V is obtained against 108 mV/V for a conventional DRAM cell transistor. There is a considerable improvement in SCE and I_{sub}/I_D characteristics; hence, with slight modification Omega MOSFETs were expected to be promising nano-era bulk CMOS devices. One of the limitations of the proposed Omega MOSFET was that simultaneous significant improvement of SCE and high driving current was not possible. Mainly to overcome this deficiency and the limitations of the conventional FinFET on SOI, a highly manufacturable fin-channel-array-transistor (FCAT) on bulk Si substrate with new STI fin formation was given [28]. FCAT showed a remarkable improvement in SCE having SS of 75 mV/dec at drain voltage $V_d = 0.1$ V and excellent DIBL behavior with

DIBL = 13 mV/V at V_d = 0.1–2.1 V, because of high controllability of the electric field. Drive current of the FCAT also improved to ~12.7 µA/cell from ~2.4 µA/cell of a planar transistor with gate and drain voltage being 2.5 V.

There are quite a few reasons for limited on-current as the technology moves below 100 nm. Channel width becomes limited and electron mobility degradation takes places due to increased channel doping. At the same time gate oxide scaling cannot be applied so as to keep off-current within standard limit (0.1 fA at 85°C). FinFETs were considered excellent alternatives for nanoscale DRAM cell transistor because of several advantages including reduction in SCE and overcoming the limitations just mentioned. A fin gate array transistor fabrication in 130 nm technology node [29], with a deep trench (DT) capacitor [30], is applied to a DRAM cell. On current of the fin-array-FET is 62 µA/cell, which is 1.7 times that of a conventional planar array cell FET, and the on-current is found to improve with shrunken gate length. At the same time off-current is 0.1 fA/cell. The fin-array-FET was designed using the 3-D process simulator (Hy Sy Pros) and device simulator (Hy DELEOS) and boron (p+) was used to dope poly gate to get positive threshold. However, the process was found to be a little complex which increased fabrication cost.

It became quite apparent that beyond 80 nm technology SCE control was difficult. Omega MOSFET, FCAT, and RCAT, etc. all tried to overcome this problem and were successful to a considerable extent, albeit with some other deficiencies. For example RCAT has much higher body-bias dependence than that of planar MOSFET. A novel body tied FinFET (but not of omega type) cell array transistor on bulk Si substrate was fabricated and claimed to be highly manufacturable for sub 60 nm technology [25]. Process flow for the proposed body-tied FinFET DRAM integration started on p-type bulk Si (100) wafer and an 80 nm Si active fin was fabricated by STI etch process and CVD thin thermal oxide and a nitride film were deposited. An HDP oxide was then deposited and planarized using CMP and the HDP oxide was selectively recessed for the DRAM cell area to define height of the fin, which was in the 80 to 120 nm range. The sub-threshold swing is remarkably better when Si active width is reduced as shown in Table 4.3; however, V_{th} starts decreasing around 100 nm technology and becomes too low and practically uncontrollable below 40 nm. The main cause of reduction in V_{th} at lower technology node is due to the out diffusion of boron of the Si fin during gate oxidation step. One suggested remedy was to use higher boron concentration but it increases junction leakage current at the storage node bottom. Some techniques were tried to overcome this problem, including use of p+ doped polysilicon gate [29]. The negative word line (NWL) scheme was found to be a better solution and additional implant dose for FinFET was not necessary. A comparison of body-bias dependence shows better performance for the bulk FinFET over planar transistor and RCAT. It also shows a better current drivability over RCAT and planar transistors. The DIBL characteristics of

TABLE 4.3

Threshold Voltage and Sub-Threshold Swing Variation with Fin Width for Body-Tied FinFET

Fin width (nm)	30	45	60	75	70	105	120
Threshold voltage (V)		0.09	0.21	0.38	0.5	0.7	0.77	0.8
Sub-threshold swing (mV/dec)	74	75	77	79	81	85	87

Source: C.H. Lee et al., "Novel Body Tied Fin FET Cell Array Transistor DRAM with Negative Word Line Operation for Sub 60 nm Technology and Beyond," *Symp. on VLSI Tech. Digest of Tech. Papers*, pp. 130–131, 2004.

FinFET (T_{si} = 80 nm, L_g = 90 nm, H = 100 nm) are compared with RCAT and planar transistors with gate length of 90 nm. DIBL of bulk FinFET was as low as 25 mV and DIBLs for the RCAT and planar transistor were 50 mV and 225 mV, respectively. Hence, a combination of NWL scheme with body tied FinFET made it very attractive [25].

Further improvement was done in body-tied FinFET DRAM through the use of damascene FinFET (D-FinFET) [31] and local-damascene (LD) FinFET DRAM [32] technologies. In conventional FinFET (C-FinFET), drain current reduces and contact resistance of the bit line to storage node increases, with decreasing fin width. Both the limitations have been overcome in D-FinFET using self-aligned local channel ion implantation (LCI), which could not be applied to the C-FinFET. Figure 4.11 shows structures of C-FinFET, D-FinFET, and LD-FinFET. After STI processes, all field areas in the cell array are etched back in C-FinFET. In the D-FinFET, only the field area under the gate line is

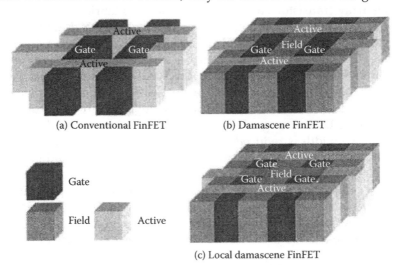

(a) Conventional FinFET   (b) Damascene FinFET

(c) Local damascene FinFET

FIGURE 4.11
Illustrations of conventional and local damascene FinFET structure. ("Local-Damascene-Fin FET DRAM integration with p+ Doped Poly-Silicon Gate Technology for sub-60 nm Device Generation," Y.S. Kim, et al., IEDM, 2005, pp. 325–328.)

etched, whereas for the LD-FinFET, the field area under the gate lines at both sides of the active fin is etched.

When the access transistor is activated in C- and D-FinFET cell array structure, the electric field at the STI sidewall increases, having damaging effects on neighboring cells. For example when negative word line is used, the access transistor gate in stand-by mode can degrade the junction leakage characteristic at the transistor drain region. In D-FinFET, the damascene area is confined to the only two sides of actives, overcoming the problem. Another advantage of LD-FinFET structure is in terms of obtaining very uniform distribution of threshold voltage. Use of p^+ boron *in situ* doped polysilicon gate gives threshold voltage of 0.75 V and its variation is about 100 mV, whereas SRCAT has higher threshold voltage of about 1.0 V and its variation range is about 300 mV.

Soon after double and triple gate body-tied FinFETs were introduced [24,27–29], and many fully integrated DRAMs were produced using such FinFETs [25]. However all of these works presented N-Channel body-tied FinFETs. To complement them, a P-channel body-tied (triple-gate) MOSFET on bulk Si wafer was also proposed [33]; here gates are formed on top and on both sidewall channel surfaces. The triple-gate PMOSFET shows much less body-bias dependence compared to planar PMOSFET at 100 nm scale but a slightly higher V_{th}. It also shows lower I_{sub}/I_D characteristic than planar PMOSFET, probably due to the drain field affected by normal gate field, and the on-current of the device is also higher. DIBL of 34 mV/V was a considerable improvement over 92 mV/V of planar pMOSFET at the same gate length of 100 nm and a sub-threshold swing of 74 mV/dec was also an improvement, in comparison to 92 mV/dec for the planar MOSFET. In addition to DRAMs, six-transistor SRAM using body-tied triple-gate MOSFETs (bulk FinFETs) has also been realized at 90 nm node technology [34].

4.6 Multichannel FET

Double gate MOSFETs and the bulk MOSFETs discussed in the last section have shown a great promise with good performance in terms of DIBL, SS, I_{sub}/I_D ratio, and so on, and also from the manufacturability point of view. Due to the use of certain processes like dry etching for trimming fin width, width of the source/drain region also narrows down and that increases their resistance [35]. In addition multi-fin layout is required in some cases for the enhancement of transistor drive [23], but due to lithographic limits, pitch is more difficult to narrow down than the design rule. So the active area is not fully utilizable for the devices using FinFETs. A highly manufacturable double FinFET on bulk Si wafer, called multichannel field effect transistor (McFET) was proposed which could do active patterning without

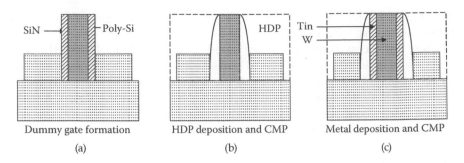

Dummy gate formation   HDP deposition and CMP   Metal deposition and CMP

(a)   (b)   (c)

FIGURE 4.12
Few steps in the process flow sequence for TiN gate McFET fabrication using replacement gate process. (Adapted from "Fully Working High Performance Multi-Channel Field Effect Transistor (McFET) SRAM Cell on Bulk Si Substrate Using TiN Single Metal Gate," S.M. Kim et al., Symp. VLSI, Techn. Dig of Tech. Papers, pp. 196–197, 2005.)

lithographical limits [36]. On a p-type (100) wafer trenches are etched which are filled with HDP oxide and first STI CMP is done. Then dummy gates are patterned and etched, HDP oxide is filled in spaces between dummy gates, and a second STI CMP follows. Si_3N_4 inner spacer is formed which decides the thickness of the channel. Then Si channels are etched and in the end channel fins are obtained on the active area. The I_D-V_D characteristics of N McFET with the gate length of 80 nm showed reduced SCE and four times more drive current than planar transistor. Source/drain resistance of McFET was found to be of the same order as that of planar transistor, but nearly 200 times less than that of FinFETs, which used active trimming. Notwithstanding good electrical characteristics, proposed McFET having twin fin channels had insufficient signal-to-noise margin (SNM) due to low value of threshold voltage. An advanced McFET using titanium nitride (TiN) single metal gate at 65 nm node technology on bulk Si wafer was proposed [37], which used the twin fin formation procedure of the McFET fabrication of ref. [36]. After the fin formation as given above, replacement gate process is used for which polysilicon dummy gate electrode is formed as shown in Figure 4.12(a). Source/drain are formed and after HDP oxide deposition and CMP, dummy gate surface is uncovered (Figure 4.12(b)). After dummy gate and dummy gate insulator removal, a second gate oxide is grown. TiN and W are deposited to form gate electrode as shown in Figure 4.12(c). This McFET showed suitable threshold and excellent characteristics of SS (70–75 mV/dec) and DIBL with $V_{thn} = 0.3$ V and $V_{thp} = -0.3$ V, and a realized McFET cell in an SRAM provided SNM of 350 mV at 1.0 V.

Initially the McFETs were used for the realization of SRAM cells. For DRAM cells (in the range of 80–60 nm) it was required that the peripheral circuits of the cell are also reduced in size so that these match the small pitches of bit lines and word lines. In addition to the necessity of having smaller peripherals from fabrication point of view, it also helped in reducing overall chip size and wire delays. The McFETs which were used for the realization of SRAM cell so far, were utilized to fabricate sense amplifiers and the sub word drivers and so on—that is, the cell periphery circuits alongside the use of FinFET for the DRAM core and planar transistors for periphery circuits [38]. The problem of integrating three different types of transistors on a single chip could be solved by using three new techniques; (1) selective STI SiN liner removal (SLR) process for different transistor profile formation, (2) selective TiN gate for McFETs, and (3) narrow active pitch patterning for current gain enhancement. Through the process flow of the SLR approach, it has clearly been shown that only one photo step and SiO spacer process selectively formed three different kinds of transistors. DIBL was not observed but SCE was shown to be suppressed down up to $L_g = 100$ nm. McFETs improved drain current by 27% and 36% for N and P types, respectively, at the 80 nm design rule based DRAM and these could be scaled down to 55 nm technology.

4.7 Saddle MOSFET

Figure 4.13 shows three-dimensional schematic and cross-sectional views of a new MOSFET structure, which is slightly different from a recess channel MOSFET as, along with channel, it has a side gate also [39]. The new transistor called saddle MOSFET has an additional side channel overlap ($L_{ov\text{-}side}$), and it is fabricated on bulk Si wafer. The lightly doped drain/source junction depth (x_j) for this saddle MOSFET is about 21 nm, and that of heavily doped drain/source is about 33 nm with gate oxide thickness of 3.5 nm. Body width is 20 nm and the overlap $L_{ov\text{-}side}$ for this width is about 10 nm. The thin body is directly connected to the substrate. The threshold voltage (V_{th}) at the side channel of the recess region is lower than that at the bottom channel due to localized channel doping (peak value 3×10^{18} cm^{-3}), in the same way as that for the conventional recess MOSFET. The I_{on}/I_{off} of the saddle MOSFET is at least 10^3 times higher in comparison with the recess MOSFET, mainly due to the 10 nm of side overlap and I_{on} is nearly 20 μA at $V_{DS} = V_{GS} = 1.5$ V; this is one order of magnitude larger than recess channel device, because of two side-gate effects. A comparison of some of the important parameters have been shown in Table 4.4, between conventional recess channel NMOS and a saddle NMOS, showing clearly the advantageous position of the saddle MOSFET. In addition to mentioned advantage of saddle MOSFET, it has excellent short channel effect (SCE) immunity but suffers the serious drawback of higher gate-induced drain leakage (GIDL) because of the wider overlap area from the gate to the source/drain when x_j is made deep to reduce junction leakage. It also has higher word-line capacitance due to the large overlap. The problems of such nature posed difficulty in practical utilization of the saddle MOSFET for DRAM technology. The saddle MOSFET, which

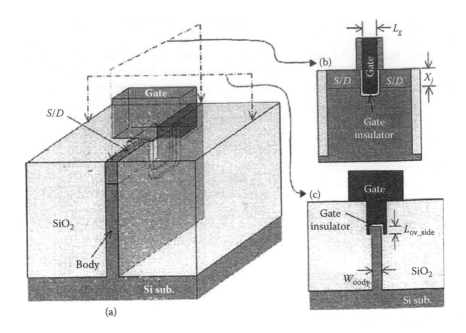

FIGURE 4.13

(a) Three-dimensional schematic view of saddle MOSFET, and its cross-sectional views across (b) the gate, and (c) the thin body. (Highly scalable Saddle MOSFET for High Density and High Performance DRAM," K.-H. Park, K-R Han and J-H. Lee, IEEE Electron. Dev. Letters, Vol. 26, No. 9, pp. 690–692, 2005.)

TABLE 4.4

Important Characteristic Comparison between Recess Channel NMOS and Saddle NMOS

Parameter	Recess Channel NMOS	Saddle NMOS
Work function to control V_{th} ($\varnothing_m$)	4.17 eV	4.71 eV
I_{off}	~0.1 pA	0.1 fA
Body width (W_{body})	≤(⅔) L_g	No constraint
I_{on} ($V_{DS} = V_{GS} = 1.5$ V)	~2 μA	20 μA
Sub-threshold swing (SS)	132 mV/dec	69.5 mV/dec
Drain Induced Barrier Leakage ($DIBL$)	246.2 mV/V	30.9 mV/V

is shown in Figure 4.14, does have a recessed channel, but with localized side-gate possesses excellent characteristics in GIDL and gate-to-S/D overlap capacitance [40]. The localized side-gate shown as a hatched region is only under a vertical position near the S/D junction depth x_j. It results in the reduction of unwanted overlap of the gate and S/D region and only necessary slight overlap remains. Hence, the only difference in the structure of the saddle MOSFET and the modified saddle MOSFET is the length of

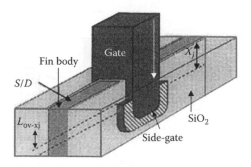

FIGURE 4.14
Three-dimensional schematic view of the modified gate of the saddle MOSFET. (Redrawn from "Simulation Study of High-Performance Modified Saddle MOSFET for Sub-50 nm DRAM Cell Transistors", K.-H. Park et al., IEEE Electron. Dev. Lett., Vol. 27, pp. 759–761, 2006.)

the S/D region from the x_j overlapped by the side-gate. At a fixed W_{body} of 40 nm, the modified saddle MOSFET has a smaller DIBL (~15 mV/dec.) and SS (~89 mV/dec.) than those of the recessed channel device (~94 mV/V and ~99 mV/dec.). However, more important is the value of GIDL which is larger than 0.1 pA at a L_{ov-xj} of 87 nm. Characteristics of modified saddle MOSFET which makes it a promising transistor for the 50 nm DRAM technology and beyond.

4.8 Saddle FinFET

Recess channel array transistors (RCATs) have the advantages of using reduced channel doping concentration, reduced leakage, and, hence, excellent refresh characteristics [4,6], and the FinFETs have high I_{on} due to considerable larger channel width. FinFETs have efficient short channel effect but require either negative word line scheme or work function engineering to control off-leakage current [25,32]. A saddle fin (S-Fin) transistor tried to combine the better aspects of the RCATs and FinFETs to make it suitable for high density DRAM. Figure 4.15 shows the simplified front and top view of the S-Fin, which is fabricated by etching both the active Si and the field oxide. Fabrication process is in line with both RCAT and damascene FinFET where only active Si and only the field oxide, respectively, were etched. As a result, a groove like RCAT in the channel length direction, and fin structure like the FinFET in the channel width direction, with a well-rounded corner profile, was formed [41].

It was observed that for normal DRAM operating condition, S-Fin current was maximum and 25% more than that of the RCAT. Although the S-Fin did not show the highest threshold, it can be adjusted in an easier way

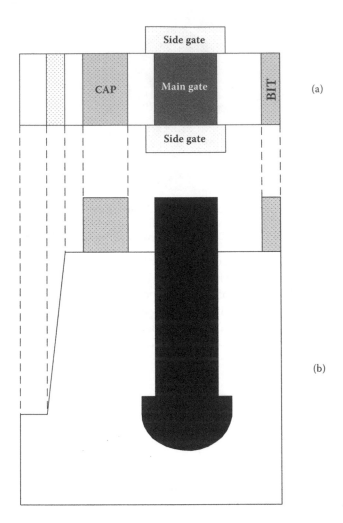

FIGURE 4.15
(a) Top view of R FinFET. (b) Front view of R FinFET. (Modified from "Highly Scalable Saddle-Fin (S-Fin) Transistor for Sub-50 nm DRAM Technology," S-W. Chung et al., Symp. VLSI Tech. Dig. of Tech. Papers, pp. 32–33, 2006.)

than damascene Fin without using any change of scheme. The S-Fin was also found to be much superior in terms of DIBL, SS, and transconductance. Investigation on the effect of variation of fin height and width of S-Fin were made on the different parameters like DIBL, V_{th}, SCE, and V_{bb} dependency, and it was found to be suitable for scalability for sub-50 nm technology [41]. However, very soon it was discovered that in spite of some excellent characteristics due to its trigate nature, S-FinFET faced certain critical problems of drain leakage and threshold voltage control [42]. Hence, an experimental analysis of the current drivability and mechanism of leakage current for the

S-Fin device as well as for RCAT, a representative recessed channel device, was taken up.

While the effective channel length becomes large, it controls SCE better, even with reduced feature size, but because of single gate, current drivability of RCAT is poor in comparison to the S-Fin, where the recessed channel is surrounded by a trigate. For the purpose of comparing the analysis of two recessed transistors, it is assumed that asymmetric channel doping profile is adopted with channel doping concentration much higher in bit line side than in the bottom and storage node side [43], so that in a DRAM cell, MOSFETs could maintain high threshold voltage and keep the leakage current low at the same time. A detailed analysis of the mechanism of leakage current concludes that the RCAT leakage mainly originates from the bottom of the channel region, because of usually thinner oxide. It creates a stronger electric field in the bottom region. Therefore, for the recessed-channel devices RCAT and S-Fin, doping level at the bottom channel region needs to be low in order to relax the electric field. However, for the S-Fin an additional reason for the larger leakage is the widened gate-drain overlap area and the strong field in that region. For the S-Fin, in spite of reduced lesser leakage from the bottom channel region, overall effectiveness in suppressing the off-stage leakage is reduced.

Another point of analysis in the recessed channel type devices was the evaluation of device degradation caused by Fowler-Nordheim and hot carrier stresses. It was observed that the interface trap density, a critical factor for leakage in recessed-channel structures, is much less in S-Fin than in RCAT. It probably happened due to silicon surface treatment used in the processing of S-Fin devices. In addition, the Fowler-Nordheim stresses create fewer traps in the S-Fin device due to thicker gate oxide in the side-gate part and the lower bottom channel doping concentration. In totality, it is the trigate channel which is helpful in suppressing the leakage, in spite of the increased total interface trap in S-Fin.

On the basis of the results obtained from the mainly two-point analysis, an improved device structure, having combined advantageous points of RCAT and S-Fin, is proposed. It is called recessed channel FinFET (R-FinFET), which has a trigate only in the bottom channel region and planar gate structure in the S/D regions. The R-FinFET does not have the side gates in S/D overlapping region, and it was designed with the same channel shape as the S-Fin. Simulation studies on the RCAT, the S-Fin and the R-FinFET show that R-FinFET has a much higher threshold voltage than the S-Fin with the same asymmetric channel doping profile of $3.5 \times 10^{18}/cm^2$ as shown in Figure 4.16(a). The R-FinFET and the S-Fin (with side gate width of 190 nm) have the same on-current because of equal side gate width as shown in Figure 4.16(b). However, the R-FinFET has lower leakage current and best I_{on}/I_{off} ratio can be obtained by the R-FinFET. In addition R-FinFET shows less leakage than the S-Fin under the off-state, even with the highest on-current level, because the side gate does not overlap with the drain region in the R-FinFET [42].

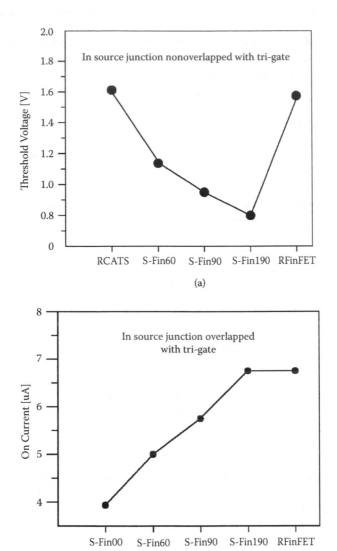

FIGURE 4.16
(a) Threshold voltage of the RCAT, S-Fin, and RFinFET structures for the case of the 160-nm source junction depth. (b) On-current of the S Fin and the RFinFET for the case of 200-nm source junction depth. (Redrawn from "A Proposal on an Optimized Device Structure with Experimental Studies on Recent Devices for the DRAM Cell Transistor," M.J. Lee et al., IEEE Trans. Electr. Dev., Vol. 34, No. 12, pp. 3325–3335, 2007.)

Finally, the retention time distribution of the RCAT, the S-Fin, and R-FinFET comparison shows that R-FinFET and RCAT give the best performance. The largest leakage in the channel structure is in the gate-drain overlap region due to high electric field profile [44] and in the S-Fin, the region is trigated, which intensifies high electric field; hence, poorer performance from retention point of view among the three devices is analyzed.

4.9 Surrounding Gate Transistor

With increasing packing density of integrated circuits, not only the channel length but channel width is also to be reduced for the planar transistors which decreases current derivability. In addition, reliability degradation due to the large electric field generation because of impurity diffusion at the LOCOS edge [45]. Shortening of channel length also causes severe problems in transistor performance, like threshold lowering (or SCE), increase of substrate bias effect due to increased impurity concentration in the channel region, and the generation of hot electrons.

At 1 Gbit DRAM level, a planar transistor can no longer be used because of the above mentioned effects, and the degradation of the sub-threshold characteristics. A three-dimensional transistor cell—that is, a trench-transistor cell (TTC), as described in Section 2.5—was realized to overcome the aforementioned problems. As the trench transistor used in the TT cell was vertical in nature, its dimensions were not constrained and it had excellent sub-threshold characteristics. However, due to the use of LOCOS process in the bit line isolation, TT cell size is limited to 6 F^2–9 F^2 (F being the minimum feature size). Surrounding gate transistor (SGT) [46–49] and SGT cell [50] were developed to overcome the limitations of the TT cell. In SGT, the transistor was formed around pillar silicon, while in the SGT cell, the capacitor is formed around the bottom of the pillar silicon island, and the transistor is also fabricated on top part of the pillar. SGT cell size could be reduced to 4 F^2, the limit of open bit line architecture, although in subsequent stacked SGT (S-SGT) structure, cell size was further reduceable.

Figure 4.17 shows the schematic view of SGT, whose source, gate, and drain are realized in a vertical fashion. Sidewalls of the pillar silicon island are used as the channel region as the sidewalls are surrounded by the gate electrode. Dimensional limitations of the planar transistor are overcome, because in SGT channel length and width depend on the pillar height and perimeter, respectively [46,47]. Moreover, in SGT structure channel region does not touch the field isolation edges; hence, LOCOS edge field enhancement effect is eliminated.

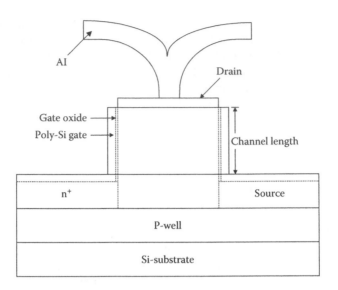

FIGURE 4.17
Side view of a surrounding gate transistor (SGT). (Redrawn from " Impact of Surrounding Gate Transistor (SGT) for Ultra High Density LSIs," H. Takato et al., IEEE Trans, Electron Devices, Vol. 38, No. 3, pp. 573–578, Mar., 1991.)

For fabricating NMOS SGT, first a p-well is formed on the Si-substrate with almost constant doping for nearly 2 μm depth for threshold voltage (V_{th}) adjustment as V_{th} adjustment is obtained by the well impurity concentration, and then a pillar Si island of 1–2 μm height is formed using Si grooving technology. Next, the surrounding gate structure is formed by depositing 20 nm thick gate oxide, n$^+$ Poly-Si is deposited and etched off, at proper places for gate electrode wiring. After that source and drain regions are formed through As implantation, [47]. Formation of contact hole at the top of the pillar is a critical step in this realization.

It is found experimentally that SGT has the same current drivability as the planar transistor of equal dimensions. However, SGT has larger effective channel width as all the sidewalls of the pillar island can be used as transistor channel. Thus SGT can reduce the occupied area easily. In addition to the high packing ability, SGT has several excellent characteristics, such as small substrate bias effects and high reliability.

A comparison of the sub-threshold characteristics of the planar transistor and those of SGT shows that threshold voltage is nearly equal for $V_{sub} = 0$ and the sub-threshold swing for the SGT is 72 mV/dec, whereas for the planar transistor it is 98 mV/dec. The value of SS is further decreased reaching toward an ideal value of 60 mV/dec when the Si island size is reduced [47]. This feature is due to the increased gate controllability with reduction in the pillar Si island's side dimension a. While studying dependence of V_{th} with respect to V_{BS} (1 V to 8 V), as a function of a, it is found that there is no

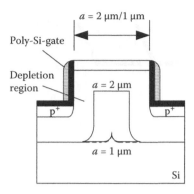

FIGURE 4.18
By decreasing the silicon pillar size of SGT from $a = 2$ to $a = 1$ the coefficient of substrate bias effect becomes much smaller, showing fullness in the pillar silicon island. (Redrawn from "Impact of Surrounding Gate Transistor (SGT) for Ultra High Density LSIs," H. Takato et al., IEEE Trans, Electron Devices, Vol. 38, No. 3, pp. 573–578, Mar., 1991.)

substrate bias effect at $a = 1.0$ μm, due to the fullness of the depletion region shown with striped line and its result of strong control of the potential of the channel region by the surrounding gate in the pillar Si island, as shown in Figure 4.18. Correspondingly, from V_{BS} changing from 1 V to 8 V V_{th} varies only a little for $a = 2$ μm in SGT but increases from 1.0 V to nearly 1.8 V for a planar transistor. It is the surrounding gate structure which gives only small depletion capacitance and the fully depleted Si pillar resulting in mentioned advantage of SGT. In addition, SGT has a small impact ionization rate, because the surrounding gate relaxes the lateral electric field at the drain edge, resulting in good reliability [47].

4.9.1 Multipillar SGT

Conventional SGT has one limitation. Its operational speed is limited due to the large gate capacitance and the large junction capacitance, which are caused by the large channel width and the large junction area at the trench bottom. To improve upon basic structure, a multipillar SGT (M-SGT) consisting of multipillar Si islands shown in Figure 4.19 can realize very high-speed operation [49]. Having very similar characteristics, it has large channel width which is a combination of circumference of all pillars, realizing high drivability than SGT.

In M-SGT, gate overlap capacitance is larger than that of a planar transistor because of larger overlap between gate and drain, but its effect is limited to an increase of nearly 20% gate electrode RC delay. At the same time gate resistance is only 20% of that of SGT or the planar transistor, because of the small occupied area, which is nearly 30% of that of planar transistor and mesh structure of gate electrode. Hence, resultant gate electrode RC delay is considerably reduced. Source junction capacitances of M-SGT and SGT

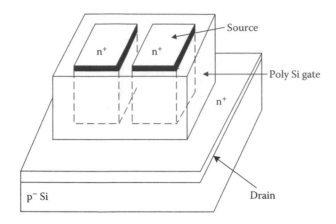

FIGURE 4.19
The gate electrode surrounds multipillar silicon island. The source, gate, and drain are arranged vertically. (Redrawn from "Multipillar Surrounding Gate Transistor (MSGT) for Compact and High-Speed Circuits," A. Nitayama et al., IEE Trans. Electron Devices, Vol. 38, No. 3, pp. 579–583, Mar., 1991.)

are much smaller than the planar transistor due to the fullness of depletion regions in the pillar Si island(s), and the drain junction capacitance of the M-SGT is 50% of that of SGT and 70% of that of the planar transistor. This also leads to the small gate electrode RC delay allowing M-SGT to operate at high speed. For example, at a gate length of 0.8 μm, propagation delay of M-SGT's comes down to nearly 20% of the delay with planar transistors. It is also observed that M-SGT has improved sub-threshold characteristics than SGT and the planar transistor. Sub-threshold swing SS is as small as 65 mV/dec, which is due to larger gate controllability of the channel. Greater gate surrounding effect takes place on the channel due to small square of pillars in comparison to a little weaker gate surrounding effect in SGT.

Significant design and optimization of vertical surrounding gate MOSFETs have been done for enhanced short channel immunity [50] and for enhanced transconductance-to-current ratio (gm/Ids) [51]. With reference to the dimensions shown in the cross-sectional view of the surrounding gate MOSFET of Figure 4.20, smaller values of both silicon film thickness, t_{si} and oxide thickness t_{ox} lead to high gate controllability, hence, enhanced short channel immunity. Once the ratio (t_{si}/t_{ox}) is reduced, short-channel immunity increases. However, to obtain smaller value of (t_{si}/t_{ox}), it should not be done by increasing t_{ox}, otherwise, gate controllability shall be reduced. As far as the (gm/Ids) value is concerned, an ideal value of 38.647 V^{-1} is obtained at channel lengths above 200 nm. An accurate model for (gm/Ids) ratio has been developed which requires accurate estimation of t_{si} and t_{ox} and the model has shown that SGTs are better than DG MOSFETs in this respect.

At reduced feature size for minimizing short channel effects it is not necessary to increase the p-doping between the source/drain region, as it creates

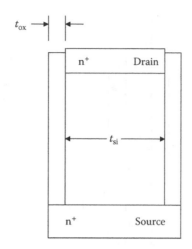

FIGURE 4.20
Basic dimensions through the cross-sectional view of the channel of a vertical surrounding gate MOSFET. (Redrawn from "Multipillar Surrounding Gate Transistor (MSGT) for Compact and High-Speed Circuits," A. Nitayama et al., IEE Trans. Electron Devices, Vol. 38, No. 3, pp. 579–583, Mar., 1991.)

the problem of increasing leakage current and deterioration in performance. It is observed that scaling down can be achieved in a fully depleted SGT structure after including some precautionary measures. It is known and can be illustrated through Figure 4.21(a) and (b) that in the fully depleted devices an active region becomes isolated from the substrate and the floating body can charge up. A transient bipolar current flows and discharges the cell capacitor [52]. The measure taken to prevent this leakage in fully depleted SGT is through the provision of a backside gate as shown in Figure 4.21(c), which allows the formation of a continuous channel and no current flows when the SGT is off [53]. Such a device was fabricated on the sidewall with a height of nearly 800 nm with such thicknesses that it reaches below 70 nm of the ridge; a 14–15 nm thick gate oxide is formed at (011) the ridge sidewall. A static retention time (bit line at 0 V) of several seconds with almost no transient leakage is observed when bit line is switched from high to low in the SGT structure with body contact. Proposed device retains the performance gain of a fully depleted body while eliminating critical floating body effects and held promise for transistors for 70 nm DRAMs and beyond.

In 2003 a study was conducted on three-dimensional transistors, double gate transistors, trench-isolated transistors (TIS) (using side-wall gate)/FinFET and SGT, from the point of view of finding pattern area reduction while realizing gate logic (NAND or NOR based), tapered buffer circuit, MPU, DRAM, and embedded DRAM. For studying typical logic gate realization, a case of four-input NAND gate was taken using each of the three-dimensional transistors mentioned above and for the purpose of comparison, area occupied by the transistors, area needed for the isolation of the transistor, well, and Al

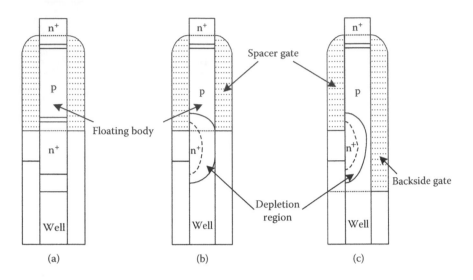

FIGURE 4.21
(a) and (b) Formation of a floating body by the lower S/D and depletion region. (c) A body contact provided by a hole accumulation layer generated by the presence of backside gate. ("Fully Depleted Surrounding Gate Transistor (SGT) for 70 nm DRAM and Beyond", B. Goebel et al., IEDM Tech. Dig., pp. 275–278, 2002.)

wiring were taken into consideration. Pattern design with SGT is markedly different because of its vertical current flow [54]. Smallest pattern area was realizable at ~47–48% through SGT or TIS/FinFET and for DRAMs minimum area of 65% was occupied by SGT and 42% for S-SGT.

In a conventional planar DRAM, chip area comprises nearly 60% of cell array, 10% tapered buffered circuit for clock series, 10% tapered buffer for I/O buffer, and 20% for sense amplifier and decoder, etc. Pattern area is reducible by 25% in double gate and FinFET case; a limit of 6 F^2 type memory cell. For SGT case and stacked-SGT case, cell area is reducible to 50% and 12%, respectively, the minimum value of 4 F^2 type memory cell or even less. These cells shall be discussed in Section 4.10.1. A summary of pattern area reduction in DRAMs is given in form of a comparative assessment in Figure 4.22(a) with the pattern area reduction in ASIC, MPU, DRAM, and eDRAM which shows the importance of SGT, though the design of its cell library was considered complicated.

The schematic cross section of an SGT DRAM cell is shown in Figure 4.22(b) [55]. Here an SGT is formed at the upper portion of the Si pillar as the fabrication procedure is described earlier and capacitor storage node is formed at the lower portion of the pillar. Since, in SGT, effective gate length is determined by the height of the upper portion, a relaxed gate length can be realized. Contact of the bit line is made at the top of the Si pillar. The cell is located in matrix-like trenches for isolation; thus, there is no inter-cell leakage even in small cell-to-cell spacing. Compared to near ideal cell-to-cell isolation characteristics of the SGT cell, punch through voltage starts

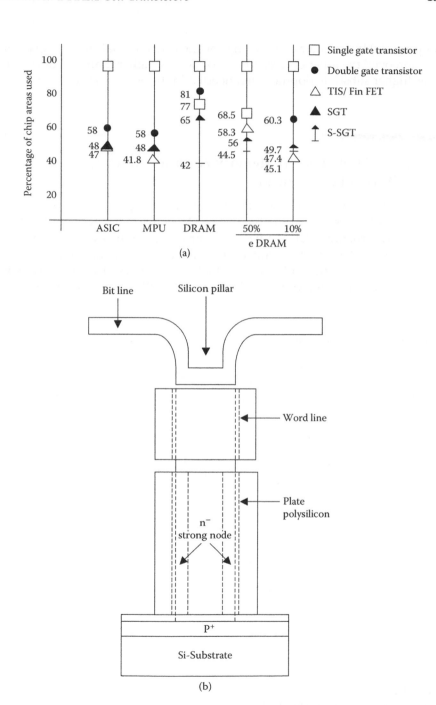

FIGURE 4.22
(a) Comparison of pattern area reduction in ASIC, MPU, DRAM and eDRAM. (b) A basic structure of an SGT cell. (Redrawn from "A Surrounding Gate Transistor (SGT) cell for 64/256 Mbit DRAMs," K. Sunouchi et al., IEDM Tech. Dig. 1989, pp. 23–26.)

dropping at 0.18 μm inter-cell spacing for the conventional trench capacitor cell. Moreover, the junction breakdown of SGT is higher than that of trench cell as higher substrate doping is not needed in the SGT structure.

4.10 Three-Dimensional Memory Architecture: Cell Area Less than 4 F^2

In trench or stacked capacitor DRAM cell, only capacitor structure was made three-dimensional. To enhance DRAM density, the SGT DRAM [55] and NAND-structured DRAM [56] were proposed, which were essentially two-dimensional memory architecture. Efforts were made to use three-dimensional space efficiently; hence, stacked-SGT DRAM technology was proposed that can realize further-high density DRAM, as the overall cell size could be less than 4 F^2 [57–60], with F being the minimum feature size.

4.10.1 Stacked-Surrounding Gate Transistor and Cell

A 3-D memory array architecture is realized by stacking a number of cells in series vertically resting on a bottom cell, which itself is connected to the bit line, as shown in Figure 4.23(a) [59]. Bottom cells are arranged in a

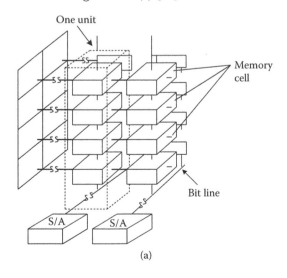

(a)

FIGURE 4.23
(a) A 3-D memory array architecture basics. (b) S-SGT DRAM based on 3-D memory array architecture: a cross-sectional view. (c) Read/write operation of S-SGT DRAM based on 3-D memory array architecture. (Redrawn from " New Three-Dimensional Memory Array Architecture for future Ultra high-Density DRAM," T. Endoh et al., IEEE J. Solid-State Circuits, Vol. 34, No. 4, pp. 476–483, 1999.)

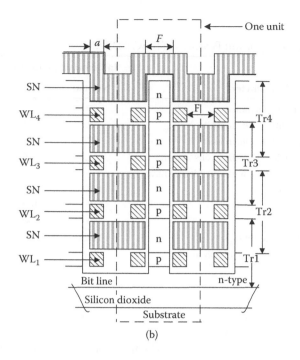

(b)

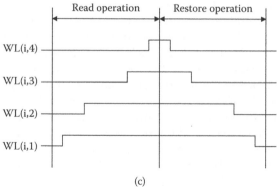

(c)

FIGURE 4.23 (continued)

conventional 2-D memory array form, which is addressed by a conventional one-dimensional column address and new two-dimension row address. In this way one cell of a conventional architecture is replaced by a unit, which is a stack of several cells. When SGT cells are stacked in such a way it is called stacked surrounding gate transistor (S-SGT) structure. Figure 4.23(b) shows S-SGT DRAM in which four SGT type cells are stacked vertically in series, containing four transistors Tr1 to Tr4 and corresponding word lines WL_1 to WL_4 and Figure 4.23(c) shows waveforms of read/restore operation of the S-SGT DRAM unit. For the read operation of a unit, after precharging

TABLE 4.5

Comparison of Important Features of S-SGT DRAM

S. No.	S-SGT Parameter	In Comparison to DRAM Type (%)		
		Normal	SGT	NAND
1.	Bit line capacitance			
	(i) with 3 cells in unit	35.7	54	—
	(ii) with 4 cells in unit	37	56	
2.	Bit line length with 4 cells in unit	11.5	25	14
3.	Signal voltage with one bit line having			
	(i) 600 cells	170	115	130
	(ii) 1000 cells	240	160	140
	(iii) 6000 cells	450	300	180
4.	RAS access time	118.5	—	—
5.	Total chip size at 1 Mbit level	46.9	67	—

of its bit line, only first word line WL is accessed and the signal in first cell from bottom is *read*. Next, after precharging the bit line, WL_2 for second cell goes high in addition to WL_1, and the signal in the next cell is *read*. The step is repeated until the signals of all cell capacitors in a vertical unit are *read*. After completing the read operation, all signals are restored in the reverse order of the read operation, as shown in the Figure 4.23(c). Important parameters of the stacked three-dimensional array that improve depend mainly on the number of cells in a vertical stack. For example, if three-dimensional array is constructed using four vertical SGT cells, its total bit line capacitance depends on (1) capacitance between substrate and bit line, (2) the capacitance of each cell at the bottom in the unselected stack, and (3) the load capacitance of the selected stack. As the numbers of cells in stack increase, components (1) and (2) decrease and (3) increases. Similarly sense signal for *high* and *low*, the RAS access time, memory array area, and the overall chip size also depend on the number of cells in a vertical array. A comparison of the stacked-SGT DRAM is shown in Table 4.5 with a normal-SGT and a NAND DRAM, which is very favorable for the S-SGT.

It may be noted that the S-SGT stack has four cells and one bit line; it can realize cell area per bit of 1.44 F^2 in comparison to 12 F^2 for normal planar DRAM, and cell area per bit becomes 2.88 F^2 when two cells are stacked [60].

Structure of a *self-aligned-type* stacked-SGT DRAM is shown in Figure 4.24 [61]. In principle it is similar to the previously reported S-SGT, that is, gate, source, drain, and capacitor storage plate are arranged on a silicon pillar vertically. However, shape of the Si pillar is like stacking one cylinder onto another with diameter of the cylinder decreasing from the bottom to the top; a steplike pillar structure is formed. Due to this steplike pillar structure, processing of cell entities can be done by a *self-aligned process* [62]. An analysis was done for the cell size dependence with respect to step widths and the

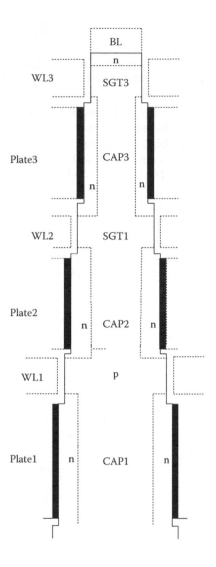

FIGURE 4.24

Basic structure of a self–aligned-type S-SGT DRAM. (Redrawn from "2.4 F2 Memory Cell Technology with Stacked-Surrounding Gate Transistor (S-SGT) DRAM," T. Endoh et al., IEEE Trans. Electron Devices, Vol. 48, No. 8, pp. 1599–1603, Aug. 2001.)

number of cells in one unit and based on analysis results a cell design procedure was adopted. By using this procedure, process simulation showed that at 0.5 μm technology levels, S-SGT DRAM can achieve a cell size of 2.4 F^2, which is half of the cell size of a conventional SGT DRAM cell of 4.8 F^2.

Another cell arrangement for stacked DRAM using vertical SGT in a simple process flow occupies 4 F^2 layout area. A practical version was realized in 0.35 μm CMOS technology with gate oxide thickness of 14 nm and channel length of 200 nm. As the body of the transistor is floating, its output characteristics show a kink effect at 1.2 V like that of a depleted SOI transistor. The transistor provides high I_{on}, but to control leakage current below 1 fA V_{th} has to be increased to 0.8 V. Another limitation of the given structure is the requirement of a higher dielectric constant material for getting sufficient storage capacitance [63].

It is known that floating body transistors are more prone to soft errors generated due to cosmic particles compared to the bulk devices; it has also come to notice that parasitic bipolar current in floating body structures also causes soft errors [64]. Although parasitic bipolar current is highly minimized in a floating channel (FC)-SGT DRAM cell compared to a planar SOI DRAM [65], it is still the major contributor toward soft errors. It is therefore essential that the floating-body effect is minimized and in this direction another floating channel (FC) SGT DRAM cell with a three-dimensional vertical storage capacitor is made available which could also go down to 4 F^2 cell area [66]. DRAM cell and array are basically same as shown in [55] and its top view is shown in Figure 4.25. It is observed that leakage current is minimized, having less heavy doping of the body, when a thin pillar is used in FC-SGT with metal gate. For reduced thickness of pillars, the amount of charge loss decreases; for example, for a 10 nm thick FC-SGT, loss of stored charge due to the parasitic bipolar current is as little as 28% of the total leakage current; it reduces soft errors effectively.

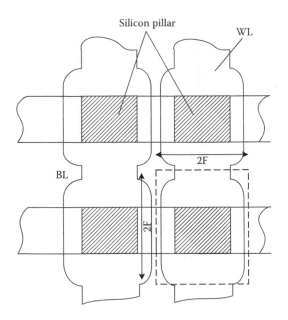

FIGURE 4.25
Top view of a floating channel–SGT DRAM cell. (Redrawn from "Device Design Guidelines for FC-SGT DRAM Cells with High Soft-Error Immunity," F. Matsuka et al., IEEE Trans. Electr. Dev., Vol. 52, pp. 1194–1199, 2005.)

4.11 BEST and VERIBEST DRAM Cells

To replace the planar access transistor by a different structure, like a vertical one, with aided improved technology, especially in the area of lithography, practical structures using vertical transistors were given in *trench transistor cell* (TTC) [67] and *trench transistor cell with self-aligned contact* (TSAC) [68]. Initial TT cell was found to be lacking not only in performance on a few counts but was suitable for an open bit-line layout only [69]; however, these cells revived interest in vertical transistor usage.

A 0.6 μm² 256 Mbit Trench DRAM at 0.25 μm design rule cell was given by IBM in 1993 with a unique feature of a self-aligned BuriEd STrap (BEST), which probably gave a real push to the vertical transistor DRAM cell technology [70]. This self-aligned buried strap was formed at the intersection of the storage trench and the junction of the array device as shown in the cross-sectional view of the buried strap in BEST cell of Figure 4.26. For the cell fabrication, first phosphorus is implanted and then the storage node trench is etched into the substrate and n⁺ capacitor plate is formed by out diffusing arsenic from the lower part of the trenches. Oxidized nitride (ON) dielectric is formed in the trench to form an oxide collar. Polysilicon trench fill is completed and recessed in three steps. Procedure of the formation of the buried strap is without using a separate mask; arsenic from the doped poly-Si in the trench diffuses into the substrate during hot process to form the strap.

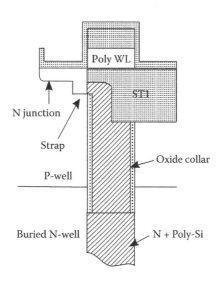

FIGURE 4.26
The self-aligned buried strap (BEST) based on merged isolation and node trench (MINT). (Redrawn from "A 0.6 μm2 256 Mb Trench DRAM Cell with Self-Aligned BuriEd Strap (BEST)," L. Nesbit et al., IEDM Tech. Dig., 1993, p. 627.)

Shallow trench isolation during the formation of buried strap isolates the buried poly-Si strap from the word line.

While fabricating a BEST cell, certain issues need to be attended to. For example, the use of As at thermal budget satisfactorily takes care of any effect of the strap out-diffusion on the device characteristic, as well as the inter-strap leakage. Buried strap junction leakage is kept below 0.1 fA/cell at 2.5 V supply by forming STI deeper than the out-diffused strap depth.

Quite a few advancements were made on the BEST cell structure to make it scalable for Gbit range. Using 0.25 µm technology, a fully planarized BEST cell was given by G. Bronner and others [71]. The cell was shown to be scalable to 0.18 µm lithography, to get a BEST cell having an area of 0.32 µm². For Gbit range and beyond not only the length but width of the channel needs to be of minimum dimensions which was relaxed earlier to get better I_{on}/I_{off} ratio. However, narrow width effect becomes dominant in the array transistor design at gigabit level. In addition control of the corner of the device associated with the shallow trench isolation, as used in the BEST cell, becomes very important. Threshold voltage may vary to ~200 mV, which requires higher doping in the well and word line boosting as well. Raised shallow trench isolation was integrated with a BEST DRAM cell which eliminates variable depletion effects around the device corner, and it was shown that a fine control of threshold voltage was achieved using this technique [72], for example the variation in the threshold voltage was reduced to ~50 mV (from 200 mV) as shown in Figure 4.27.

Fully planarized BEST cell along with raised shallow trench isolation lead to a VERIBEST DRAM cell at 4 Gbit/1 Gbit level [73], and it used a vertical

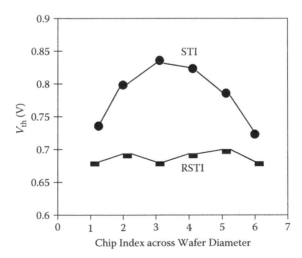

FIGURE 4.27
Mean threshold-voltage variations across a wafer processed with conventional STI and RSTI. ("A Novel 1 Gb Trench DRAM Cell with Raised Shallow Trench Isolation (RSTI)," J. Alsmeier et al., Symp. VLSI Tech. Dig. of Tech. Papers, 1997, p. 19–20.)

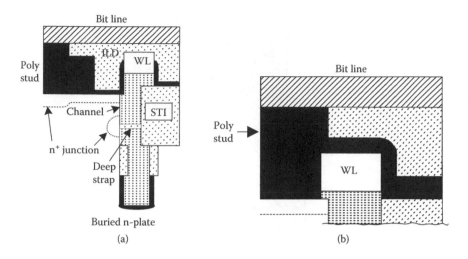

FIGURE 4.28

(a) 8 F^2 VERIBEST DRAM cell with a vertical transistor channel, and (b) possibility to shrink to sub-8 F^2 cell. (Redrawn from "A Novel Trench DRAM Cell with a VERtIcal Access Transistor and BuriEd S Trap (VERY BEST) for 4Gb/16Gb," U. Gruening et al., IEDM 1999, pp. 25–28.)

pass transistor along the deep trench sidewall [67] and self-aligned to its buried strap. Figure 4.28 shows the VERIBEST cell, as well as a possible sub-8 F^2 cell. The vertical array transistor integration was able to meet DRAM retention time requirement quite satisfactorily as channel length was relaxed, resulting in less doping concentration and controlled leakage current showing its ability for further scaling down.

A fully planarized process sequence with a VERIBEST raised shallow trench isolation is used to form vertical array as well as the planar support device in a functional 0.135 μm², 6 F^2 DRAM cell [74]. It used a trench capacitor sidewall vertical access transistor at 150 nm technology with a herringbone active area layout along with two levels of bit line wiring. The herringbone active area pattern maintains a $2F$ minimum-pitch word line even with maximum separation between each pair of devices. Figure 4.29 shows the cell cross section, where transistor channel self-aligned buried strap node contact is formed at the intersection of the trench and the active area pattern. Two levels of bit line interconnect using a W dual-damascene process make it possible to twist them for folded sensing leading to reduced interference. Use of raised shallow trench isolation and capacitor formation using LOCOS collar [30] helps in possible scalability to 4 Gbit/16 Gbit DRAM density level. In the trench sidewall vertical pass transistor, channel length is not constrained by the minimum feature size lithographic schemes and it is set by the recess step, which can be selected to achieve better threshold voltage control. However, in the 6 F^2 cell of ref. [74], back-to-back sharing of bit contact does not allow the cells to be fabricated close enough, thereby limiting scalability. To overcome the limitation, a different layout was employed

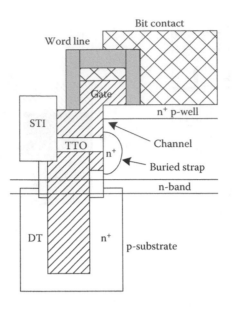

FIGURE 4.29

6 F^2 vertical cell schematic cross-section using VERIBEST structure. (Adapted from "A 0.135 μm² 6 F^2 Trench-Sidewall Vertical Device Cell for 4Gb/16Gb DRAM," C.J. Radens et al., Symp. VLSI Tech. Dig. of Tech. Papers, 2000, pp. 80–81.)

in which adjacent devices were isolated and bit-line contacts were provided for each device. Figure 4.30 shows schematic cross-section of such an orthogonal 6 F^2 DRAM cell [75]. An n⁺ buried strap is used to connect the vertical channel to the deep trench storage capacitor node. Figure 4.30 clearly shows the isolation of the channel and the storage capacitor. It also shows the isolation provided to the transistor channel and the buried strap by the n-band buried well, n⁺ source/drain regions, and the p-well. Adjacent devices are isolated toward bit line using an extended collar. Combination of constructional features makes the scheme scalable up to 0.1 μm and beyond.

4.12 Vertical Transistors

In vertical transistors, channel length can be chosen independently of the ground rules, allowing it to fulfill the requirement of $I_{off} < 1$ fA/cell. However, longer channel results in lower I_{on} with minimum size width. A double gate vertical transistor DRAM cell having trench capacitor and double buried node contact has been given for 100 nm and beyond, in which gate width doubles compared to the conventional transistor (with same ground rules) and it overcomes the limitation of lower I_{on} [3]. It is always preferable to use

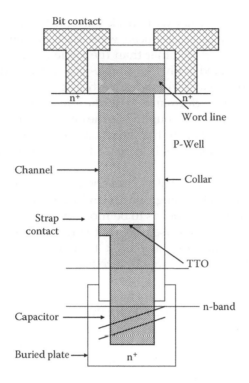

FIGURE 4.30
Schematic cell cross-section across the word line for an orthogonal 6 F^2 trench side wall vertical device cell. (Redrawn from "A Orthogonal 6 F2 Trench-Sidewall Vertical Device Cell for 4 Gb/16 Gb DRAM", C.J. Radens et al., IEDM Tech. Dig., pp. 349–352, 2000.)

simple geometric patterns for the layouts of the different masks from the lithographic point of view. The double-gate cell using line masks, robust self-aligned trench top structure for borderless bit lines, in addition to the self-aligned word line, makes the structure very lithography-friendly. In addition to overcoming the limitation of I_{on}, I_{off} of the planar and single gate vertical transistor, the double gate vertical transistor has significantly better threshold voltage control through major reduction of short-channel effects using channel dopant optimization by defining the location of peak doping density through varying energy level of the implant [76]. In a comparison between planar and vertical pass transistors, I_{on} is considerably increased to 80 μA for the vertical double gate transistor against 35 μA for the planar transistor. Instead of vertical implant, an angled implant provides a nonuniform lateral doping profile into the deep trench sidewalls. This improves the body effect from 150 mV to 120 mV, which can further be improved to 100 mV by working on low thermal budget [76].

In double-gate device channel if implant was done vertically, the body effect was high, which leads to higher array V_{th} during write-back of a *high* and reduces its drive/lowers write-back current. In fact it is well known that large body effect causes reduced write-back current in DRAMs [77]. To minimize the body effect, other than the angled implant scheme, super-halo design is integrated with the double-gate vertical transistor DRAM cell [78]. Using super-halo asymmetric vertical pass transistor designs can double the write-back performance without affecting other performances. The super-halo also allows decoupling of the channel and buried-strap doping which minimizes impact to the retention time. Super-halo could not be applied in planar technology as it increased either junction leakage or process complexity.

Some of the challenges faced at 110 nm DRAM technology level are as follows. Cell design should be lithography friendly so that self-alignment could be used. Storage capacitance needs to be more than 30 fF per cell preferably with conventional dielectrics used till that stage. It should have sufficient I_{on} without word line boosting and $I_{off} < 1$ fA/cell. It is observed that the aforementioned challenges are met in a report [79] that deployed a vertical transistor trench cell of R. Weis [3], having channel length of 200 nm and a bottle shaped trench capacitor with HSG.

Vertical transistor DRAM cells could further be scaled down to 70 nm scale once some important technological requirements were met. To understand that, one needs to know the limitations while scaling the vertical transistor and remedies for them as well. Deep isolation is required for the vertical transistor for suppressing interaction of adjacent devices. Conventional process used at 110 nm technology nodes used isolation as deep as ~550 nm, which did not suffice at 70 nm. Spin-on-glass (SOG) and HDP oxide cap isolation fill process [80] is good enough for continued scaling down. Another important issue is that of trench series resistance of polysilicon with 80:1 aspect ratio which lowers signal at sensing time. Use of polysilicon liner and Ti N fill reduces the trench parasitic resistance by a factor of four at 110 nm. However, it increases at lower technology node levels and some layout modifications to reduce parasitic bit-line capacitance and cell-to-cell interaction are also required [81]. The vertical transistor cell [3] with a double contact has increased bit line capacitance, which requires larger value of storage capacitance. Vertical transistor is modified in the form of a single-bordered contact as shown in Figure 4.31 and cell size is increased to 9 F^2. To reduce cell-to-cell interaction and bordered contact, spacing between trenches was increased if wiggled word lines were offset from the trenches as shown in Figure 4.31. It helps in having larger trenches and allows application of bordered contact.

Another route to scaling down of vertical transistor DRAM cell to 70 nm and beyond was made available by J. Beintner and others [82] through the use of self-aligned buried strap (SABS) process, which significantly improved retention characteristics. In vertical trench DRAM technology the buried strap is directly linked to the highly doped region under the gate, which could lead to high GIDL, and therefore degrades retention characteristics [83]. Scaling down of the vertical DRAM cell also results in a floating

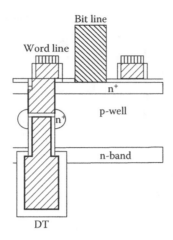

FIGURE 4.31
Cross-section of a vertical transistor cell. The trenches are offset from the wiggled word line. The active area and bit lines run vertically. (Modified from R. Divakaruni et al., " Technologies for Scaling Vertical Transistor DRAM Cell to 70 nm," Symp. VLSI Tech. Dig. of Tech. Papers, pp. 59–60, 2003.)

body-activated leakage current because of the pinching off the upper part of the p-well from its lower part [77]. So far, the buried strap was formed by diffusion from arsenic-doped polysilicon. In SABS, undoped trench-top oxide (TTO) is replaced with a phosphorus-doped oxide, which acts as a second source of solid phase out-diffusion to connect the transfer device and the buried strap. Two shallow junctions are formed—(1) the original buried strap, and (2) the doped TTO solid phase out-diffusion—and each junction can independently optimize the device performance. Combined with drastically reduced thermal budget, proposed technology enables the scalability of the vertical transistor technology down to 70 nm and beyond.

Some problems like buried out-diffusion and electrical cross-talk have been reported for the combination of trench capacitor and vertical channel access transistor (VCAT) arrangements [82,84], whereas stack capacitor and surrounding gate VCAT is a good combination to extract advantages of advanced capacitor process technologies such as mechanically robust and metal-insulator-metal (MIM) storage node structures. On these lines, a new core design that could accommodate 4 F^2 cell at 80 nm design rule has been proposed for manufacturability [85–87]. Arrangement of a bulk-Si based 4 F^2 cell array has proposed vertical access transistors located at cross points of word lines and bit lines. The cell array consists of a stack of capacitors and surrounding gate VCATS on buried bit lines [87]. The vertical pillars have 1.0 F space along word lines and 0.5 F in bit line direction and the bit line is self-aligned to pillars with 0.25 F oxide spaces as shown in Figure 4.32. The bit lines are made by n+ doping and by using 0.25 F oxide spacer and 0.5 F pillar space. The bottoms of the pillar patterns are connected in bit line direction

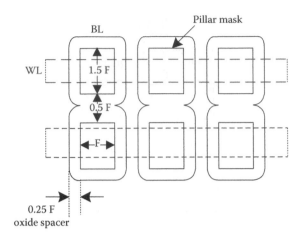

FIGURE 4.32
Differential spaces for a VPT cell array as BL is self-assigned along BL direction by 0.25 F oxide spacer formation. (Adapted from "A Novel Low Leakage Current VPT (Vertical Pillar Transistor) Integeration for 4F2 DRAM Cell Array with sub 40 nm Technology," J-M. Moon et al., *DRC Tech. Dig.*, pp. 259–260, 2006.)

with n$^+$ diffused layer while isolated in word line direction. The word lines are integrated by the self-align process for gate formation and follow damascene process for connecting inter gate poly-Si.

For the characterization of the parameters of the vertical pillar transistor (VPT) structure, simulations were done with and without offset Si structure. To evaluate effect of offset Si, 20 nm thick pillar transistor and two offsets of 30 nm and 60 nm were used. The 60 nm offset Si shows about one order lower GIDL than that of 30 nm and excellent DIBL of 25 mV/V, which is essential for stable retention characteristics during dynamic mode operation (for 30 nm case DIBL is 190 mV/V). GIDL characteristic can also be improved with the help of better controllability of phosphorus doping concentration in the protruded storage node.

Surrounding-gate VCAT has some excellent characteristics like much better short-channel immunity and higher current-driving capability and possibility of achieving sub-femto ampere level of off state channel leakage and the junction leakage. The fabricated VCAT has excellent I_{on}-I_{off} characteristics due to the small sub-threshold swing of <80 mV/dec., showing more than twice I_{on} in VCAT than RCAT [4] at similar I_{off} level [87].

Functional 4 F^2 DRAM based on VCAT is proposed having a new core design to accommodate 2 F-pitched bit line. When the primitive core array for the conventional 6 F^2 DRAM and the proposed 4 F^2 DRAM are compared, it is observed that pure cell area can be decreased only as much as effective while transition from 6 F^2 to 4 F^2 takes place; planar cell CMOS core region is hard to shrink. To enhance cell efficiency, sense amplifier blocks were reconstructed and distributed. For the fine pitch bit-line sense amplifier, rotation technique was used, avoiding the violation of CMOS design rules by releasing the sense amplifier block width where metal 1 line connects buried bit line to rotated sense amplifier. Another area of special consideration was the word line and bit line resistance. The dominant degradation factor for the read latency tRCD can be damascene word line resistance which is limited by the thickness of the surrounding gate material. For the improvement of tRCD, basic cell mat was subdivided into four parts and damascene WL was strapped to the M1 global word line. Using 80 nm design rule, measured random cycle time (tRC) and tRCD were 31 ns and 8 ns, respectively. The $4 F^2$ DRAM cell continues its advancement using VPT at 30 nm technology level. Its current drivability improved to 33 µA at $V_{DS} = V_{GS} = 1.2V$ which is nearly 3 times compared to RCAT for same level of I_{off}. Sub threshold slope also improved to 77 mV/dec with 30 nm diameter of pillars used. Matter of concern for the scaled VPT cell is the dynamic retention characteristic. However, the problem could be solved by using lightly doped drain structure [88]. Vertical transistor research is ongoing [89–91] elsewhere and more is expected in the future.

4.13 Advanced Recessed FinFETs

CMOS FinFETs fabricated on bulk Si substrate have long been considered viable economic alternatives, offering lower wafer cost compared to FinFET on SOI substrate. Efforts have been successful in meeting performance requirements of DRAM cells employing FinFETs in terms of I_{off}, I_{on}, and so on, while scaling down to even sub-50 nm technology [31]. Having n^+ poly-Si gate realizes low threshold voltage in bulk FinFETs, which results in higher I_{off} [25]. Hence, higher work function material, say p^+ poly-Si, was used to increase the threshold voltage, but I_{off} still remained high on account of greater GIDL, hence an n^+/p^+ poly-Si gate FinFET with n^+ gate toward the drain was used. The scheme works well in reducing I_{off} below 1 fA/cell, with a constraint that fin body width W_{fin} shall be less than half of the gate length to meet an important requirement of the suppression of short channel effect (SCE) [92]. However, keeping a uniform fin width throughout the wafer at sub 50 nm is challenging while using simple trimming process. A new locally separated channel (LSC) structure has been given in an n^+/p^+ poly-Si gate bulk FinFET so that W_{fin} is uniformly trimmed [93]. Fins with good uniformity were realized using spacer hard mask [94] and the spacers were formed using LPCVD and etch back. For the purpose of comparing the electrical characteristics with a conventional p^+/n^+ poly-Si gate FinFET [93], bulk FinFET using LSC structure with oxide as a channel separation material was designed. It was observed that I_{off} increases with increasing subsidiary n^+ poly-Si gate length L_s due to decreasing V_{th} and at the same time I_{off} increases with decreasing

L_s due to GIDL. Bulk FinFET with p^+/n^+ gate having p^+ poly-Si for channel separation improves SCE, SS and I_{on} considerably. However, an important condition for meeting the requirement of keeping I_{off} <1 fA requires that p^+ poly-Si filled channel separation region should not touch the drain region.

RCAT, SRCAT, and some of their variations have been used widely in forming DRAM cells at 80 nm node and 60 nm, respectively [95]. Improvements have been done to get better V_{th}, SS, SCE, and so on. A major consideration is the requirement of further reduction in GIDL. Moreover, as fabrication is done at sub-50 nm, constraints become more stringent in terms of channel doping concentration, body effect, cell drivability, and scalability for sustained improved performance. A partially insulated RCAT (PiRCAT) using partially insulating oxide layers in bulk Si substrate has been fabricated to match the mentioned requirements. In PiRCAT even with reduced channel doping, higher V_{th} comparable with that of SRCAT is achieved and extension of depletion charge is suppressed due to the presence of partially insulated oxide (PiOx) layer that with its lesser permittivity reduces electric field from gate. PiRCAT is fabricated using RCAT process in conjunction with Si-epi growth process. The rest of the fabrication of PiRCAT is the same as that used in the fabrication of partially insulated cell array transistor (PiCAT) [96]. It is observed that with the increase in the thickness of the epitaxial layer of PiRCAT, substrate doping concentration is increased under the recessed channel, resulting in higher threshold and lower sub-threshold leakage. Body-bias effect was improved by about 80 mV/V and sub-threshold slope by about 10 mV/dec, in comparison to the conventional RCAT. Current drivability is more than 30% that of conventional RCAT. A new isolation technology Pi-STI, which is a necessary requirement at sub-50 nm technologies formed with PiRCAT, is deployed for both core and peripheral transistors. Breakdown voltage remains constant even at a distance of 0.12 μm between N-well to N^+ in P-well, whereas it deteriorates around 0.29 μm in a conventional STI structure.

In spite of the FinFET being a device of strong choice for DRAMs, only a few successful fabrications were available in sub-50 nm technology. The most likely reasons are the difficulty in obtaining uniform fin shape as mentioned above and a proper design of impurity profile in the fin area. In practice, lithography was not good enough to provide fin patterns with 10 nm width and some trimming method was essential. However, conventional techniques of oxidation (and its stripping) caused void formation in HDP SiO_2 fill and extra parasitic gate capacitance [96]. Simultaneous oxidation of silicon fin and SiN cap was done using a new plasma oxidation technique to overcome the mentioned problems. Using this technique sub-10 nm fin widths have been fabricated [97]. Punch-through stopper (PTS) structure has been used at the channel bottom to suppress leakage currents in bulk. In a new PTS scheme lateral doping by straggling ions from the isolation region was done at 0° tilt angle with considerable lower implantation energy. Laterally scattered implanted ions reach the fin region forming PTS. The novel PTS structure has shown good characteristics in terms of V_{th} roll-off, junction capacitance, and punch-through immunity for CMOS bulk FinFET with less than 30 nm gate length.

In the bulk FinFET of Okano and others [97], the subchannel that is formed at the bottom part of the gate, as a result of the vertically directed implantation, cannot be controlled by the electrostatic potential of the gate. In some other reported FinFETs [25] GIDL is increased either due to the DRAM transistor with p+ *in situ* doped poly-Si gate or the use of negative word line. A new bulk FinFET in a partially insulated structure (Pi-FinFET) was given that could control the subchannel, reducing GIDL and hot carrier effect (HCE) [98]. The main features of the Pi-FinFET are the use of partially insulating oxide layer under source and drain and a novel structure *pad-polysilicon side contact* (PSC) to reduce GIDL and improve I_{on} and SCE. In this DRAM structure cell array transistors are Pi-FinFETs where peripheral and core transistors are fabricated on epi-Si and partial insulation of drain and source is done using the process used in ref. [95]. DIBL is decreased by nearly 25% and the sub-threshold swing is decreased nearly 20% compared to FinFETs and junction leakage is decreased by about 50% due to the blocking of the vertical leakage path by partially-insulating oxide; however GIDL remains same without any improvement. For improving GIDL, it was found that "Pi-FinFETs with gradual pad-polysilicon doping concentration, which have multilayer with undoped polysilicon and highly doped polysilicon," show that GIDL is decreased about two orders with the same I_{off} as in bulk FinFETs and I_{on} improved by about 30%.

Further improvement was made in the Pi-FinFET with a PSC where fabrication process of the Pi-FinFET was same using local damascene and a self-aligned contact (SAC) process to form a PSC structure [99]. Three kinds of Pi-FinFETs—normal Pi-FinFETs, Pi-FinFET with PSC (PSC-1), and Pi-FinFET with undoped silicon selective epitaxial growth (SEG) buffered PSC (PSC-2)—were fabricated and their characteristics were compared with bulk FinFETs as shown in Table 4.6, PSC-2 is found to be the most promising structure.

Figure 4.33 shows a fin and recessed channel MOSFET (FiReFET), which tries to improve the driving capability of RCAT that was affected because of longer effective channel length [96]. The main features of the FiReFET are (1) trench gates and (2) thin fin channel and wide source/drain region. It results in the reduction in I_{off} without using PTS implantation and gives improvement in I_{on}. Simulation based analysis was done on a FiReFET of gate length and fin width of 30 nm and 20 nm, respectively with t_{ox} of 2 nm on (100) surface and 2.5 nm on (110) surface and total active width being 60 nm. Junction depth was fixed at 50 nm and recess depth at 150 nm. Initially I_{off} is more than a corresponding RCAT because of GIDL that was a result of enhancement of field at the fin corner. To minimize GIDL, underlapped gate-to-source/drain structure was adopted in line with a concept of the non-overlapped gate-to-source/drain region [101]. DIBL, SS, and I_{on} are improved as the recessed depth increased from 50 nm to 200 nm. The

TABLE 4.6

Comparison of Electrical Characteristics of PiFETs

		Pi-FinFET		
Median Value @ 85°C	FinFET	Pi-FinFET	PSC-1	PSC-2
Fin height (nm)	150	80	80	80
V_{th} (V) in %	100	86	86	86
I_{on} @ 1.2 V (µA/cell) in %	100	110	117	123
Normalized I_{on} with width (µA/µm) in %	100	170	210	200
Leakage current reduction%	—	50	15	50
DIBL (mV/V) in %	100	100	96	86
Sub-threshold swing (mV/dec) in %	100	100	104	104
g_{max} (µs) in %	100	—	112	112
Word line capacitance (C_{WL}) in %	100	75	75	75

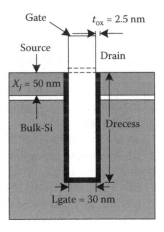

FIGURE 4.33
Front view of FiReFET which was used for simulation. (Adapted from "Fin and Recess Channel MOSFET (FiReFET) for Performance Enhancement of Sub-50 nm DRAM Cell," J.Y. Song et al., ISDRS, 2007.)

structure needed optimization of the underlap of the gate to drain in order to achieve high current drivability as well as low leakage.

FinFETs with short fin width are preferred for achieving better SCE [102]; however, it leads to larger parasitic series resistance. Another well-known problem is large GIDL due to enhancement of electric field at the fin corners. Source/drain regions were extended outside the gate region using silicon epitaxial regrowth [103]; however, the technique suffers from process complexity, hence cost. A buried-gate fin and recess channel MOSFET (BG-FiReFET) has been given in which source/drain region become wider without using the silicon epitaxial regrowth. It overcomes the problem of parasitic series resistance. Buried gate structure helps in minimizing the GIDL current [104] also. Fin width as small as 20 nm with gate length of 30 nm was used. DIBL and SS are 8.4 mV/V and 73 mV/dec, respectively; both being small even with short channel length as the fin channel was deeply recessed up to 150 nm.

References

1. Kinam Kim, "Future Memory Technology: Challenges and Opportunities" *Intern. Symp. on VLSI—TSA*, pp. 5–9, 2008.
2. Y.K. Park et al., "Fully Integrated 56 nm DRAM Technology for 1 Gb DRAM," *VLSI Tech. Digest*, pp. 190–191, 2007.
3. R. Weis et al., "A Highly Cost Efficient 8 F² DRAM Cell with a Double Gate Vertical Transistor Device for 100 nm and Beyond," *Tech. Digest of IEDM*, pp. 415–418, 2001.
4. J.Y. Kim et al., "The Breakthrough in Data Retention Time of DRAM Using Recess-Channel-Array Transistor (RCAT) for 88 nm Feature Size and Beyond," *Symp. on VLSI Tech. Digest of Tech. Papers*, pp 11–12, 2003.
5. J.Y. Kim et al., "The Excellent Scalability of the RCAT (Recess-Channel-Array-Transistor) Technology for Sub-70nm DRAM Feature Size and Beyond," *Int. Symp. on VLSI—TSA*, pp. 33–34, 2005.
6. J.Y. Kim et al., "S-RCAT (Sphere-shaped-Recess-Channel-Array Transistor) Technology for 70 nm DRAM Feature Size and Beyond." *Symp. on VLSI Tech. Digest of Tech. Papers*, pp. 34–35, 2005.
7. K. Kim, et al., "Body-bias on drain-induced barrier lowering in sphere-shaped recess cell-array transistor," *Electronic Letters*, Vol. 48, No. 7, March 2012.
8. W. Wakamiya et al., "Novel Stacked Capacitor Cell for 64 Mb DRAM," *Symp. Tech. Dig. Tech. Papers*, p. 31, 1989.
9. H. Takahashi et al., "High Performance CMOS Surrounding Gate Transistor (SGT) for Ultra High Density LSIs," *IEDM Tech. Dig.*, pp. 222–225, 1986.
10. D. Hisamoto, T. Kaga, and E. Takeda, "Impact of Vertical SOI 'DELTA' Structure on Planar Device Technology," *IEEE Trans. Electr. Devices*, Vol. 38, pp. 1419–1424, June, 1991.
11. H.S. Wong et al., "Design and Performance Considerations for Sub-0.1 μm Double-Gate SOI MOSFETs," *IEDM Tech. Dig.*, pp. 747–750, 1994.
12. M. Ieong et al., "High Performance Double Gate Device Technology Challenges and Opportunities," *Proc. Int. Symp. Quality Electr. Design* (ISQED), 2002.
13. D. Hisamoto et al., "Metalized Ultra-Shallow-Junction Device Technology for Sub-0.1 μm Gate MOSFETs," *IEEE Trans. Elect. Dev.*, Vol. 41, pp. 745–750, 1994.
14. D. Hisamoto et al., "Fin FET—A Self-Aligned Double-Gate MOSFET Scalable to 20 nm," *IEEE Trans. Electron Devices*, Vol. 47, no. 12, pp. 2320–2325, 2000.

15. S. Kimura et al., "A 0.1 µm-Gate Elevated Source and Drain MOSFET Fabricated by Phase-Shifted Lithography," IEDM Tech. Dig., pp. 117–120, 1996.
16. T.J. King et al., "Electrical Properties of Heavily Doped Polycrystalline Silicon Germanium Films," *IEEE Trans. Electr. Dev.*, Vol. 41, pp. 228–232, 1994.
17. X. Huang et al., "Sub 50-nm Fin FET: PMOS," *IEDM*, pp. 67–70, 1999.
18. Y.K. Choi et al., "Sub 20-nm CMOS Fin FET Technologies," *IEDM*, pp. 421–424, 2001.
19. K.G. Anil et al., "CMP-less Integration of Fully Ni-Salicided Metal Gates in FinFETs by Simultaneous Salicidation of the Source, Drain, and the Gate Using a Novel Dual Hard Mask Approach," *VLSI Tech. Symp.*, p. 198–199, 2005.
20. Y.X. Liu et al., "Flexible Threshold Voltage FinFETs with Independent Double Gates and an Ideal Rectangular Cross-Section Si-Fin Channel," IEDM Tech. Dig., pp. 986–988, 2003.
21. H. Shang et al., "Investigation of Fin FET Devices for 32 nm Technologies and Beyond," *Symp. on VLSI Tech. Dig. of Tech. Papers*, p. 2006.
22. B. Yu et al., "FinFET Scaling to 10 nm Gate Length," IEDM Tech. Dig., pp. 251–254, 2002.
23. F-L. Yang et al., "25 nm CMOS Omega FETs," IEDM Tech. Dig., pp. 255–258, 2002.
24. T. Park et al., "Fabrication of Body-Tied FinFET (Omega MOSFETs) Using Bulk Si Wafers," *Symp. VLSI Tech. Dig., Tech. Papers*, pp. 135–136, 2003.
25. C.H. Lee et al., "Novel Body Tied Fin FET Cell Array Transistor DRAM with Negative Word Line Operation for Sub 60 nm Technology and Beyond," *Symp. on VLSI Tech. Digest of Tech. Papers*, pp. 130–131, 2004.
26. J.-H. Lee et al., "Simulation Study of a New Body Tied FinFETs (Omega MOSFETs) Using Bulk Si Wafers," *Si-Nanoelectron. Tech. Dig.*, pp. 102–103, 2003.
27. T.-S. Park et al., "Characteristics of the Full CMOS SRAM Cell Using Body-Tied TG MOSFETs (Bulk FinFETs)," *IEEE Trans. Electr. Dev.*, Vol. 53, no. 3, pp. 481–487, 2006.
28. D.-H. Lee et al., "Fin-Channel-Array Transistor (FCAT) Featuring Sub-70 nm Low Power and High Performance DRAM," *IEDM*, pp. 407–410, 2003.
29. R. Katsumata et al., "Fin-Array-FET on Bulk Silicon for Sub-10 nm Trench Capacitor DRAM," *Symp. VLSI Tech. Dig. Tech. Papers*, pp. 61–62, 2003.
30. T. Rupp et al., "Extending Trench DRAM Technology to 0.15 µm Ground Rule and Beyond," IEDM Tech. Dig., pp. 33–36, 1999.
31. Chul Lee et al., "Enhanced Data Retention of Damascene-Fin FET DRAM with local Channel Implantation and < 100 > Fin Surface-Orientation Engineering," *IEDM*, pp. 61–64, 2004.
32. Y.S. Kim et al., "Local-Damascene-Fin FET DRAM Integration with p+ Doped Poly-Silicon Gate Technology for Sub-60 nm Device Generation," *IEDM*, pp. 325–328, 2005.
33. T.-S. Park et al., "Characteristics of Body-Tied Triple-Gate pMOSFET," *IEEE Electr. Dev.*, Vol. 25, no. 12, pp. 798–800, 2004.
34. T.-S. Park et al., "Characteristics of the Full CMOS SRAM Cell Using Body-Tied TG MOSFETs (Bulk FinFETs)," *IEEE Trans. Electr. Dev.*, Vol. 53, no. 3, pp. 481–487, 2006.
35. T. Park et al., "Body-Tied Double-Gate SONOS Flash (Omega Flash) Memory on Bulk Si," *Tech. Dig. of DRC*, p. 33, 2003.

36. S.M. Kim et al., "A Novel Multi-Channel Field Effect Transistor (McFET) on Bulk Si for High Performance Sub-80nm Applications," IEDM Tech. Dig., pp. 639–642, 2004.
37. S.M. Kim et al., "Fully Working High Performance Multi-Channel Field Effect Transistor (McFET) SRAM Cell on Bulk Si Substrate Using TiN Single Metal Gate," *Symp. VLSI Tech. Dig. Tech. Papers*, pp. 196–197, 2005.
38. M. Yoshida et al., "A Full Fin FET DRAM Core Integration Technology Using a Simple Selective Fin Formation Technique," *Symp. VLSI Tech. Dig. Tech. Papers*, pp. 34–35, 2006.
39. K.-H. Park, K.-R. Han, and J.-H. Lee, "Highly Scalable Saddle MOSFET for High Density and High Performance DRAM," *IEEE Electron. Dev. Letters*, Vol. 26, no. 9, pp. 690–692, 2005.
40. K.-H. Park et al., "Simulation Study of High-Performance Modified Saddle MOSFET for Sub-50 nm DRAM Cell Transistors," *IEEE Electron. Dev. Lett.*, Vol. 27, pp. 759–761, 2006.
41. S.-W. Chung et al., "Highly Scalable Saddle-Fin (S-Fin) Transistor for Sub-50 nm DRAM Technology," *Symp. VLSI Tech. Dig. of Tech. Papers*, pp. 32–33, 2006.
42. M.J. Lee et al., "A Proposal on an Optimized Device Structure with Experimental Studies on Recent Devices for the DRAM Cell Transistor," *IEEE Trans. Electr. Dev.*, Vol. 34, no. 12, pp. 3325–3335, 2007.
43. J.W. Lee et al., "Improvement of the Data Retention Time in DRAM Using Recessed Channel Array Transistors with Asymmetric Channel Doping for 80 nm Feature Size and Beyond," *Proc. S.S. Dev. Res. Conf.*, pp. 449–452, 2004.
44. W.-S. Lee et al., "Analysis on Data Retention Time of Nano-scale DRAM and Its Prediction by Indirectly Probing the Tail Cell Leakage Current," IEDM Tech. Dig., pp. 395–398, 2004.
45. S. Swada et al., "Effects of Field Boron Dose on Substrate Current in Narrow Channel LDD MOSFET," IEDM Tech. Dig., pp. 778–781, 1984.
46. H. Takato et al., "High Performance CMOS Surrounding Gate Transistor (SGT) for Ultra High Density LSIs," *IEDM*, pp. 222–225, 1988.
47. H. Takato et al., "Impact of Surrounding Gate Transistor (SGT) for Ultra High Density LSIs," *IEEE Trans. Electron Devices*, Vol. 38, no. 3, pp. 573–578, Mar., 1991.
48. S. Miyano et al., "Numerical Analysis of a Cylindrical Thin-Pillar Transistor (CYNTHIA)," *IEEE Trans. Electron Dev.*, Vol. 39, no. 8, pp. 1976–1981, Aug., 1992.
49. A. Nitayama et al., "Multipillar Surrounding Gate Transistor (MSGT) for Compact and High-Speed Circuits," *IEEE Trans. Electron Dev.*, Vol. 38, no. 3, pp. 579–583, Mar., 1991.
50. A. Karanti et al., "Design and Optimization of Thin Film Fully Depleted Vertical Surrounding Gate (VSG) MOSFETs for Enhanced Short Channel Immunity," *Solid State Electronics*, Vol. 46, pp. 1333–1338, 2002.
51. A. Karanti et al., "Design and Optimization of Vertical Surrounding Gate MOSFETs for Enhanced Transconductance-to-Current Ratio (gm/Ids)," *Solid State Electronics*, Vol. 47, pp. 155–159, 2003.
52. A. Wei and D.A. Antoniadis, "Measurement of Transient Effects in SOI DRAM/SRAM Access Transistors," *IEEE Electron. Dev. Lett.*, Vol. 17, pp. 193–195, 1996.
53. B. Goebel et al., "Fully Depleted Surrounding Gate Transistor (SGT) for 70 nm DRAM and Beyond," IEDM Tech. Dig., pp. 275–278, 2002.

54. S. Watanabe, "Impact of Three-Dimensional Transistor on the Pattern Area Reduction for ULSI," *IEEE Trans. Electr. Dev.*, Vol. 50, pp. 2073–2080, 2003.
55. K. Sunouchi et al., "A Surrounding Gate Transistor (SGT) Cell for 64/256 Mbit DRAMs," IEDM Tech. Dig., pp. 23–26, 1989.
56. T. Hasegawa et al., "An Experimental DRAM with a NAND-Structured Cell," *IEEE J. Solid-State Circuits*, Vol. 28, pp. 1099–1104, 1993.
57. T. Endoh et al., "The Analysis of the Stacked Surrounding Gate Transistor (S-SGT) DRAM for the High Speed and Low Voltage Operation," *IEICE Trans. Electron*, Vol. E81-C, no. 9, p. 1491, 1998.
58. T. Endoh et al., "The Stacked-SGT DRAM Using 3D-Building Memory Array Technology," *Trans. Inst. Electron, Inf. Commun., Eng. Jpn.*, J 81-C-1, p. 288, 1998.
59. T. Endoh et al., "New Three-Dimensional Memory Array Architecture for Future Ultra High-Density DRAM," *IEEE J. Solid-State Circuits*, Vol. 34, no. 4, pp. 476–483, 1999.
60. T. Endoh et al., "The 1.44 F² Memory Cell Technology with the Stacked-Surrounding Gate Transistor (S-SGT) DRAM," *Proc. 22nd Int. Conf. on Microelectronics (MIEL'2000)*, Vol. 2, NIS, SERBIA, May 2000.
61. T. Endoh et al., "2.4 F² Memory Cell Technology with Stacked-Surrounding Gate Transistor (S-SGT) DRAM," *IEEE Trans. Electron Dev.*, Vol. 48, no. 8, pp. 1599–1603, 2001.
62. M. Suzuki et al., "The 2.4 F² Memory Cell Technology with Stacked-Surrounding Gate Transistor (S-SGT) DRAM," *Trans. Inst. Electron Inf. Commun. Eng. Jpn.*, Vol. J83-C, p. 92, 2000.
63. F. Hofmann and W. Rosner, "Surrounding Gate Select Transistor for 4 F² Stacked Gbit DRAM," *SSDRC*, pp. 131–134, 2001.
64. H. Iwata and T. Ohzone, "Numerical Analysis of Alpha-Particle-Induced Soft Errors in SOI Devices," *IEEE Trans. Electron. Dev.*, Vol. 39, pp. 1184–1190, 1992.
65. F. Matsouka and F. Masouka, "Numerical Analysis of Alpha-Particle-Induced Soft Errors in Floating Type Surrounding Gate Transistor (FC-SGT) DRAM Cell," *IEEE Trans. Elect. Dev.*, Vol. 50, pp. 1638–1644, 2003.
66. F. Matsuoka, H. Sakuraba, and F. Masuoka, "Device Design Guidelines for FC-SGT DRAM Cells with High Soft-Error Immunity," *IEEE Trans. Electr. Dev.*, Vol. 52, pp. 1194–1199, 2005.
67. W.F. Richardson et al., "A Trench Transistor Cross-Point Cell," *IEDM Dig. Tech. Papers*, pp. 714–717, 1985.
68. M. Yanagisawa et al., "Trench Transistor Cell with Self-Aligned Contact (TSAC) for Megabit MOS DRAM," *IEDM Dig. Tech. Papers*, pp. 132–135, 1986.
69. N.C.C. Lu, "Advanced Cell Structures for Dynamic RAMs," *IEEE Circuit and Device Magazine*, pp. 27–36, Jan. 1989.
70. L. Nesbit et al., "A 0.6 µm² 256 Mb Trench DRAM Cell with Self-Aligned BuriEd Strap (BEST)," IEDM Tech. Dig., p. 627, 1993.
71. G. Bronner et al., "A Fully Planarized 0.25 µm CMOS Technology for 256 Mbit DRAM and Beyond," *Symp. VLSI Tech. Dig. of Tech. Papers*, pp. 15–16, 1995.
72. J. Alsmeier et al., "A Novel 1 Gb Trench DRAM Cell with Raised Shallow Trench Isolation (RSTI)," *Symp. VLSI Tech. Dig. of Tech. Papers*, pp. 19–20, 1997.
73. U. Gruening et al., "A Novel Trench DRAM Cell with a VERtical Access Transistor and BuriEd STrap (VERY BEST) for 4Gb/16Gb," *IEDM*, pp. 25–28, 1999.

74. C.J. Radens et al., "A 0.135 µm^2 6F^2 Trench-Sidewall Vertical Device Cell for 4Gb/16Gb DRAM," *Symp. VLSI Tech. Dig. of Tech. Papers*, pp. 80–81, 2000.

75. C.J. Radens et al., "An Orthogonal 6 F^2 Trench-Sidewall Vertical Device Cell for 4 Gb/16 Gb DRAM," IEDM Tech. Dig., pp. 349–352, 2000.

76. K. McStay et al., "Vertical Pass Transistor Design for Sub-100 µm DRAM Technologies," *Symp. VLSI Tech. Dig. of Tech. Papers*, pp. 180–181, 2002.

77. J. Mandelman et al., "Challenges and Future Directions for the Scaling of Dynamic Random Access Memory (DRAM)," *IBM J. Res. & Dev.*, Vol. 46, no. 2–3, pp. 187–212, 2002.

78. D. Chidambarrao et al., "Super-Halo Asymmetric Vertical Pass Transistor Design for Ultra-Dense DRAM Technologies," *VLSI Tech, Sys. and Appl.*, pp. 25–28, 2003.

79. H. Akatsu et al., "A Highly Manufacturable 110 nm DRAM Technology with 8 F^2 Vertical Transistor Cell for 1 Gb and Beyond," *VLSI Tech. Dig. Tech. Papers*, pp. 52–53, 2002.

80. J-H. Heo et al., "Void Free and Low Stress Shallow Trench Isolation," *Symp. on VLSI Tech. Dig. of Tech. Papers*, pp. 132–133, 2002.

81. R. Divakaruni et al., "Technologies for Scaling Vertical Transistor DRAM Cell to 70 nm," *Symp. VLSI Tech. Dig. of Tech. Papers*, pp. 59–60, 2003.

82. J. Beintner et al., "Vertically Self-Aligned Buried Junction Formation for Ultra High-Density DRAM Applications," *IEEE Electr. Dev. Letters*, Vol. 25, no. 5, pp. 259–261, 2004.

83. K. Saino et al., "Impact of Gate Induced Drain Leakage Current on the Tail Distribution of DRAM Data Retention Time," IEDM Tech. Dig., pp. 837–840, 2002.

84. J. Beintner et al., "On the Retention Time Distribution of Dual Channel Vertical DRAM Technologies," *VLSI Tech. Dig.*, pp. 243–246, 2003.

85. J-M. Yoon et al., "A Novel Low Leakage Current VPT (Vertical Pillar Transistor) Integration for 4 F^2 DRAM Cell Array with Sub 40 nm Technology," *DRC Tech. Digest*, pp. 259–260, 2006.

86. K.-W. Song et al., "A 31 ns Random Cycle VCAT-Based 4 F^2 DRAM with Enhanced Cell Efficiency," *Symp. on VLSI Circuits, Dig. Tech. Papers*, pp. 132–133, 2009.

87. K.-W. Song et al., "A 31 ns Random Cycle VCAT-Based 4 F^2 DRAM with Manufacturability and Enhanced Cell Efficiency," *IEEE J. Solid State Circuits*, Vol. 45, pp. 880–887, 2010.

88. H. Chung, et al., "Novel 4F^2 DRAM cell with Vertical Pillar Transistor (UPT)," *Proc. Europian S.S. Dev. Res. Conf.*, pp. 211–214, 2011.

89. T. Endoh, et al., "Study of Self-Heating in Si Nano Structure for Floating Body-Surround Gate Transistor with High-k Dielectric Films," *Jpn. J. Appl. Phys.*, Vol. 46, No. 5B, pp. 3189–3192, 2007.

90. K. Sakui and T. Endoh, "A New Vertical MOSFET Vertical Logic Circuit (VLC) MOSFET" Suppressing Asymmetric Characteristics and Realising an Ultra Compact and Robust Logic Circuit," *ISDRS*, DOI: 10.1109/ISDRS,2009.5378136, pp. 1–2, 2009.

91. K. Sakui and T. Endoh, "A Compact, High-Speed, and Low-Power Design for Multi-Pillar Vertical MOSFET's Suppressing Characteristic Influence by Process Fluctuations," VLSI Technology Systems and Applications (VLSI-TSA), Int. Symp. pp. 30–31, 2010.

92. K.H. Park and J.H. Lee, "Gate Workfunction Engineering in Bulk FinFets for Sub-50 nm DRAM Cell Transistors," *IEEE Electr. Dev. Letters*, Vol. 28, pp. 148–150, 2007.
93. H-A-R. Jung et al., "n⁺/p⁺ Gate Bulk FinFETs with Locally Separated Channel Structure for Sub-50-nm DRAM Cell Transistors," *IEEE Elect. Dev. Lett.*, Vol. 28, pp. 1126–1128, 2007.
94. S.P. Kim et al., "Paired FinFET Charge Trap Flash Memory for Vertical High Density Storage," *VLSI Symp. Tech. Dig.*, pp. 84–85, 2006.
95. J.-M. Park et al., "A Novel Body Effect Reduction Technique to Recessed Channel Transistor Featuring Partially Insulating Layer under Source and Drain: Application to Sub-50 nm DRAM Cell," *VLSI Symp. Tech. Dig.*, pp. 903–906, 2007.
96. K.H. Yeo et al., "80 nm 512 M DRAM with Enhanced Data Retention Time Using Partially-Insulated Cell Array Transistor (PiCAT)," *Symp. VLSI Tech. Dig. Tech. Papers*, pp. 30–31, 2004.
97. K. Okano et al., "Process Integration Technology and Device Characteristics of CMOS FinFET on Bulk Silicon Substrate with Sub-10 nm Fin Width and 20 nm Gate Length," *IEDM Tech. Dig.*, pp. 721–724, 2005.
98. J-M. Park et al., "Fully Integrated Advanced Bulk FinFETs Architecture Featuring Partially-Insulating Technique for DRAM Cell Application of 40 nm Generation and Beyond," *IEDM Tech. Dig.*, pp. 889–892, 2006.
99. S.Y. Han et al., "A Novel DRAM Cell Transistor Featuring a Partially-Insulated Bulk FinFET (Pi-FinFET) with a Pad-Polysilicon Side Contact (PSC)," *Symp. VLSI Tech. Dig. Tech. Papers*, pp. 166–167, 2007.
100. J-Y. Song et al., "Fin and Recess Channel MOSFET (FiReFET) for Performance Enhancement of Sub-50 nm DRAM Cell," *Int. Symp. ISDRS*, pp. 1–2, 2007.
101. V. Trivedi et al., "Nanoscale FinFETs with Gate-Source/Drain under Lap," *IEEE Trans. Elect. Dev.*, Vol. 52, pp. 56–62, 2005.
102. K. Suzuki et al., "Scaling Theory for Double-Gate SOI MOSFET," *IEEE Trans. Elect. Dev.*, Vol. 40, pp. 2326–2329, 2003.
103. J. Kedzierski et al., "Extension and Source/Drain Design for High-Performance FinFET Devices," *IEEE Trans. Elect. Dev.*, Vol. 50, pp. 952–958, 2003.
104. J.Y. Song et al., "Fabrication and Characterization of Buried-Gate Fin and Recess Channel MOSFET for High Performance and Low GIDL Current," *ISDRS*, 2009.

5

Storage Capacitor Enhancement Techniques

5.1 Introduction

One of the most critical issues in DRAM cell designs is the fabrication of sufficient value reliable storage capacitance while cell size is continuously decreasing. In the beginning, SiO_2 was the dielectric used in planar forms of capacitors and reduction in its thickness was the main option for maintaining realized capacitance value with increasing DRAM density. However, beyond 4 Mbit, high leakage prevented further reduction in dielectric thickness and alternatives in the form of the use of materials other than (only) SiO_2, structural innovations in the cell, and modification in electrode surface were applied.

In the discussion of the development of DRAMs from 16 Mbit onwards to around 1 Gbit in Section 3.4 of Chapter 3, problems associated with the necessity of fabrication of storage capacitor in ever-decreasing chip area were clearly emphasized. In that period of DRAM development SiO_2 and its combination with silicon nitride was replaced with higher dielectric constant Ta_2O_5. In addition, the capacitor was realized in three dimensions instead of being planar; the crown-shaped capacitor was a preferred shape. At 256 Mbit and 1 Gbit level, high dielectric constant material like BST film was also used. Storage nodes other than polysilicon, like RuO_2/TiN, were employed. In addition, texturing of the bottom plate of the capacitor through formation of hemispherical grain (HSG) was used to increase the effective capacitor area (hence its value). In Section 5.2 generation of HSG on the storage node shall be discussed along with the variations in the properties of HSGs on account of process and processing temperature used. Properties of higher permittivity and layered dielectric involving SiO_2 and Si_3N_4, using slightly varying fabrication processes is then taken up. It is followed up with description of capacitors formed using Ta_2O_5 as dielectric, whose properties get changed with the method of deposition of Ta_2O_5, annealing temperature and metal used for the electrodes. Since in embedded DRAMs processing temperature should be lower, formation of HSG in such cases is also discussed briefly in Section 5.4.

A different kind of technology is used for storage capacitance fabrication in trench DRAMs for sub-100 nm design rules. Bottle-shaped trenches with Al_2O_3 film for an aspect ratio more than 60 have been realized. Application of growth of HSG in such capacitors is given in Section 5.5. Metal-insulator-metal (MIM) capacitor structures using materials with increasing value of dielectric strength material like Ta_2O_5, BST, HfO_2, ZrO_2, and layered combinations of some of these materials are taken up next. Use of different metals for the top and bottom electrodes while using mentioned dielectrics has also been described. A common theme is that it is always important to monitor the reliability of the fabrication capacitor especially in terms of its leakage current density and the compatibility of the all the materials used so that their combinations do not degrade the quality of the capacitor and the DRAM cell.

5.2 Hemispherical Grain Storage Node

Before depositing thin dielectric layer, texturing of the bottom polysilicon electrode of poly-to-poly capacitor was initially realized by Fazan and Lee [1] to increase its capacitance without increasing its projected area. The effective storage surface area was increased by more than 30% due to the creation of asperities on the polysilicon surface and the technique was later found to be applicable in all stacked or trench capacitor cells which had a bottom polysilicon electrode. Increase in capacitor area due to the creation of the texture depended strongly on the polysilicon deposition temperature, polysilicon thickness, doping process, and oxidation temperature [2]. For measuring electrical characteristics in such schemes, after texturing, polysilicon, a highly reliable composite oxide/nitride dielectric was formed by depositing a 10 nm thick Si_3N_4 using LPCVD and slight reoxidation [3]. Top capacitor plate of ~200 nm thick polysilicon was then deposited to form the capacitor. Maximum capacitance increased to 6.7 from 5 fF/μm^2 with the textured bottom plate. Leakage current was measured through the textured structure and slight increase was observed at high fields; however, at low or operating fields no difference was observed. This increase in leakage can be understood by following the difference between electrical conduction mechanism through SiO_2 and Si_3N_4. In SiO_2 conduction is through Fowler-Nordheim emission and leakage current depends on the electric field near the injecting electrode and increases considerably due to the presence of asperities. In Si_3N_4 conduction is bulk-limited and governed by Frenkel-Pool emission which is due to field-enhanced thermal excitation of trapped electrons [4]. Other important electrical characteristics such as the time-dependent dielectric breakdown (TDDB) measurement at the operating field of 3.6 MV/cm

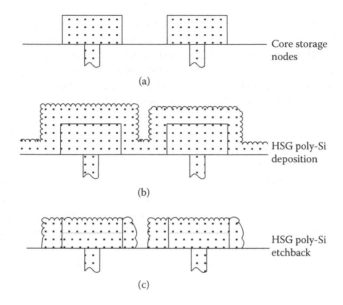

(a)

(b)

(c)

Core storage
nodes

HSG poly-Si
deposition

HSG poly-Si
etchback

FIGURE 5.1
Fabrication steps for HSG poly-Si storage nodes for enlargement of storage node area by the HSG poly-Si deposition and etch back. ("A Capacitor-over-Bitline (COB) Cell with a Hemispherical Grain Storage Node for 64 Mb DRAMs", M. Sakao, et al., IEDM, pp. 655–658, 1990.)

showed that textured capacitors have much higher lifetime values than the non-textured or smooth capacitors [1].

A hemispherical grain (HSG) poly-Si storage node has been developed at 64 Mbit DRAM density level for a capacitor-over-bit-line (COB) cell structure [5]. Figure 3.4 showed a perspective view for the cell with HSG poly-Si storage node, which doubled the effective surface area in comparison to that of conventional smooth poly-Si and hence helped in eliminating the need of using a complex storage node structure and reduced storage node height. The HSG storage node fabrication requires two additional process steps, that is, HSG poly-Si deposition and an etch back as shown in Figure 5.1. After the formation of capacitor contact and conventional storage node, HSG poly-Si is deposited on core storage node. The HSG poly-Si is etched back by reactive ion etching with HBr gas. For a 1.8 μm^2 memory cell at 0.4 μm minimum feature size, a 0.5 μm high poly-Si storage node with 0.1 μm spacing covered with hemispherical grain was formed after the etch back, which provided storage capacitance of 30 fF, an increase of more than 100% in value without surface texturing.

Since the dielectric films with less than 3 nm SiO_2 equivalent thickness are required, stacked capacitor cells using oxide-nitride-oxide (ONO) layers can hardly satisfy the requirements of cell area and storage capacitance beyond 64 Mbit DRAM level. To overcome this problem engraved storage electrode poly-Si film (Figure 5.1), which has a rugged surface, was used

[6–8]. However, it was observed that the surface morphology of poly-Si film deposited by LPCVD changes with the deposition temperature [9]. It was also observed that poly-Si film deposited at 560°C has both amorphous area and polycrystalline area with hemispherical shape grains. The grain diameter increased and grain density also decreased with rise in working temperature and at 565°C grain shape changed from hemispheric to cylindrical and amorphous area disappeared. At still higher temperature of 580°C small grains combine together and form smooth polycrystalline film [10]. Poly-Si storage elctrode deposited at about 570°C has rugged surface and Si_3N_4 film deposited at 620°C. A 0.1 μm poly-Si film achieved 2.5 times more surface area than the storage electrode deposited at 620°C, which made it suitable for 64 Mbit DRAM and beyond. It is necessary to improve oxidation resistance of thin Si_3N_4 film if it is to be used as ONO composite film. It is observed that deposition temperature considerably affects the oxidation resistance of the Si_3N_4 film; when deposited at 600°C it has higher oxidation resistance than the film deposited at conventional temperature of 900°C.

Ultra-thin oxide/nitride dielectric layer is needed for 64 and 256 Mbit stacked DRAMs, but when the thickness of the nitride film is reduced to less than 4 nm range, an abrupt drop in capacitance appears. This sharp transition occurs when the ultra-thin nitride film is unable to withstand the following reoxidation step [11], which grows a thick film on the heavily doped bottom polysilicon electrode. Because of the possibility of the oxidation punch-through mechanism it imposes an upper limit to the maximum capacitance of 6.9 and 12.3 fF/μm² for smooth and rugged structures, respectively, which can be realized with ON dielectric. The capacitance increase of 80% using ON films on rugged polysilicon was as high as the maximum values reported for Ta_2O_5 high dielectric constant-material used with smooth surface in the years 1990–91 [12].

A different surface modifying method was developed by Jun and others in 1992 to obtain a modulated stacked (MOST) capacitor, in which surface area of the storage electrode was increased up to eight times that of an ordinary stacked capacitor [13]. Usual process steps are taken to form structure below the stacked capacitor and to form its first electrode; an amorphous silicon layer is deposited. An oxide layer of approximately 100 nm is deposited and then top poly-Si layer is deposited under controlled pressure and temperature conditions. This layer has rough surface consisting of hills and valleys. Subsequent processing steps leave the storage electrode poly-Si in pillar form. Then a composite ON layer of dielectric is formed and finally a 200 nm thick poly-Si layer to form capacitor plate electrode. Electrical properties such as leakage current, breakdown voltage, and TDDB are comparable to those of conventional stacked capacitors.

The conventional oxide/nitride (ON) dielectric used in 4 Mbit and 16 Mbit DRAMs can provide capacitance values on the order of 6–7 fF/μm² for 3.3 V supply [14]. For 64 and 256 Mbit stacked memory chips high dielectric constant Ta_2O_5 films on rugged poly-Si bottom electrode have been suggested

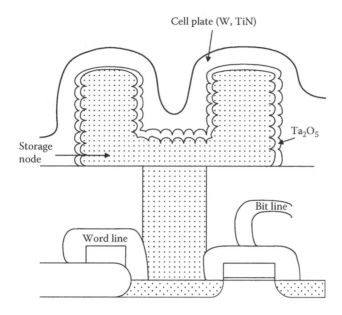

FIGURE 5.2

Cross-sectional view of a typical capacitor structure combining HSG and Ta_2O_5 for 256 Mb DRAMs. (Redrawn from "A High-C capacitor (20.4 fF/µm2) with ultrathin CVD Ta_2O_5 Films Deposited on Rugged Poly-Si for High Density DRAMs", P.C. Fazan, et al., IEDM, pp. 263–266, 1992.)

which are able to provide capacitance values as high as 12 fF/µm² [5,12,14]. However, a chemical vapor deposited (CVD) thin Ta_2O_5 film with rugged poly-Si and a high capacitance of 20.4 fF/µm² without increasing projected area can be achieved, without degrading the leakage current, and TDDB characteristics, as shown in a simple stacked cell of Figure 5.2 [15]. A 150 mm <100> silicon wafer is covered with a 200 nm thick SiO_2 layer and 100 nm thick poly-Si is deposited at 550°C to form rugged hemispherical grained bottom electrode structure. A 15 nm thick Ta_2O_5 film is then deposited at 400 or 450°C, which is annealed by rapid thermal annealing (RTA) process to reduce leakage current. Finally the top electrode is deposited. It is observed that when compared to standard ON/smooth structures, Ta_2O_5/rugged poly-Si capacitor fabrication in core or crown-shaped capacitor for 0.5 µm² 256 Mbit DRAM cell, the node height is reduced by a factor of three or more, which considerably reduces topography issues. Excellent leakage current, TDDB characteristics, and capacitor values of more than 20 fF/µm² make the given structure very suitable for 256 Mbit DRAM applications. Another cylindrical capacitor structure using hemispherical grained-Si developed for 256 Mbit DRAM by Watanabe and others [16] achieves a cell capacitance of 30 fF with 0.4 µm high storage electrodes in a 0.72 µm² cell area. The HSG-Si cylindrical capacitor was developed by applying the seeding method [17] in which Si molecular beam deposition and subsequent annealing are done to

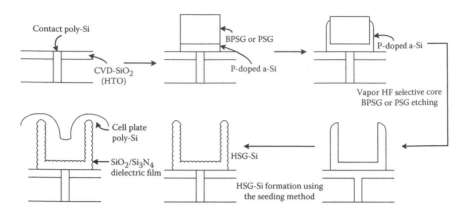

FIGURE 5.3
Process flow of an HSG-Si cylindrical capacitor. ("A New Cylindrical Capacitor Using Hemispherical Grained Si (HSG-Si) for 256 Mb DRAMs", H. Watanabe et al., IEDM, pp. 259–262, 1992.)

form Si microcrystals on undoped amorphous Si surface, along with the use of a vapor hydrogen fluoride (HF) selective etching. Under the low-pressure vapor HF etching, high selective etching of borophosphosilicate glass (BPSG) to SiO_2 is obtained. The cylindrical electrode is simply formed by this selective etching method. Figure 5.3 shows the process flow of a HSG-Si cylindrical capacitor. Grain size was controlled through adjusting annealing time. Both sides of the electrode walls could be covered with HSGs.

BPSG film was used as the core to form cylinders. BPSG and p-doped amorphous-silicon films for the bottom electrode was patterned by lithography and reactive ion etching. A cylindrical wall electrode was formed by deposition and successive etch back of the p-doped amorphous silicon film. The core BPSG was selectively removed by using LP HF vapor etching. Native oxide on the cylindrical electrode was removed and HSG-Si appeared on p-doped amorphous Si surface using the *seeding method*, Si_2H_6 molecule irradiation, and annealing at 580°C in ultra-high vacuum chamber. A 5.0 nm SiO_2 equivalent-thick SiO_2/Si_3N_4 film was formed on the HSG-Si cylindrical electrode.

The influence of the temperature during Si_2H_6 molecule irradiation and subsequent annealing was investigated both for undoped and p-doped amorphous Si [18]. It is observed that small and large HSG-Si grains cover the whole surface of an electrode formed on an undoped amorphous Si at 600°C and that HSG-Si partially covers the surface of the electrode formed on a p-doped amorphous Si film at the same temperature, whereas uniform size HSG-Si covers the electrode surface formed on p-doped amorphous Si at 580°C. It was also observed that grain-size of HSG-Si increased and the thickness of cylindrical wall decreased with increasing annealing time. Realized capacitance increased 1.6 times after 10 second annealing, 2.0 times after 30 seconds and 2.3 times after 55 seconds; however, if annealing is continued even after that the grains become almost spherical and the cylindrical wall disappears [18]. Similar attempt was made to obtain rugged poly-Si cell structure at 256 Mbit DRAMs density level having low aspect storage node without any complex process [19]. In this scheme double-sided rugged poly-Si fin STC technology is used in which the shape of the rugged poly-Si is controlled by ion implantation. The separated-grain poly-Si is deposited by LPCVD. The shapes of the grains are changed by ion implantation with arsenic dose more than 5E15 per cm^2 and neighboring grains get connected to each other. Cell capacitance reaches 25 fF/cell in STC with 2 fins having 4.5 nm thick ONO film and the storage node height is reducible to 300 nm.

Above examples show the formation of HSG-Si using different forms of capacitor structures like fin, stacked, cylindrical, or trench type, with different dielectrics like ON, ONO, Ta$_2$O$_5$. Methods of formation of grains like seeding/annealing or ion implantation and variation in temperatures were used and slightly varying results were obtained. Investigation was made by A. Ils and others [20] on the formation of HSG-Si on undoped and doped amorphous silicon by the seed and anneal techniques; effects of dopants on grain formation were also studied. For undoped amorphous Si, it was not only confirmed that Si atoms near the seed migrated more toward the seed to form HSG and the depression around [21], but the original silicon surface moves progressively toward the substrate with longer annealing due to nucleation. It was assured that in this case HSG-Si is formed through nucleation and by mass transport, which results in mushroom-shaped grain. For the p-doped amorphous Si case no depletion was found around the grains and the annealing time had little effect on grain size, which was much smaller. It was concluded that grains were formed mainly from the epitaxial growth, which takes place during seeding.

While forming grained polycrystalline Si film for DRAM storage capacitor, it is important to determine its electrical properties. Study shows that a strong correlation exists between the surface roughness measured in terms of reflectance and the electrical area enhancement factor (AEF) of the film. Leakage current density of the device with rough electrode is less than the AEF times that of the device with controlled poly-Si electrode. It is shown that the morphology roughness and doping of the film strongly affect the electrical performance of the capacitor [22].

5.3 Higher Permittivity and Layered Dielectrics

For 1 Gbit DRAM, the cell area is typically 0.24 μm^2. To realize 30 fF in this small area high permittivity materials like BaSr, TiO$_3$, Pb, and lead, zirconium, and titanium oxides are recommended. However, ultra-thin Ta_2O_5 capacitors can also give desirable results with low cost of ownership, though they require some key technologies, which are as follows [23]:

1. Use of tungsten for the storage node surface suppresses native oxide growth on the electrode, resulting in 40% reduction of the SiO_2 equivalent thickness in comparison with n^+ poly-Si electrode [24].

2. New oxygen plasma annealing for Ta_2O_5 film reduces leakage current.

3. The tungsten film is to be prepared by LPCVD on the HSG-Si structure.

By adapting the aforementioned technologies, the storage capacitance increased 2.6 times over that of a conventional stacked capacitor, reaching a value of 29.6 fF/μm^2 (against 20.4 fF/μm^2 for Ta_2O_5/HSG-Si capacitor structure). This cell capacitance is obtained with 0.6 μm-high HSG-Si storage-node in a 0.24 μm^2 cell area. Leakage current characteristic is almost the same as that of no-HSG structure. It indicates that the 1.6 nm equivalent thick Ta_2O_5 capacitor on HSG structure with 10^{-8} A/cm^2 at half of 1.5 V operating voltage can be achieved.

Even with the use of high dielectric constant films, a simpler planar stacked capacitor cannot provide a cell capacitance with such a small area available for it beyond 1 Gbit DRAMs. Hence a three-dimensional capacitor is developed in which sputtered TiN of the TiN/poly-Si electrode is replaced with the plasma enhanced CVD-WN [25].

Before the development of 256 kb DRAMs, only SiO_2 was used as the dielectric, however, with the development of 256 Kbit DRAM oxide-nitride-oxide/oxide-nitride (ONO/ON), multilayered dielectric has been used [5,10,12,16], which resulted in higher effective dielectric constant and lower defect density. Further, when DRAMs moved over to 1 Gbit range, complex cell structure using ONO/ON dielectric became no longer feasible and dielectrics with higher dielectric constant were required as their use simplified the cell structure and simple structure was translated to simpler processes, leading to low cost of production. Use of HSG-poly-Si while using different dielectrics and different structures was also very attractive.

Use of three-dimensional structures in Mbit DRAMs, SiO_2 ($e_s = 3.9$), faced a reliability problem since locally reduced oxidation rates caused defects and reduced breakdown fields; hence, Si_3N_4 ($e_r = 7.2$) became the most practical dielectric when combined with rugged poly-Si electrode such as HSG poly-Si and/or three-dimensional capacitor structure. The development of ONO/ON capacitor film structure is shown in Figure 5.4 [26].

The ONO film structure of Figure 5.4(a) begins with thermal oxidation of bottom poly-Si, LP CVD deposition of Si_3N_4 and thermal oxidation of the Si_3N_4 film. Once the bottom oxide layer was removed for reducing the SiO_2 equivalent thickness (tox_{eq}) as shown in Figure 5.4(b), it is called ON. When tox_{eq} becomes less than 5 nm, leakage current is dominated by direct

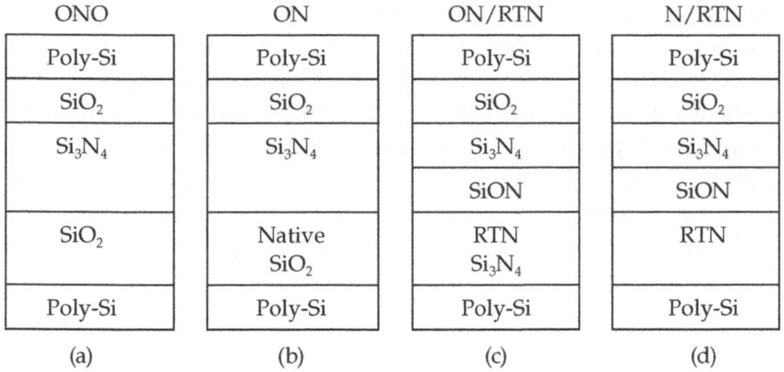

ONO	ON	ON/RTN	N/RTN
Poly-Si	Poly-Si	Poly-Si	Poly-Si
SiO$_2$	SiO$_2$	SiO$_2$	SiO$_2$
Si$_3$N$_4$	Si$_3$N$_4$	Si$_3$N$_4$	Si$_3$N$_4$
		SiON	SiON
SiO$_2$	Native SiO$_2$	RTN Si$_3$N$_4$	RTN
Poly-Si	Poly-Si	Poly-Si	Poly-Si
(a)	(b)	(c)	(d)

FIGURE 5.4
Film structure formation trends in Si$_3$N$_4$ capacitors. ("Trends in DRAM Dielectrics", K.S. Tang, et al., *Circuit and Devices*, pp. 27–34, 1997.)

tunneling, thus putting a limit to tox_{eq} to 5 nm. A transition layer is formed when an initial layer of Si$_3$N$_4$ is deposited on the native oxide. As the Si$_3$N$_4$ film thickness is reduced, the tunneling leakage current rapidly increases. To overcome this problem, the native oxide was changed to a nitride layer through rapid thermal nitridation (RTN) on the polysilicon electrode surface before CVD of Si$_3$N$_4$ [26], as shown in Figure 5.4(c). For further improvement to N/RTN structure of Figure 5.4(d), processing steps are to be increased. However, there is general agreement that making use of ONO/NO for giga-bit and beyond requires too many complex processes, hence higher dielectric constant materials are preferred [27].

5.3.1 Ta$_2$O$_5$–Based Capacitors

Ta$_2$O$_5$ (e_r ~25) was found to be a very suitable dielectric for 256 Mbit and 1 Gbit DRAM capacitors as it could replace Si$_3$N$_4$ without major process integration problems. Many methods for depositing Ta$_2$O$_5$ film have been proposed, for example, CVD methods such as LP CVD, plasma-enhanced CVD (PECVD) and electron cyclotron resonance plasma-enhanced ECR PECVD. In the first half of 1980s Ta$_2$O$_5$ film formation using CVD method was widely adapted. The films were deposited on Si or metal-coated substrates by the thermal decomposition of tantalum ethylate. Film deposition was done between 450°C and 700°C, but high dielectric strength of nearly 26 was obtained at about 600°C. Dielectric constant of the Ta$_2$O$_5$ film was independent of its thickness for MIM structures, but reduces in value as the thickness decreases when film is grown on Si substrate [28].

High-temperature annealing is usually necessary to reduce the excessive leakage current of as-deposited Ta$_2$O$_5$ films probably due to the deficiency of oxygen in it. To reduce leakage current and to improve the electrical properties of CVD Ta$_2$O$_5$ capacitors, various post-deposition annealing techniques are proposed to fill vacancies with oxygen. Out of these, rapid thermal annealing (RTA) in N_2O is found to be the most effective in suppressing leakage current. Another problem faced is that post-anneal Ta_2O_5 capacitors in DRAM cells have to withstand high-temperature processes such as boro-phosphosilicate glass (BPSG) reflow after capacitor formation and Ta_2O_5-based capacitors degrade seriously after BPSG reflow, which is caused by TiN oxidation at the TiN/BPSG interface [29]. However, high-temperature annealing is not needed in the PECVD and ECR-PECVD processes due to the crystallization of the Ta_2O_5 film and growth of interfacial SiO_2 layer [27]. The PECVD processes are based on the utilization of a precursor activated through high-energy plasma, whereas the ECR-PECVD process creates plasma at low operating pressure using microwave as well as high magnetic field. The plate electrode in Ta_2O_5 film based capacitor is generally a stack of highly doped poly-Si/TiN or WN/Ta_2O_5/highly doped poly-Si. The stack of doped poly-Si/TiN or WN is used so that TiN or WN layer prevents the escaping of oxygen from the Ta_2O_5 film. Then doped poly-Si is used on top of the TiN or WN layer so that the metal electrodes are protected in subsequent process steps.

It is important to note that an interfacial oxide layer always grows during Ta_2O_5 film deposition and during any annealing process. Since this oxide has low dielectric constant, effective capacitance is reduced. Therefore, thickness of the interfacial oxide should be as small as possible. Bottom electrode material affects the interfacial oxide thickness. Use of platinum (Pt) bottom electrode completely eliminates the interfacial oxide layer, but it can be deposited only by physical vapor deposition. Other bottom electrode materials are TiN and W, which can be deposited by CVD. TiN was the most widely used material because of its stability, even at high temperatures. It may be noted that Ru bottom electrodes having Ta_2O_5 film were used only at the end of 1990s and shall be discussed at a later stage.

The downscaling limit of Ta_2O_5 capacitors is estimated to be around an equivalent oxide thickness of 1.5 nm [26]. Combined with poly-Si electrode and capacitance enhancement techniques such as HSG, Ta_2O_5 capacitors are practical up to 1 Gbit DRAMs; the effective capacitance is difficult to maintain/increase further. While successfully fabricating a 1 GB DRAM with Ta_2O_5 capacitor, slightly different reports were available about the possible thinning of Ta_2O_5 film. Accordingly the leakage current shall increase drastically if thickness of the film is less than 3.5 nm [25]. Of course, procedures have been used to reduce the leakage current but going below 3 nm was not advisable. Hence, to ensure a capacitance density of 145 fF/μm^2 for a 1 Gbit DRAM, a three-dimensional capacitor was fabricated with 150 nm wide undercut having top electrode of TiN/poly-Si and WN/poly-Si as shown in Figure 5.5. It was observed that for a Ta_2O_5 capacitor, WN is comparable with TiN as electrode material in terms of leakage before BPSG flow; however after BPSG flow at 850°C for 30 minutes, leakage current in WN/poly-Si top electrode became one-order less (compared to the TiN/polySi electrode), which was less than 2×10^{-15} A/cell at 1.6 V.

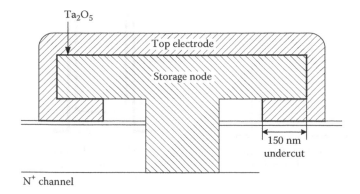

FIGURE 5.5
A stacked capacitor with 150 nm-wide undercut. (Redrawn from "Ta$_2$O$_5$ Capacitors for 1 G bit DRAM and Beyond", K.W. Kwon et al., IEDM, pp. 835–838, 1994.)

Another 1 GB DRAM using a CROWN cell was given with CVD TiN top electrode and Ta$_2$O$_5$ film having SiO$_2$-equivalent oxide thickness of nearly 1.6 nm. As three-dimensional bottom metal electrode could not be fabricated easily using conventional method of CVD or sputtering, a substituted tungsten (SW) electrode technology was developed in which W was substituted for poly-Si so that shape and thickness of the electrode could be processed easily [30]. Measured capacitance was nearly 17.1 fF/cell, but 20 fF/cell was reachable and the leakage current was nearly 0.1 µA/cm^2 at a working voltage of 0.75 V. In spite of good electrical characteristics, a CVD-W-based storage node could not be used at 0.12 µm technology because of its surface morphology. Hence, a low-pressure CVD-WN technique was used for surface smoothing and an MIM cylinder storage node capacitor with an ultra-thin Ta$_2$O$_5$ was realized [31]. Improved technology and processes made it possible to realize 30 fF/cell capacitance in 0.25 × 0.5 µm^2 and leakage was only 10^{-15} A/cell at 0.75 V.

5.4 Low-Temperature HSG

In large-scale system integration, high-speed logic device processing and high-density embedded DRAM cell formation steps should be compatible and capacitor fabrication temperature should be lower than 700°C to suppress the degradation of CoSi$_2$ in cobalt-salicided p$^+$ diffusion layers. However, with lowering the process temperature, HSG grains are depleted which reduces the storage capacitance. A new low-temperature HSG cylinder ON capacitor process was developed by Yamamoto and others [32] as shown in Figure 5.6. Anisotropic silicon deposition is done and a cylindrical

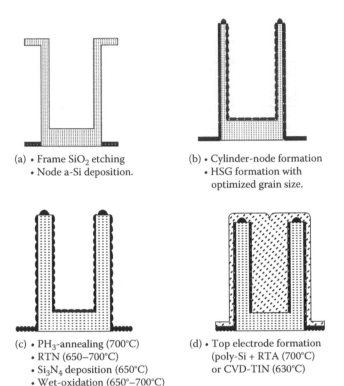

(a) • Frame SiO$_2$ etching
    • Node a-Si deposition.

(b) • Cylinder-node formation
    • HSG formation with
     optimized grain size.

(c) • PH$_3$-annealing (700°C)
    • RTN (650–700°C)
    • Si$_3$N$_4$ deposition (650°C)
    • Wet-oxidation (650°–700°C)

(d) • Top electrode formation
    (poly-Si + RTA (700°C)
    or CVD-TIN (630°C)

FIGURE 5.6
Low-temperature HSG cylinder ON capacitor process. (Redrawn from "Low-Temperature Metal/ON/HSG-Cylinder Capacitor Process for High Density Embedded DRAMs", I. Yamamoto, et al., Symp. on VLSI Tech., pp. 157–158, 1999.)

node is formed (Figure 5.6(a), (b)). After forming HSG grains on inner-type cylinders, ON dielectric film of 4–5 nm (equivalent SiO$_2$ thickness) and top-plate-electrode were grown. For the development of low-temperature process, phosphorus doping to HSG grain with oxide-free surface by PH$_3$ annealing was done at 700°C as shown in Figure 5.6(c) and TiN-plate electrode was used to suppress the depletion in the HSG grains and in the top electrode. Figure 5.6(c) shows the structure after top electrode formation. The process temperature was held below 700°C in the entire capacitor formation process. As an application, a graphic accelerator using 0.24 µm design rule embedded DRAM was fabricated at low temperature without degrading the performance of the logic devices.

To reduce the cost of ownership (COO) on account of using high-k dielectric a trench capacitor DRAM is realized [33] for embedded system using the trench DRAM as it appears to be more suitable for the logic LSI than the stacked DRAM [34]. To ensure enough capacitance, the scheme used gas phase doping (GPD) bottle-shape trench (BT) [35] and then HSG technique was applied for 0.08 μm trench capacitor. The BT technique increases the trench diameter and the HSG technique increases the sidewall area and realized capacitance is increased by nearly 5% with GPD, 10% with BT, and 50% with HSG, respectively. As the trench is deep, it is important to have uniform formation of HSG.

5.5 Sub-100 nm Trench Capacitor DRAMs

Important work has been done for capacitance enhancement in sub-100 nm trench DRAMs [36,37,38]. Use of bottle-shaped trench, integration of HSG, and replacement of ON dielectric by Al_2O_3 form a common theme with some variations in the process of the schemes. In fact, Al_2O_3 was used for the first time as high-k node dielectric in silicon-insulator-silicon (SIS) trench capacitor in 2001. However, deposition of Al_2O_3 film was done using atomic layer deposition (ALD) to achieve excellent step coverage at aspect ratios of up to ≈60, an essential requirement at 90 nm trench level [36].

Successful implementation of HSG on fully integrated trench wafer was achieved for the first time in 2001, and the thermal stability of Al_2O_3 in trench capacitor was also shown. Both selective and nonselective HSG processes have been developed for trench application. In the selective HSG process grains are formed selectively below the collar. Leakage versus t_{ox} data given in Table 5.1 for planar SIS Al_2O_3 capacitor (and oxide nitride capacitor) after annealing at 1050°C clearly demonstrates thermal integrity of Al_2O_3. A capacitance increase of up to 30% was achieved with leakage current remaining well below 1×10^{-8} A/Cm² or 1 fA/cell. In a similar approach a manufacturable integration scheme for Al_2O_3 as node dielectric in trench capacitor was implemented in 170 nm technology, but claimed to be scalable below 100 nm [37]. Once a deep trench is formed, ALD process is used for Al_2O_3 film deposition and post-deposition annealing is done using rapid thermal process (RTP). First stage arsenic doped poly-Si deposition is done for the top electrode, then a recess is formed in it and Al_2O_3 is selectively removed as shown in Figure 5.7. At the final stage oxide collar is formed and the upper trench region is filled by a second poly-Si deposition. Obviously with decreasing minimum feature size and aspect ratio of trench being high (say >50),

TABLE 5.1

Variation of Leakage with t_{ox} for Planar Capacitor

t_{ox} (nm)		3.5	3.75	4.0	4.25	4.5
Leakage current	Oxide nitride	5×10^{-8}	3×10^{-8}	1×10^{-8}	9×10^{-9}	
(A/cm²) at 1.0 V	Al_2O_3		2×10^{-9}	1×10^{-9}	7×10^{-10}	—

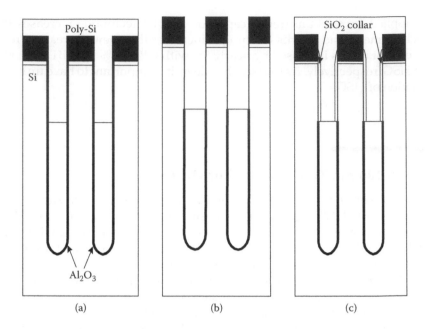

FIGURE 5.7
Schematic representation of the high-k integration flow. (a) After Al_2O_3 and poly-Si deposition, (b) poly-Si recess and selective Al_2O_3 removal, and (c) SiO_2 collar formation and poly-Si fill. (Redrawn from "A Fully Integerated Al2O3 Trench Capacitor DRAM for Sub-100nm Technology", H. Seidl, et al., IEDM, pp. 839–842, 2002.)

integration of thin Al_2O_3 film becomes demanding. Measured capacitance, in this report, was 36 fF/cell with 45 Å (angstrom) thick Al_2O_3 and leakage was 0.1 fA/cell at ± 1.0 V.

J. Lützen also used bottle-shaped trench capacitor with HSG deposition and Al_2O_3 dielectric for sub-100 nm DRAM. In trench technology for DRAMs, dielectric liner of sufficient thickness is required in the upper part of the trench to turn off the vertical parasitic device between the drain/buried strap region and the buried electrode of the capacitor. Conventional schemes reduce the trench diameter using oxide layer which increases the resistance of trench fill, thereby increasing RC delay, a constraint for future generations. A buried collar concept was implemented to increase enough space for the connection from the array device to inner electrode, as shown in Figure 5.8. Not only does the method result in lower node resistance, it does not require any sacrificial fills (resist or poly-Si) to define the depth of different features [38]. In the proposed scheme the integration of nonselective or selective HSG is straightforward [36,37]. Widening of the trench by wet etches and use of HSG deposition gives capacitance enhancement factor (CEF) of up to two. Proposed technology is compatible with high-k dielectric materials to be used with trench technology, which guaranteed values

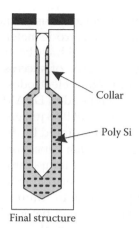

Final structure

FIGURE 5.8
Buried collar concept. (Adapted from "Integration of Capacitor for Sub-100-nm DRAM Trench Technology", Jorn Lützen et al., Symposium on VLSI Tech. Dig. Technical Papers, pp. 178–179, 2002.)

of more than 30 fF/cell at sub-100 nm ground rules. IBM Microelectronics fabricated a 512 Mbit prototype DRAM with a 84.6 mm² die size using ArF lithography, utilizing an $8F^2$ vertical bottle-shaped, 6.5 nm deep trench, and hemispherical grains [39]. The technology was shown to be scalable without changing the basic architecture or cell layout.

5.6 Metal Insulator Metal Structure

In Section 5.3 while progressing through the use of dielectric from SiO_2 to its combination with silicon nitride and then application of Ta_2O_5, structure of the capacitor also changed from SIS (silicon-insulator-silicon) to MIS (metal-insulator-silicon) and then to MIM (metal-insulator-metal). The main aim was to realize increasing capacitance density while keeping leakage under control. Different metals were being used for the capacitor electrodes like TiN, WN, or Ru. Selection of the electrode metal influences the obtained capacitance density, but a critical requirement is that it should be compatible with the other materials on the chip, especially at different working temperatures, and it should not degrade the performance of the realized capacitor. As the DRAM density went on increasing, requirement of still higher capacitance density led to searches for different capacitor structures and dielectric materials with large value of dielectric constant; of course, as mentioned, the new dielectrics must be compatible with new electrode materials. Following sections are based on the type of dielectric used like Ta_2O_5, Al_2O_5, HfO_2, ZrO_2, BST series, and Ti based. Different electrode materials like TiN, platinum (Pt), ruthenium (Ru), RuO_2, and Ti/Sr based have been used with the above mentioned dielectrics with varying degree of success and performance. Choice of dielectric material, top and bottom electrodes, and the process temperature became more stringent for embedded DRAMs for downward scaling of minimum feature size. For example, at 0.18 μm generation large-scale DRAM and high-performance logic was integrated by developing MIS (TiN/Ta_2O_5/poly-Si) capacitor technology in which capacitor element was formed after the transistor formation, at a process temperature of 800°C [40]. However, beyond 0.15 μm generation capacitor formation temperature has to be lower than 600°C, the reason being that due to thinner gate oxide and lower heat immunity of salicide, transistors formed before the formation of capacitors became susceptible to the higher thermal budget. However, reduction of temperature leads to larger capacitor leakage current and higher capacitor contact resistance, which makes it almost impracticable. One practical solution is the changeover from MIS to MIM. Without adopting any new material for the sake of continuity reason, an MIM capacitor at 0.15 μm technology node was fabricated at nearly 500°C process temperature in which TiN was used both for the top and bottom electrodes and Ta_2O_5 was used as the insulator [41]. Figure 5.9 shows schematic view of the fabricated MIM capacitor in which leakage current could be contained at 8E-15 A/μm² at 125°C. The MIM capacitor having equivalent oxide thickness of 17 Å was integrated and used in a 4 Mbit test chip, its cell size being 0.425 μm².

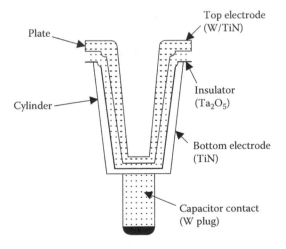

FIGURE 5.9
Schematic cross-sectional view of an MIM capacitor at 0.15 μm technology. (Adapted from "A 0.15 μm Logic Based Embedded DRAM Technology Featuring 0.425 μm² Stacked Cell Using MIM (Metal-Insulator-Metal) Capacitor," M. Takeuchi et al., Symp. on VLSI Techn, Dig. Tech. Papers, pp. 29–30, 2001.)

5.6.1 Ru/Ta$_2$O$_5$/Ru Capacitor Technology

Ta$_2$O$_5$ film has widely been used in SIS and MIS DRAM capacitors. Effort was also on to find a suitable metal to be used for both electrodes with Ta$_2$O$_5$ as dielectric. Reasons for keeping Ta$_2$O$_5$ in continued use were many, for example the Ta$_2$O$_5$ film deposition is an easier process, even at lower process temperature and has excellent step coverage. Moreover, the equipment used for Ta$_2$O$_5$ film deposition could also be used for economic reasons. While selecting other materials for electrodes several reports have been made available about the use of ruthenium (Ru) as electrode along with its possible drawbacks. Use of Ru electrodes has shown a distinct lowering of equivalent oxide thickness of Ta$_2$O$_5$ films. To study the electrical properties of a Ru/Ta$_2$O$_5$/Ru capacitor a sample parallel plate structure was fabricated in which Ru was deposited by sputtering [42]. Dielectric constant of the crystalline Ta$_2$O$_5$ in the sample structure was found to be about 65 and that was independent of the film thickness; hence, equivalent oxide thickness decreased linearly with film thickness. Leakage current density of the Ru/crystalline-Ta$_2$O$_5$/Ru capacitor deteriorated after forming gas annealing (FGA) process. To minimize this effect, an encapsulating barrier layer of 150 Å Al$_2$O$_3$ was used and found to be effective. As the storage node can be of different structures, a cylinder-type storage node shape was selected because Ru metal is easily etched by reactive ion etching (RIE). Hence, an Ru cylinder-type structure was fabricated with its storage node height being 3500 Å and the use of 120 Å thick film of crystalline Ta$_2$O$_5$ gave capacitance of 39 fF/cell [42].

Figure 5.10 shows the fabrication process of CVD-Ru/Ta$_2$O$_5$/CVD-TiN capacitor with a cylinder-type storage node for a DRAM at 0.13 μm technology level. Bottom electrode was formed using CVD process for TiN, and then Ta$_2$O$_5$ was deposited followed by annealing. For top electrode different materials, like sputtered-TiN, CVD-TiN, and CVD-Ru films, were used for comparison. The storage of node height was nearly 1 μm and step coverage

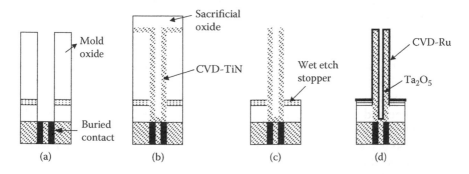

(a) (b) (c) (d)

FIGURE 5.10
Process sequence to fabricate CVD-Ru/Ta$_2$O$_5$/CVD–TiN capacitor with cylinder-type storage node. ("Development of CVD-Ru/Ta$_2$O$_5$/CVD-TiN Capacitor for Multi Gigabit-Scale DRAM Generation," W.D. Kim et al., Symp. on VLSI Tech. Dig. Tech. Papers, pp. 100–101, 2000.)

of the Ru film was approximately 80%, covering all the exposed area of the bottom electrode resulting in maximum capacitance of 40 fF/cell. Out of the three materials used for the top electrode, CVD-Ru film was found to give the best performance in terms of leakage current stability and capacitance density [43]. A satisfactory value of 25 fF/cell was achievable even at 0.1 μm node technology with a 1 μm high storage node and tox_{eq} = 18 Å. Hence, given capacitor structure was usable at several generations of DRAMs.

For gigabit-level DRAM capacitance, Ru has been found to be the most promising metal for electrodes because of its property of easy etching by oxygen plasma and good electrical properties. To get a good conformal CVS-Ru film and smooth morphology a novel process was developed by Won and others [44]. The technique was adapted to fabricate a concave-type of storage node of 1 μm height in 0.13 μm technology that provides a capacitance of 33.5 fF/cell with leakage current of 1 fA/cell at +1.4 V and –1.3 V.

5.6.2 (Ba$_x$Sr$_x$) TiO$_3$ (BST)–Based Storage Capacitor

There is a practically obtainable dielectric constant limit of Ta_2O_5 films, hence materials with even higher dielectric constants are required. One of the better options came in the form of the use of ferroelectric materials. These ferroelectric materials have a characteristic temperature, the transition temperature T_o, at which the material makes a structural phase change from a polar phase (ferroelectric) to a nonpolar phase (paraelectric). The ferroelectric phase possesses spontaneous polarization, which can be reversed by an applied electric field. Because of the larger variations in the material properties that take place at the transition temperature, it is desirable to use a material whose transition temperature is outside the working temperature range of DRAMs (say 0–100°C on chip) [27].

Among the great variety of ferroelectrics, two families of materials are most promising: the lead titanate family, comprising ferroelectric phase materials (e_r > 100) such as $PbTiO_3$, lead zirconate titanate (PZT) and PLZT, and the barium strontium family comprising paraelectric phase materials such as $SrTiO_3$ (e_r ≈ 200) and (BaSr) TiO_3 (BST, e_r ≈ 400) [45].

$SrTiO_3$ and BST have given better promise for DRAM applications because they exhibit a negligible frequency dependence of the dielectric constant. In addition both BST and $SrTiO_3$ are paraelectric at operating temperatures. There is a variety of methods of depositing BST and $SrTiO_3$ such as sputtering, CVD, pulsed laser ablation, or metal organic deposition [27]. Among the different techniques, CVD is often considered most interesting. It has several advantages such as excellent step coverage, high deposition rate, an easy control of composition, and good thickness and composition uniformities over large-size wafers.

($Ba_{0.5}Sr_{0.5}$) TiO_3 was first utilized in a practical stacked capacitor for a 256 Mbit DRAM in 1991 [45]. One of the reasons for BST usage was that its composition was easy; hence, it was expected that its electrical characteristics would be stable. Selected electrode material was Pt mainly because of its resistance to oxidation. However, it also needed a Ta film under Pt to prevent its contamination by Si at the electrode contact. Hence 50 nm layers each of Ta and Pt were sputtered and then a 100 nm thick layer of BST was deposited at 650°C by RF-magnetron sputtering. Along with a sidewall of SiO_2, 100 nm thick TiN top electrode was sputtered. For a stacked cell area of 0.5 μm^2, calculated value of the capacitor was 20 fF with dielectric thickness of 100 nm. A large number of reports are now available which use BST as dielectric with decreasing oxide equivalent thickness. A few of these have been summarized in Table 5.2, containing BST film thickness, electrodes metals used, leakage current and capacitance density realized [46–53] and some of the sketches of the realizations are shown in Figure 5.11(a–f). At 0.8 μm technology a 4 Mbit DRAM was fabricated, in which BST film thickness was between 50 to 100 nm, which was deposited through sputtering providing tox_{eq} between 0.67 to 1.2 nm. Storage node material was again Pt, of course with a diffusion barrier on a p-doped poly-Si plug [47]. Soon advances were made in DRAM density in gigabit range using BST. In one such report tox_{eq} of 0.59 nm was obtained by ECR MOCVD in BST-based stacked capacitor [48]. It employed bottom electrode of 50 nm Ti and 50 nm TiN through dc-sputtering, and RTA treatment converted Ti to $TiSi_x$ salicide layer, which reduced contact resistance between TiN and poly-Si. Layers of Ru (50–100 nm) and RuO_2 (350–500 nm) were formed so that TiN surface is not oxidized. $(Ba_{0.4}Sr_{0.6})$ TiO_3 thin layer was then laid on the storage node. Cell capacitance of 25 fF was obtained while using only 0.125 μm^2 area. At 0.14 μm technology level combination of BST with Ru storage nodes became attractive and a number of gigabit DRAM were realized [49,50].

Use of BST for 4 Gbit DRAM required still lower tox_{eq}, which was possible at higher BST deposition temperature. However, it caused lateral oxidation of TiN increasing its resistance. To prevent the lateral oxidation RuO_2/Ru storage node was constructed on a $TiN/TiSi_x$/poly-Si contact plug. An 80 nm thick BST was deposited at 550°C, though thickness of BST film on RuO_2/Ru storage node sidewall was 40 nm providing tox_{eq} = 0.4 nm. Storage capacitance of 25 fF was obtained in chip area of 0.065 μm^2 with 0.25 μm high storage node [51].

It may be noted that initially Pt was used for capacitor electrode but it was replaced by Ru/RuO_x electrodes because Pt was difficult to etch and oxygen can readily diffuse through it; hence it needs an oxidation resistance barrier. However, with Ru electrodes leakage current density is increased and thermal stability is reduced. Hence, for multi-gigabit density level once again Pt electrode or Ru electrode with Pt encapsulation was used.

Properties of the storage electrode metal are extremely important for the proper functioning of the BST capacitor. It is already mentioned that for Pt electrode etching is difficult as it is done mostly by physical sputtering procedure. For storage nodes below 0.22 μm minimum feature size use of Pt becomes inefficient and results in loss of electrode area, and, hence, it is no

TABLE 5.2

Brief Descriptions of BST Capacitors

Dielectric Combination	Electrode Material	DRAM Density	Dielectric Thickness (nm)	Equivalent Oxide Thickness (nm)	Leakage Current (A/cm²)	Realized Capacitance (fF/cm²)	References
$(Ba_xSr_{1-x})TiO_3$	Pt and TiN	256 MB	100	—	$<10^{-7}$	40	K. Koyama [45]
$(Ba_0Sr_{0-3}) TiO_3$	Pt/Ti or Pt/TiN/Ti (Storage Node)	—	140	1.3	2×10^{-9} at 3.3 V	32	E. Fujii [46]
$(Ba_xSr_{1-x}) TiO_3$	Pt/diffusion barrier/P-doped Poly-Si	4 MB	50–100 (R.F. Sputtering)	0.67–1.2	1×10^{-7} at 1.65 V	25	Y. Ohno [47]
$(Ba_{0.4}Sr_{0.6}) TiO_3$	$RuO_2/Ru/TiN/TiSi_x$ (Storage Node)	1 GB	100 (ECR-MOCVD)	0.59	1×10^{-6} at 1.0 V	—	S. Yamamichi [48]
$(BaSr) TiO_3$	Ru/BST/Ru	1 GB	25 (MOCVD)	0.56	7×10^{-9} at 1.1 V	30	Y. Nishioka [49]
$(BaSr) TiO_3$	Ru (Storage Node)	1 GB	25 (CVD)	0.5	1.5×10^{-8} at 0.6 V	30	A. Yuuki [50]
$(BaSr) TiO_3$	Ru O_2/Ru/storage node on TiN capped plug	4 GB	30 (ECR plasma MOCVD)	0.4	8.5×10^{-7} at 1.0 V	25	H. Yamaguchi [51]
$(Ba_{0.5}Sr_{0.5}) TiO_3$	Pt electrode	Gigabit	40 (magnetron sputtering)	—	1.2×10^{-7} at 1.0 V	40	R.B. Khamankar [52]
$(BaSr) TiO_3$	Ru + Pt encapsulation	—	MOCVD	—	—	—	K.N. Kim [53]

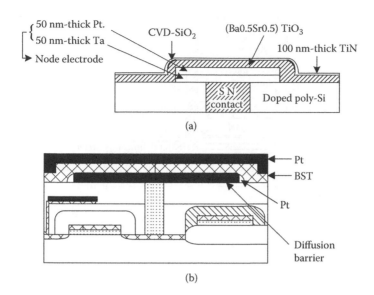

(a)

(b)

FIGURE 5.11

(a) BST based stacked capacitor fabrication. (Redrawn from K. Koyama et al., "A Stacked Capacitor with (Ba_xSr_{1-x}) TiO_3 for 256 M DRAM," IEDM Tech. Dig., pp. 823–826, 1991.) (b) Part of the cross-section of the memory cell having storage capacitor with BST Film. (Modified from Y. Ohno et al., "A Memory Cell Capacitor with Ba_xSr_{1-x} TiO_3 (BST) Film for Advanced DRAMs," Symp. VLSI Tech. Dig. Tech. Papers, pp. 149–150, 1994.) (c) Schematic view of the capacitor structure with $RuO_2/Ru/TiN/TiSi_x$ storage node. (From S. Yamamichi et al., "An ECR MOCVD (Ba, Sr) TiO_3 Based Stacked Capacitor Technology with $RuO_2/Ru/TiN/TiSi_x$ Storage Nodes for Gbit-Scale DRAMs," IEDM Tech. Dig., pp. 119–122, 1995.) (d) Schematic cross-sectional view of DRAM memory cells with Ru/BST/Ru stacked capacitors. (Redrawn from Y. Nishioka et al., "Giga-bit Scale DRAM Cell with New Simple Ru/(Ba, Sr) TiO_3/Ru Stacked Capacitor Using X-ray Lithography," IEDM Tech. Dig., pp. 903–906, 1995.) (e) Schematic cross-section of a BST capacitor structure with Pt electrode. (Redrawn from R.B. Khamankar et al., "A Novel BST Storage Capacitor Node Technology Using Platinum Electrode for Gigabit DRAMs," IEDM Tech. Dig., pp. 245–248, 1997.) (f) Illustration of an MIM BST capacitor with Pt-encapsulated Ru storage node. (Redrawn from K.N. Kim et al., "A DRAM Technology Using MIM BST Capacitor for 0.15 μm DRAM Generation and Beyond," Symp. on VLSI Tech. Dig. Tech. Papers, pp. 33–34, 1999.) Pt plate and Tin electrode are deposited over the BST film.

more suitable for the realization of memory cell capacitance [53]. Another serious issue is the barrier layer between poly-silicon plug and metal storage electrode. Sometimes misalignment leads to a direct touch between BST layer and the barrier layer such as TiN or TiSiN, and it causes low Schottky barrier-generated large dielectric leakage current; it needs fixation. As shown in Figure 5.11(f) the problem is solved through the development of a recessed barrier layer with SiN spacer and Pt-encapsulated Ru storage node. Since etching of Pt electrode is almost impractical beyond 0.13 μm, Ru electrode is used often but rough surface of Ru electrode leads to larger leakage current. It is for this reason that the Ru electrode is encapsulated by a Pt layer [53].

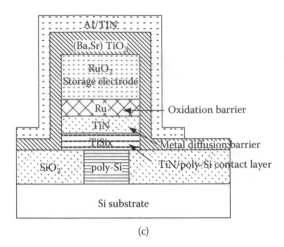

(c)

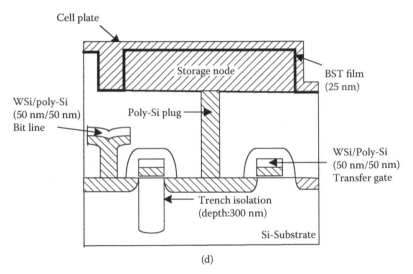

(d)

FIGURE 5.11 (continued)

Another paraelectric dielectric $SrTiO_3$ has been widely believed to be a suitable candidate for fabricating DRAM capacitor in gigabit range [45,46]. A new capacitor structure using MOCVD $SrTiO_3$ thin film on the sidewall of an RuO_2/TiN storage electrode has been fabricated. A reason for selecting such a structure was that at gigabit scale sidewall surface area became larger than top surface of the storage electrode in a stacked capacitor, whereas in earlier cases only top surface of the Pt electrode was used for capacitance purpose. RIE patterning of RuO_2/TiN storage node and low temperature deposition of $SrTiO_3$, of 40 nm thick layer on 0.5 μm high node provided capacitance of 25 fF. Leakage current density was 8×10^{-7} A/cm^2 at half-V_{cc} of +1 V [54].

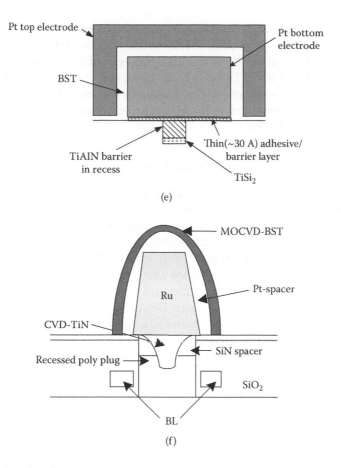

FIGURE 5.11 (continued)

TABLE 5.3

Estimation of Storage Node Height for Different Feature Sizes

Feature size (nm)		70	80	90	100	110	120	130
Storage node height (nm)								
t_{eq} = 0.6 nm (SrTiO$_3$), t_{eq} = 1.5 nm (Ta$_2$O$_5$)	SrTiO$_3$ cylinder	500	400	375	350	325	300	250
c_s = 30 fF/cell	SrTiO$_3$ concave	1550	1300	1175	1050	980	910	860
Ru Thickness = 25 nm	Ta$_2$O$_5$ cylinder	1575	1325	1225	1110	1050	990	950

Height of storage node becomes important especially at gigabit density level DRAMs from the fabrication point of view. An estimate of the same has been obtained based on data shown in Table 5.3 for an ($F * 3F$) storage node, to obtain capacitance of 30 fF/cell [55] for (1) cylindrical structure with Ta$_2$O$_5$ (t_{eq} ~1.5 nm), (2) concave structure with SrTiO$_3$ (t_{eq} ~0.6 nm), and

(3) cylindrical structure with $SrTiO_3$ (t_{eq}~0.6 nm) to get a capacitance of 30 fF/ cell. As obvious from the data, $SrTiO_3$ cylindrical structure is preferable as its storage node height is lower than 500 nm even at 0.1 µm technology level. A cylindrical $CVD-Ru/SrTiO_3/CVD-Ru$ capacitor was successfully realized with storage node height of 300 nm and using cell area of 0.31 µm². A 22 nm thick $SrTiO_3$ layer provided 18 fF/cell capacitance with leakage current of < 0.1 fA/cell at ± 0.7 V [55].

5.6.3 HfO₂-Based Capacitor Technology

In the search for high-k materials for memories, although having different characteristics, HfO_2-based materials and Al_2O_3 dielectrics have also been used. ALO dielectrics have very high conduction band offset (CBO) value (2.5 eV) but relative low dielectric constant (~9), whereas HfO_2 dielectric have high dielectric constant (~25) but a little lower CBO value (1.5 eV). ALO has been used mostly in trench capacitor as given in Section 5.5 or in conjunction with HfO_2; only a few examples are available using ALO other than for MIM capacitors for high-density DRAMs because of its low dielectric constant in comparison to other materials. A few examples using HfO_2 and AHO (Al_2O_3/HfO_2) are discussed below in brief.

An MIM capacitor using HfO_2 as dielectric was developed for the first time within a maximum temperature of 200°C (the maximum temperature of 200°C is very compatible with Cu and low permittivity interconnect technology) [56]. An HfO_2 film of 56 nm was deposited on substrate by ablating Hf target in O_3 ambient and then Al was deposited as the top electrode. Dielectric constant was around 18.5 and high capacitance density of 3.0 fF/µm² with a leakage current of 2×10^{-9} A/Cm² at 3.0 V. Many of the MIM capacitors using HfO_2 have been used in silicon RF and mixed-signal IC applications. For such applications capacitor voltage linearity is very important; for the 56 nm HfO_2 MIM capacitor, obtained parameters meet the requirements of the ITRS Roadmap 2001 [57].

An MIM capacitor compatible with back end of the line was fabricated in which an HfO_2 film was deposited using ALD at 300°C and annealing was done at 400°C. Samples of the capacitor were fabricated with varying thickness to study its effect on capacitance density, frequency dispersion, voltage coefficient of capacitance (VCC), and leakage current [58]. The results are shown in Table 5.4 (VCC for 1 MHz only). The fact that HfO_2 capacitor has nearly the same and small frequency dispersion with thickness variation of HfO_2, it makes it suitable dielectric for DRAMs.

A cylindrical $TiN/HfO_2/TiN$ capacitor applicable to 70 nm generation DRAM was developed for the first time in which HfO_2 film was deposited by ALD using Hf (NEtMe) as precursor and O_2 plasma as a reactant [59]. MOCVD TiN was used as top electrode. It was observed that leakage current strongly depended on the film deposition temperature. With 13 Å tox_{eq},

TABLE 5.4

Dependence of Capacitance Parameters on HfO_2 Film Thickness

HfO_2 Film Thickness (nm)	Capacitance Density (fF/μm^2)	Leakage Current at Room Temp. (10^{-8} A/Cm^2)	For V_{cc}, $C(V) = (\alpha V^2 + \beta V + 1)$	
			α ppm/V^2	β ppm/V
30	5.0	2.1	238	206
20	~9.0	3.1	450	260
10	13.0	5.95	831	607

leakage current was stable at 0.1 μA/cm^2 at 1.5 V and 85°C when the film was deposited at 300°C.

HSG-merged Al_2O_3/HfO_2 (AHO) capacitor technique was introduced in 2005 [60]; however, getting uniform capacitance over full area of wafer, without any capacitor-related failure, proved to be difficult. Reliable mass production was not possible using this technique. S.G. Kim and others reported a fully integrated 512 Mbit DRAM with HSG-merged-AHO cylinder capacitor in 2006 [61]. The technique used reverse HSG one cylinder storage node (RHOCS) and AHO capacitor process as RHOCS process did not decrease the space between the storage nodes. A comparison is illustrated in Figure 5.12 between RHOCS and conventional double HSG one-cylinder storage node

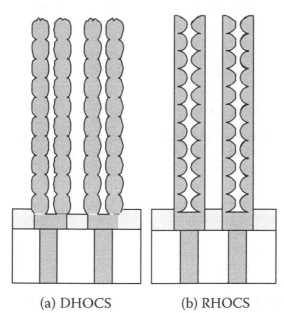

(a) DHOCS   (b) RHOCS

FIGURE 5.12

Comparison between (a) DHOCS, and (b) RHOCS. (Adapted from "Fully Integrated 512 Mb DRAMs with HSG-Merged-AHO Cylinder Capacitor", S.G. Kim et al., Solid-State Electronics, 50, pp. 1030–1034, 2006.)

(DHOCS) in which HSG is grown on both inner and outer sidewalls of the cylinder by forming HSG after lift-off process. In RHOCS scheme HSG is grown on only the inner sidewall of the cylinder by forming HSG after deposition of storage node material. In the proposed scheme AHO bilayer is implemented to combine the good points of the two dielectrics, that is, high dielectric constant of HfO_2 and higher conduction band offset (CBO) of ALO.

As predicted in earlier works, improved electrical characteristics and retention time were achieved in the fabricated DRAM in 110 nm technology without any capacitor-related leakage current failure. It was expected that the technology would be used in sub-100 nm scale.

Leakage current of the DRAM capacitor has to be equal to or less than 0.1 $\mu A/cm^2$, irrespective of the equivalent oxide thickness of the capacitor. It was expected that for keeping constant EOT the use of higher permittivity materials will lead to a physically thicker dielectric film, and it would result in lower leakage current; however, that is not so. Investigation was done in terms of tunneling barrier, or in other words leakage current, work function, and applied voltage at a constant equivalent oxide thickness. It was concluded from the study that high permittivity value is effective only with large work function or at small voltage [62]. Experimental verification was done on a TiN/Hf/ALO/TiN capacitor, which with an EOT of 0.7 nm was having a leakage current of 0.08 $\mu A/cm^2$ at 1 V. However, level of leakage was achieved only when Al_2O_3 was inserted at the bottom of the dielectric to suppress the interfacial increase in EOT. In addition Al_2O_3 was also inserted in the bulk HfO_2 to increase the bulk permittivity. Hence a MIM capacitor with HfO_2 and TiN as dielectric material combination was claimed to be one of the most promising combinations for DRAM.

5.6.4 ZrO_2-Based DRAM Capacitor

As the minimum feature size comes down with each DRAM generation, fabrication of the storage capacitor for the realization of 25 fF/cell capacitance faces new challenges. As illustrated in Figure 5.13, for the design rules between 45 nm to 50 nm, required equivalent oxide thickness has to be between ~6 to 8 Å with a storage node height of 1.3 μm [63]. It was not possible to get required capacitance density at 45 nm technology level with already in-use materials like HfO_2/Al_2O_3 stack and $Hf_xAl_yO_z$ [64]. ZrO_2 has been investigated in this respect and results have been very encouraging. Dielectric constant in the range of 15 to 18 has been achieved with good thermal stability, excellent conformality, and large band-gap [65]. ZrO_2 films deposited on planar Si (100) wafers and patterned amorphous Si cylinders by RTCVD process give good conformal coverage over high aspect ratio. Properties of ZrO_2 were utilized in the form of a single ZrO_2 layer using atomic layer deposition in a 50 nm technology DRAM in a planar type of capacitor [66]. This simple TiN/ZrO_2/TiN structure had a limitation of poor

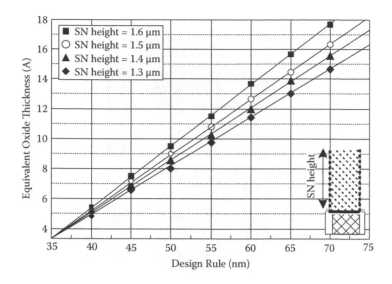

FIGURE 5.13

Required equivalent oxide thickness values for few design rules. ("Development of New TiN/ $ZrO_2/Al_2O_3/ZrO_2$/TiN Capacitors Extendable to 45 nm Generation DRAMs Replacing HfO_2 based Dielectric," D-S. Kil et al., Symp. on VLSI Techn. Dig. Tech. Papers, pp. 38–39, 2006.)

leakage current for negative bias and hence could not be easily deployed in DRAM structure. Dielectric film structure from ZrO_2 and Al_2O_3 was then fabricated which is easily extendable to 45 nm node level [63]. A $ZrO_2/Al_2O_3/ZrO_2$ dielectric contains tetragonal ZrO_2 and amorphous Al_2O_3, where ZrO_2 film is deposited from $Zr(NEtMe)_4$ as precursor and ozone as oxidant by ALD process. It was observed that compared to $HfO_2/Al_2O_3/HfO_2$, its capacitor value was nearly 25% higher. Equivalent oxide thickness (EOT) of ~6.3 Å could be obtained when the dielectric thickness was 32 angstroms, while keeping the leakage within bounds of 1 fA/cell at ± 1 V.

For stacked capacitors one manufacturability requirement is to keep aspect ratio (AR) for contact and vias nearly the same at all generations. For a selected AR of 7, required EOT and height of the cylindrical capacitor for different technology levels between 45 nm and 90 nm have been shown [67], which suggests that for a 45 nm eDRAM EOT below 1 nm is needed with capacitor height being 300 nm. The basic structure of TiN/ZrO_2/TiN eDRAM fabricated at 65 nm design rule is shown in Figure 5.14. The ZrO_2 film was deposited using ALD process at 275°C and 10 nm thick TiN films were deposited at 400°C by ALD to form top and bottom electrodes. It is observed that even with an EOT less than 8 Å, leakage was less than 15 nA/cm² at ± 1.0 volt (within 1 fA/cell at 125°C), which is better than that of EOT = 19 Å of Al_2O_3. Results shown by the process predict good manufacturability for 45 nm eDRAM technology.

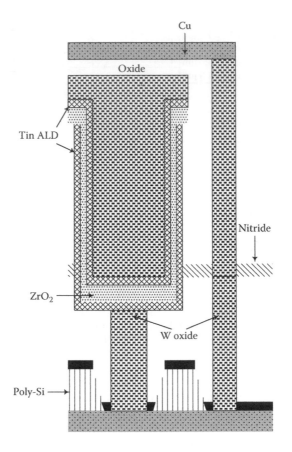

FIGURE 5.14
Schematic of a CUB embedded DRAM with TZT stacked capacitor. (Modifed from "Highly
Reliable TiN/ZrO$_2$/TiN 3D Stacked Capacitors for 45 nm Embedded DRAM Technologies,"
A. Barthelot et al., Solid State Research Conf. (ESSDERC), pp. 343–346, 2006.)

MIM capacitors having dielectrics of capacitance equivalent thickness
(CET) ~ 0.9 nm were manufactured for DRAMs in 2009, and further reduction in the physical thickness of the oxide and reduction in CET, while keeping the leakage current density of <1 fA/cell or 1×10^{-8} to 1×10^{-7} A/Cm^2
was needed [68]. A TiN/ZrO$_2$/TiAl MIM capacitor using a new zirconium
precursor ZrD-04 was developed to satisfy the requirement of ref. [68] and
[69]. On an n$^+$ type Si substrate, a layer of TiN metal of nearly 10 nm is formed
through CVD and ~7–8 nm thick ZrO$_2$ film is deposited using ALD process at
275°C. For increasing the dielectric value without significantly increasing the
leakage current premetallization rapid thermal annealing (RTP) is done at
either 400°C, 500°C, or 600°C. Obtained dielectric value after RTP process is
22.5 ± 1.5 to 32 ± 5 while leakage current is still below $1 \times 10^{-8} - 1 \times 10^{-7}$ A/cm^2
at 1 V.

5.6.5 Advanced MIM Capacitors

HfO_2- and ZrO_2-based high-permittivity dielectrics have been used for the fabrication of MIM capacitors with satisfactory electrical performance [70]. Formation of $ZrO_2/Al_2O_3/ZrO_2$ prevents crystallization during deposition helping in reducing leakage [63]. However, even these high dielectric value materials shall not be adequate to the meet the challenge of fabricating DRAM capacitor with further reduced features, mainly on account of not meeting the leakage current specifications.

To maintain the downward scaling of DRAM capacitor, Table 5.5 shows the requirements/suggestions as given by the International Technology Roadmaps for Semiconductors (ITRS). Use of suggested dielectrics with permittivity more than 100 becomes necessary. However, these materials have lower band gaps (~3.2 eV), which may affect leakage current. TiO_2 as dielectric material and RuO_2 for the electrodes have been selected in two reports on the fabrication of advance MIM capacitors [71,72]. TiO_2 has been chosen because of its very high permittivity of 90 to 170 in its rutile phase, though high-temperature post-deposition annealing at nearly 800°C is needed to obtain pure rutile. Pretreatment of the Ru electrode by O_3 results in a thin RuO_2 film which works as a seed layer for TiO_2 rutile phase growth at only 250°C. In these two reports, obtained characteristics of the fabricated MIM capacitors are given in Table 5.6. Unfortunately leakage current density is

TABLE 5.5

Requirement/Suggestions for Advance DRAMs

ITRS	For Likely Technology Node	Capacitance Density (fF/μm²)	EOT (nm)	Leakage Current (A/cm²)	Suggested Materials for	
					Electrodes	Dielectric
2007	≤32	60	<0.4	3×10^{-7} at 1.1 V	Ru,RuO$_2$,Pt, IrO$_2$,SrRuO$_3$	TiO$_2$,STO, BST
2009	<22 nm	70	<0.5	2×10^{-7} at 0.8 V	Ru,RuO$_2$,Ir IrO$_2$	TiO$_2$,SrTiO$_3$ BaSrTiO$_3$

TABLE 5.6

Obtained Characteristics for Two RuO_2/TiO_3 MIM Capacitors

Technology Node	TiO$_2$ Film Thickness (nm)	EOT (nm)	Dielectric Constant	Capacitance Density (fF/μm²)	Leakage Current Density	Reference
32	14 nm and 13 nm	0.39	130	90	Higher than ITRS 2007	B. Hudec [70]
<32	20 nm	0.5	140	~65	1.5×10^{-6} A/cm² at 0.8 V with Pt top electrode	B. Hudec [71]

higher than desired. It comes under some control with Pt as top electrode instead of RuO_2. $SrTiO_3$ (STO), mentioned as an alternative in Table 5.6 is also a promising material for DRAM capacitor. However, much work has not been done on such capacitor fabrications. ALD STO using Sr $(thd)_2$ as precursor for Sr has been reported using metal electrodes of Ru and Pt with promising results [73,74]. Reported process could not use TiN for electrodes, as high temperature deposition or annealing was necessary in oxidizing ambient. Using another precursor and controlling the deposition variables, composition, and post-deposition processing, combination of STO and TiN has also been given [75]. New precursor was chosen as Sr $(t-Bu_3C_p)_2$, H_2O, and Ti $(OCH_3)_4$, and by changing the Sr and Ti precursors' pulse sequence, STO layer of wide compositional variety was deposited by ALD. Conformal deposition of STO in high aspect ratio is guaranteed in the process. Crystallization of sub-10 nm standard composition and Sr-rich films after 1 minute anneal in N_2 at 550°C was practically available as observed in the cross-section TEM figure in Ref. [75]. For Pt/STO/ALD/TiN MIM capacitors, where STO was annealed for crystallization, provided EOT ~0.49 nm for 7.5 nm Sr-rich film and EOT = 0.69 nm for a 9 nm standard film. Leakage current was 3.5×10^{-7} A/cm^2 at + 1 V and 8.9×10^{-7} A/cm^2 at –1 V.

STO layers have been deposited by ALD using new precursors like Sr $(^ipr_3C_p)_2$ (air liquid) as Sr precursors [76] and (C_pMe_5) Ti $(OMe)_3$ ($C_p = C_5H_5$, Me = CH_3) (air liquid) as Ti precursors [77]. High aspect ratio STO layers were grown for MIM capacitor with thickness of layer in the range of 18–30 nm [78]. The growth temperature was 250°C or 300°C and annealing was done in nitrogen at 650°C or 750°C for 10 minutes. The film composition was close to the stoichiometric Sr TiO_3 films with Sr to Ti atomic ratio in the range of 0.82 to 1.10. Up to $k = 210$ was obtained for film thickness between 7 and 20 nm, but the leakage current was high at ~ 1×10^{-5} A/cm^2. If Sr content was reduced, leakage current could be reduced, but the permittivity came down to 30. MIM capacitors using TiN and Pt as electrode and STO as dielectric were fabricated. Dielectric films grown at 300°C using ozone provided permittivity in range of 60–65. However, leakage current does not meet the set conditions and needs further study.

References

1. P.C. Fazan and R.R. Lee, "Thin Nitride Films on Textured Polysilicon to Increase Multimegabit DRAM Cell Charge Capacity," *IEEE Electron Device Letters*, Vol. 11, no. 7, pp. 279–281, 1990.
2. L. Faraone, "Thermal SiO_2 Films on n^+ Polycrystalline Silicon: Electrical Conduction and Breakdown," *IEEE Trans. Electron Devices*, Vol. ED-33, p. 1785, 1986.

3. Y. Ohji et al., "Reliability of Nanometer Thick Multi-layer Dielectric Films on Polycrystalline Silicon," *Proc. Int. Rel. Phys. Symp.*, p. 55, 1985.

4. S.M. Sze, *Physics of Semiconductor Devices*, 2nd ed., Wiley, 1981.

5. M. Sakao et al., "A Capacitor-Over-Bitline (COB) Cell with a Hemispherical Grain Storage Node for 64 Mb DRAMs," *IEDM*, pp. 655–658, 1990.

6. T. Mine et al., "Capacitance-Enhanced Stacked Capacitor with Engraved Storage Electrode for Deep Submicron DRAMs," *Ext. Abs. 21st SSDM*, p. 137, 1989.

7. Y. Hayshide et al., "Fabrication of Storage Capacitance-Enhanced Capacitors with a Rough Electrode," *Ext. Abs. 22nd SSDM*, p. 869, 1990.

8. H. Watanabe et al., "A New Stacked Capacitor Structure Using Hemispherical Grain Polysilicon Electrodes," *Ext. Abs., 22nd SSDM*, p. 873, 1990.

9. P. Joubert, "Pressure Dependence of *in-situ* Boron-Doped Silicon Films Prepared by Low Pressure Chemical Vapor Deposition," *J. Appl. Physics*, Vol. 66, no. 10, pp. 4806–4811, 1989.

10. M. Yoshimaru et al., "Rugged Surface Poly-Si Electrode and Low Temperature Deposited Si_3N_4 for 64 Mbit and Beyond STC DRAM Cell," *IEDM*, pp. 659–661, 1990.

11. Z.A. Weinburg et al., "Ultrathin Oxide-Nitride-Oxide Films," *Appl. Phys. Lett.*, Vol. 57, p. 1248, 1990.

12. P.C. Fazan et al., "Ultra-thin Oxide/Nitride Dielectrics for Rugged Stacked DRAM Capacitors," *IEEE Electron Device Letters*, Vol. 13, no. 2, pp. 86–88, 1992.

13. Y.K. Jun et al., "The Fabrication and Electrical Properties of Modulated Stacked Capacitor for Advanced DRAM Applications," *IEEE Electron Device Letters*, Vol. 13, no. 8, pp. 430–432, Aug. 1992.

14. H. Shinriki and M. Nakata, "UV-O, and Dry-O_2: Two Step Annealed Chemical Vapour-Deposited Ta_2O_5 Films for Storage Dielectrics of 64 Mb DRAMs," *IEEE Trans. Electron Devices*, Vol. 38, p. 455, 1991.

15. P.C. Fazan et al., "A High-C Capacitor (20.4 fF/μm^2) with Ultrathin CVD-Ta_2O_5 Films Deposited on Rugged Poly-Si for High Density DRAMs," *IEDM*, pp. 263–266, 1992.

16. H. Watanabe et al., "A New Cylindrical Capacitor Using Hemispherical Grained Si (HSG-Si) for 256 Mb DRAMs," *IEDM*, pp. 259–262, 1992.

17. A. Sakai et al., "Novel Seeding Method for the Growth of Polycrystalline Si Film with Hemispherical Grains," *App. Phys. Letters*, Vol. 61, p. 159, 1992.

18. H. Watanabe et al., "Hemispherical Grained Si Formation on In-Situ Phosphorus Doped Amorphous-Si Electrode for 256 Mb DRAMs Capacitor," *IEEE Trans. Electron Devices*, Vol. 42, no. 7, pp. 1247–1253, 1995.

19. H. Ogihara et al., "Double-Sided Rugged Poly-Si FIN STC (Stacked Capacitor Cell) Technology for High Density DRAMs," *52nd Device Research Conference*, pp. 53–54, 1994.

20. A. Ils et al., "Investigation of the Formation Mechanism of Hemispherical Grained Silicon (HSG-Si) on Undoped and Doped Amorphous Silicon for DRAM Applications," *Proc. 29th European Solid State Device Research Conf.*, pp. 232–235, 1999.

21. Akira Sakai et al., "Growth Kinetics of Si Hemispherical Grains on Clean Amorphous-Si Surface," *J. Vacc. Sci. Technol. A*, Vol. 11, no. 6, Nov./Dec., 1993.

22. A. Banerjee et al., "Morphology and Integration of Rough Polycrystalline Silicon Films for DRAM Storage Cell Applications," *J. Electrochem. Soc.*, Vol. 146, no. 6, pp. 2289–2293, 1999.

23. S. Kamiyama et al., "Ultra-Thin TiN/Ta$_2$O$_5$/W Capacitor Technology for 1 G Bit DRAM," *IEDM*, pp. 49–52, 1993.
24. K. Kamiyama et al., "High Reliable 2.5 nm Ta$_2$O$_5$ Capacitor Process Technology for 256 Mbit DRAMs," *IEDM, Tech. Digest*, p. 455, 1991.
25. K.W. Kwon et al., "Ta$_2$O$_5$ Capacitors for 1 G bit DRAM and Beyond," *IEDM*, pp. 835–838, 1994.
26. A. Ishitanai et al., "Trends in Capacitor Dielectrics for DRAMs," *IEICE Trans. of Electronics*, Vol. E.76-C, no. 11, pp. 1564–1581, 1983.
27. K.S. Tang, W.S. Lau, and G.S. Samudra, "Trends in DRAM Dielectrics," *Circuit and Devices*, pp. 27–34, May, 1997.
28. M. Saitoh et al., "Electrical Properties of Thin Ta$_2$O$_5$ Films Grown by Chemical Vapour Deposition," *IEDM*, pp. 680–683, 1986.
29. K. Kwon et al., "Thermally Robust Ta$_2$O$_5$ Capacitor for the 256-Mbit DRAM," *IEEE Trans. Electron Devices*, Vol. 43, no. 6, pp. 919–923, 1996.
30. T. Kaga et al., "A 0.29 µm^2 MIM-CROWN Cell and Process Technologies for 1-Gigabit DRAMs," *IEDM*, pp. 927–929, 1994.
31. S. Kamiyamama et al., "Highly Reliable MIM Capacitor Technology Using Low Pressure CVD-WN Cylinder Storage-Node for 0.12 µm-Scale Embedded DRAM," *VLSI Techn. Dig. Tech. Papers*, pp. 39–40, 1999.
32. I. Yamamoto et al., "Low-Temperature Metal/ON/HSG-Cylinder Capacitor Process for High Density Embedded DRAMs," *Symp. on VLSI Tech.*, pp. 157–158, 1999.
33. S. Saida et al., "Embedded Trench DRAMs for Sub-0.10 µm Generation by Using Hemispherical-Grain Technique and LOCOS Collar Process," *Proc. Int. Symp. on Semiconductor Manufacturing*, pp. 177–180, 2000.
34. H. Ishiuchi et al., "Embedded DRAM Technologies," *Proc. of the IEDM*, pp. 33–36, 1997.
35. T. Ozaki et al., "0.228 µm^2 Trench Cell Technologies with Bottle-shaped Capacitor for 1 G bit DRAMs," *Proc. of the IEDM*, pp. 661–664, 1995.
36. M. Gutsche et al., "Capacitance Enhancement Techniques for Sub-100 nm Trench DRAMs," *IEDM*, pp. 411–414, 2001.
37. H. Seidl et al., "A Fully Integrated Al$_2$O$_3$ Trench Capacitor DRAM for Sub-100 nm Technology," *IEDM*, pp. 839–842, 2002.
38. Jörn Lützen et al., "Integration of Capacitor for Sub-100-nm DRAM Trench Technology," *Symposium on VLSI Tech. Dig. Technical Papers*, pp. 178–179, 2002.
39. H. Akatsu et al., "A Highly Manufacturable 110 nm DRAM Technology with 8 F^2 vertical Transistor Cell for 1 Gb and Beyond," *Symp. VLSI Techn. Dig. of Technical Papers*, pp. 52–53, 2002.
40. M. Hamada et al., "A High-Performance 0.18 µm Merged DRAM/Logic Technology Featuring 0.45 µm^2 Stacked Capacitor Cell," *IEDM Tech. Dig.*, pp. 45–48, 1999.
41. M. Takeuchi et al., "A 0.15 µm Logic Based Embedded DRAM Technology Featuring 0.425 µm^2 Stacked Cell Using MIM (Metal-Insulator-Metal) Capacitor," *Symp. VLSI Techn, Dig. Tech. Papers*, pp. 29–30, 2001.
42. J-W. Kim et al., "Development of Ru/Ta$_2$O$_5$/Ru Capacitor Technology for Giga-Scale DRAMs," *IEDM Tech. Dig.*, pp. 793–796, 1999.
43. W.D. Kim et al., "Development of CVD-Ru/Ta$_2$O$_5$/CVD-TiN Capacitor for Multi Gigabit-Scale DRAM Generation," *Symp. VLSI Tech. Dig. Tech. Papers*, pp. 100–101, 2000.

44. S-J. Won et al., "Conformal CVD-Ruthenium Process for MIM Capacitor in Giga-bit DRAMs," IEDM Tech. Dig., pp. 789–792, 2000.
45. K. Koyama et al., "A Stacked Capacitor with (Ba_xSr_{i-x}) TiO_3 for 256 M DRAM," IEDM Tech. Dig., pp. 823–826, 1991.
46. E. Fujii et al., "ULSI DRAM Technology with $Ba_{0.7}Sr_{0.3}$ TiO_3 Film of 1.3 nm Equivalent SiO_2 Thickness and $10^{-9}A/Cm^2$ Leakage Current," IEDM Tech. Dig., pp. 267–269, 1992.
47. Y. Ohno et al., "A Memory Cell Capacitor with Ba_xSr_{1-x} TiO_3 (BST) Film for Advanced DRAMs," Symp. VLSI Tech. Dig. Tech. Papers, pp. 149–150, 1994.
48. S. Yamamichi et al., "An ECR MOCVD (Ba, Sr) TiO_3 Based Stacked Capacitor Technology with $RuO_2/Ru/TiN/TiSi_x$ Storage Nodes for Gbit-Scale DRAMs," IEDM Tech. Dig., pp. 119–122, 1995.
49. Y. Nishioka et al., "Giga-bit Scale DRAM Cell with New Simple Ru/(Ba, Sr) TiO_3/ Ru Stacked Capacitor Using X-ray Lithography," IEDM Tech. Dig., pp. 903–906, 1995.
50. A. Yuuki et al., "Novel Stacked Capacitor Technology for 1 Gbit DRAMs with CVD-(Ba, Sr) TiO_3 Thin Films on a Thick Storage Node of Ru," IEDM Tech. Dig., pp. 115–118, 1995.
51. H. Yamaguchi et al., "A Stacked Capacitor with an MOCVD-(Ba, Sr) TiO_3 Film and a RuO_2/Ru Storage Node on a TiN-Capped Plug for 4 Gbit DRAMs and Beyond," IEDM Tech. Dig., pp. 675–678, 1996.
52. R.B. Khamankar et al., "A Novel BST Storage Capacitor Node Technology Using Platinum Electrode for Gigabit DRAMs," IEDM Tech. Dig., pp. 245–248, 1997.
53. K.N. Kim et al., "A DRAM Technology Using MIM BST Capacitor for 0.15 μm DRAM Generation and Beyond," Symp. on VLSI Tech. Dig. Tech. Papers, pp. 33–34, 1999.
54. P-Y. Lesaicherro et al., "A Gbit-Scale DRAM Stacked Capacitor Technology with ECR MOCVD $SrTiO_3$ and RIE Patterned RuO_2/TiN Storage Nodes," IEDM Tech. Dig., pp. 831–834, 1994.
55. C.M. Chu et al., "Cylindrical $Ru/SrTiO_3/Ru$ Capacitor Technology for 0.11 μm Generation DRAM," Symp. VLSI Techn. Dig. Tech. Papers, pp. 43–44, 2001
56. H. Hu et al., "A High Performance MIM Capacitor Using HfO_2 Dielectrics," IEEE Electr. Dev. Letters, Vol. 23, pp. 514–516, 2002.
57. The International Technology Roadmap for Semiconductors, Semicond. Ind. Assoc., 2001.
58. X. Yu et al., "A High-Density MIM Capacitor (13 fF/μm²) Using ALD HfO_2 Dielectrics," IEEE Electron Device Letters, Vol. 24, pp. 63–65, 2003.
59. S-H. Oh et al., "$TiN/HfO_2/Tin$ Capacitor Technology Applicable to 70 nm Generation DRAMs," Symp. on VLSI Techn. Dig. Tech. Papers, pp. 73–74, 2003.
60. S.G. Kim et al., "Improved Electrical Characteristics and Retention Time of DRAMs Using HSG-Merged," AHO Cylinder Capacitor," Int. Semiconductor Dev. Research Symp., pp. 113–114, 2005.
61. S.G. Kim et al., "Fully Integrated 512 Mb DRAMs with HSG-Merged-AHO Cylinder Capacitor," Solid-State Electronics, no. 50, pp. 1030–1034, 2006.
62. N. Mise et al., "Theoretical Screening of Candidate Materials for DRAM Capacitors and Experimental Demonstration of a Cubic Hafnia MIM Capacitor," IEEE Trans. Electr. Dev., Vol. 57, pp. 2080–2082, 2010.

63. D-S. Kil et al., "Development of New TiN/ZrO_2/Al_2O_3/ZrO_2/TiN Capacitors Extendable to 45 nm Generation DRAMs Replacing HfO_2 Based Dielectric," *Symp. VLSI Techn. Dig. Tech. Papers*, pp. 38–39, 2006.

64. D-S. Kil et al., "Development of Highly Robust Nano-mixed $Hf_x Al_y O_z$ Dielectrics for TiN/$Hf_x Al_y O_z$/TiN Capacitor Applicable to 65 nm Generation DRAM," *VLSI Tech. Dig. Tech. Papers*, pp. 126–127, 2004.

65. J.P. Chang and Y-S. Lin, "Highly Confirmal ZrO_2 Deposition for Dynamic Random Access Memory Application," *Journal of Applied Physics*, Vol. 90, pp. 2964–2969, 2001.

66. K.R. Yoon et al., "Extended Abstracts," *SSDM*, p. 188, 2005.

67. A. Barthelot et al., "Highly Reliable TiN/ZrO_2/TiN 3D Stacked Capacitors for 45 nm Embedded DRAM Technologies," *Solid State Research Conf. (ESSDERC)*, pp. 343–346, 2006.

68. "The International Technology Roadmap for Semiconductors," *Process Integration, Devices and Structures*, pp. 29–35, 2007.

69. S. Mongham et al., "TiN/ZrO_2/Ti/Al Metal-Insulator-Metal Capacitors with Subnanometer CET Using ALD-Deposited ZrO_2 for DRAM Applications," *IEEE Electr. Dev. Letters*, Vol. 30, pp. 219–221, 2009.

70. T.S. Böseke et al., "Stabilization of Higher-k Tetragonal HfO_2 by SiO_2 Admixture Enabling Thermally Stable Metal-Insulator-Metal Capacitors," *Appl. Phys. Lett.*, Vol. 91, pp. 072902 (1–3), 2007.

71. B. Hudec et al., "Low Equivalent Oxide Thickness Metal/Insulator/Metal Structures for DRAM Applications," *7th Int. Conf. on Adv. Semiconductor Dev. and μ-Systems*, pp. 123–124, 2008.

72. B. Hudec et al., "RuO_2/TiO_2 Based MIM Capacitor for DRAM Applications," *Adv. Semiconductor Devi. and Microsystems*, pp. 341–344, 2010.

73. S-W. Lee et al., "Enhanced Electrical Property of $SrTiO_3$ Thin Film Grown by Atomic Layer Deposition at High Temperature for Dynamic Random Access Memory Applications," *Appl. Phys. Lett.*, Vol. 29, p. 222903-3, 2008.

74. O.S. Kwon et al., "Atomic Layer Deposition and Electrical Properties of $SrTiO_3$ Thin Films Grown Using Sr $(C_{11}H_{19}O_2)_2$, Ti $(OC_3H_7)_4$ and H_2O_7," *J. Electron. Soc.*, Vol. 154, p. 127, 2007.

75. N. Menou et al., "0.5 μm EOT Low Leakage ALD $SrTiO_3$ on TiN MIM Capacitors for DRAM Applications," *IEDM*, pp. 1–4, 2008.

76. M. Rose et al., "Atomic Layer Deposition of Titanium Dioxide Thin Films from C_p*Ti (OMe) and Ozone," *J. Phys.-Chem, C.*, Vol. 113, pp. 21825–21830, 2009.

77. M. Vehkamaki et al., "Growth of $SrTiO_3$ and $BaTiO_3$ Thin Films by Atomic Layer Deposition," *Electrochem. Solid-State Lett.*, Vol. 2, pp. 504–506, 1999.

78. H. Garcia et al., "Characterization of $SrTiO_3$-based MIM Capacitors Grown by Using Different Precursors and Growth Temperature," *Proc. 8th Spanish Conf. Electr. Dev. (CDE'2011)*, 2011.

6

Advanced DRAM Technologies

6.1 Introduction

Shrinking technology has remained the most important priority for DRAM development. At the same time it is responsible for most of the challenges. When the requirement of small area enforced adoption of the three-dimensional cell instead of the planar cell, technological advancements followed in parallel for the trench cells as well as for the stacked cells (and their variations), having their own advantages and limitations. From a topological point of view, the trench cell DRAM was found to be better as its capacitor was below the silicon surface. However, increase in DRAM density required a deeper trench for realizing sufficient storage capacitance; a bit impractical. The feature of low thermal budget fabrication owing to capacitor formation at early stage is also an advantage; hence, trench cell became more appropriate for merged DRAM where density might not be the first priority. However, not altogether replacing trench cells, stacked cell structures have found favor at the gigabit level. But even these kinds of cells are facing many challenges with decreases in technology scale. One of the significant technology changes came in the form of capacitor-over-bit-line (COB) structure from the capacitor-under-bit-line (CUB) structure, to get certain minimum cell capacitance in the stacked capacitor cells. Though to get advantages offered by the COB structure, certain difficulties in fabrication had to be solved. Steps taken in this direction that enhance its suitability for high-density DRAMs shall be discussed.

To sustain shrinking, technology's first requirement was having the basic cell components (the access transistor and the storage capacitor) confined to ever-decreasing projected chip area, while keeping/improving performance that satisfied set standards in terms of speed, power consumption, data retention time, and so on. At the same time complexities in process fabrication were also to be eased and overall chip size was to remain within practical/economical/yield limits. The following simple relation can work as a guideline:

$$\text{Chip area} = \frac{\text{Cell size} \times \text{DRAM density}}{\text{Array Efficiency}} = \frac{8\,F^2 \times N}{0.65}$$

where F and N are the minimum feature size and number of cells, respectively, for a folded bit-line array structure and the array efficiency is taken as 65%. In addition to pursuing open bit-line structure (with minimum cell size of 6 F^2), technological advancements are also being continuously made for realizing < 4 F^2 cell size either using vertical [1] or stacked surrounding gate (S-SGT) transistor [2] or three-dimensional stacking of DRAMs [3]. Smaller cell size directly requires small size access transistor while maintaining its drive, keeping I_{off}/I_{on} within prescribed limits and overcoming secondary effects (becoming more dominant) with F becoming smaller like short channel effect (SCE), DIBL, GIDL, and so on. The storage capacitor is the second most important component directly affected by the shrinking technology and needs adequate attention. Though capacitor in planar form is always preferred, other forms like trench, stacked, or vertical (with variations) have been investigated. SiO_2, which was the basic dielectric, could not remain viable and hence was replaced by Si_3N_4 (or layered with it), Ta_2O_5, and other high-k dielectrics at increasing levels of DRAM density. Enhancement in storage capacitance value was commonly achieved using grained or textured surface of storage node. Efforts were always made to reduce leakages to a minimum as it affected cell and chip design considerably.

Quadrupling of DRAM density every three years could become practically possible only through technological advancements in the area of lithography, isolation, and etch process. Better resolution, suppression of proximity effects using optimization technique and minimization of mask alignment errors became essential. Solutions were needed for reducing depth of focus for resolution enhancements. Wavelength in optical lithography was reduced from 468 nm to 365 nm to 248 nm and then to 193 nm. Efforts are now on to see how 13.5 nm lithography could be put in practical use. Advancements in lithography were not only essential for the realization of small size transistor and storage capacitor but their contacts with word line and bit line through tiny holes/vias became extremely important. Thin word lines and bit lines and other connecting wire strips of ever-decreasing thickness with small pitch were not possible without lithographic and etching process advancements. Isolation technique and lithography advancements are related and their combination helps reducing cell size, as isolation technique controls the process margin and compensates for alignment errors. At small technology nodes those margins become extremely important and sometimes result in open circuiting thin wires or short circuiting small gaps.

With advancements in device technology and cell capacitor, performance started to be affected with bottlenecks in the support circuit technology. Chip areas occupied by the support transistors in CMOS technology and its leakage current and power consumption became important. All parasitic capacitance and resistance of the WLs and BLs and other interconnects needed to be reduced, which meant usage of better conductivity wires and low-dialectic inter-layer material development. Better connection technology became essential for all such wires. Hence, discussion in the following sections is mainly on the development in the area of memory cell transistor and storage capacitor technology, lithography technology including isolation techniques, cell connection technology including landing pad formation and storage node contact and cell contact. Interconnect technology also forms an important component of the discussion.

6.2 Advanced Cell Structures

Advanced cell structure era started with the realization of trench capacitor cells in which highly doped polysilicon was filled in a trench below the planar surface of the substrate, which served as one plate of storage capacitor. Thin SiO_2 layer (or a composite layer with Si_3N_4) was thermally grown on the trench walls to act as dielectric and the storage electrode was in the substrate in the first generation of such cells. These initial attempts in three-dimensional cell design revealed quite a few limitations. It became difficult to increase DRAM density further with reduced projected area, as a deeper trench became necessary in order to realize a minimum of (say 25 fF) capacitance. Moreover as the storage electrode was a part of the substrate, chances of higher inter-cell leakage current increased in closely spaced cells. Higher level of substrate doping did reduce the thickness of depletion region, reducing inter-cell leakage current, but cells could have avalanche breakdown at in-between spacing of 0.8 μm. Starting from the corrugated capacitor cell of Sunami [4] several structures like folded capacitor cell [5], vertical capacitor cell [6], isolation merged vertical capacitor cell [7], buried isolation capacitor cell [8], and FASIC cell [9] have appeared. The aims behind all these design modifications included enhancement of capacitance with shallower trench and/or less size, reduction of leakage, and making the cell less susceptible to soft errors due to alpha particles. Description of these first-generation cells is given in Section 2.3.1 of Chapter 2.

To overcome the major problem of punch-through among closely placed cells and minimization of soft errors in increasing density DRAMs, second generation trench capacitor cell structures were developed in which the storage electrode was inside the trench and the plate electrode was on the outside of the trench. Examples of such cells, which were also called *inverted trench cells*, are buried storage electrode cell [10] substrate plate trench capacitor cell [11], and the stacked transistor cell [12]. These are discussed in Sections 2.6, 2.3.2.1, and 2.7, respectively. Another important development in the inverted type of trench cell came in the form of the fabrication of access transistor above the trench capacitor so as to save precious chip area. Important advance in the category is the trench transistor cell (TTC) [13] discussed in Section 2.5.

Progress of trench capacitor DRAM structures slowed down on account of difficulty in realizing deeper trench for getting sufficient capacitance in extremely reduced area. Technological advancement in fabricating deep trenches, having aspect ratio of the order of 70, was made at sub-100 nm technology nodes. In the meantime different versions of 3-D cells in the form of stacked capacitor cells (STCs) in which capacitor was fabricated above the access transistor instead of being in a trench, as described in Section 2.8, appeared to be better placed in using high permittivity dielectric in metal-insulator-metal (MIM) capacitors. The STC structure also helped in mini-mizing cell signal charge reduction effect which was present in earlier planar type capacitors due to the minority carriers generated in the silicon substrate. Different versions of STC cells in the form of horizontal and verti-cal fin structures, with varying degrees of complexity and ability to real-ize higher capacitance density, were fabricated. To get more capacitance per unit area STC structure was changed from capacitor-under-bit-line (CUB) to capacitor-over-bit-line (COB) for gigabit-density DRAMs, in spite of hav-ing few initial stage technological constraints. Use of self-aligned technology needed to be deployed in COB structures for fabricating landing pads and contact holes. Aspect ratio of the storage node was also to be reduced and reduction of parasitic resistance and capacitance needed sufficient attention. A few early COB structures like straight-line trench isolation and trench-gate transistor (SLIT) [14] and related technology were discussed in Section 2.8, whereas advanced COB cell structures are given in Section 3.5. MIM capaci-tors with different materials for top and bottom electrodes are to be used beyond 100 nm to get sufficient storage capacitance. Obtaining mechanical stability for the storage node is another important requirement for gigabit DRAMs. Two mechanically robust COB structures, LERI (learning extermi-nated ring-type insulator), and MESH (mechanically enhanced storage node for virtually unlimited height), are discussed in the following section.

6.3 Robust Memory Cell—Mechanical Stability of Storage Node

COB DRAM cells have proved to be suitable for DRAMs from 16 Mbit den-sity level onwards. Different approaches have been used for enhancing cell capacitor/area and the use of cylindrical capacitor has been very prominent since the early 1990s [15]. To satisfy the requirement of at least 25 fF/cell, first MIS capacitors were used in which Ta_2O_5 dielectric film was introduced at 0.18 µm technology. From 0.13 µm technology node, both MIS capacitor with

TABLE 6.1

Capacitance of a Cylindrical Capacitor
as a Function of Film Thickness

Dielectric thickness (nm)	30	25	20	15
Capacitance (fF/μm^2)	42	48	61	80

Ta_2O_5 as dielectric layer and MIM capacitor with different materials (but mostly Ru), for top and bottom electrode were used and beyond 100-nm the Ta_2O_5 capacitor no longer remained useful as its dielectric constant was not sufficiently high. For 0.13 μm technology, in general, cylinder-shaped storage node structure replaced BOX-shape storage node to increase the surface area of the storage node. When DRAM level improved to 0.11 μm technology, height of the cylindrical node was to be increased and its shape was also modified. The idea behind change of shape was to provide mechanical stability to the storage node to avoid cell-to-cell short or twin-bit failure.

At 0.13 μm design rule stage metal/Ta_2O_5/metal capacitor was realized in which CVD TiN film was used for cylindrical node [16]. Cell capacitance of 40 fF/cell with leakage current of 0.1 fA/cell at $\pm$ 1 V applied voltage was achieved. Another MIM DRAM cell at 0.11 μm technology level having cylindrical electrodes with Ru was realized [17]. A two-step CVD-SrTiO$_3$ (ST) was employed with SiO$_2$ equivalent thickness of 0.6 nm. Table 6.1 shows the capacitance of a cylindrical capacitor as a function of SrTiO$_3$ film thickness having storage node height of 300 nm and cell area of 0.13 μm^2 with a dielectric constant of ~135.

When Ta_2O_5 is used in MIM capacitors, best performance is obtained when electrodes are realized with Ru as its dielectric constant rises to 70 or SiO$_2$ equivalent of 0.7 nm (Ta_2O_5 ~10 nm). However, there is a major practical problem of integrating three-dimensional electrodes with Ru. Formation of cylindrical bottom electrode with Ru faces two serious problems: (1) adhesion between Ru and SiO$_2$ is very poor, causing seepage of wet etching at the interface during sacrificial SiO$_2$ strip process and effects the collapse of cylindrical electrodes (Figure 6.1(a)), and (2) oxygen easily penetrates through Ru film, oxidizing the surface of the contact plug material (Figure 6.1(b)) resulting in some contact failures. These problems have been solved by a liner-supported cylinder (LSC) technology and a robust cylindrical electrode with Ru, which has been realized at 0.13 μm technology [18]. TiN liner is formed in the shape of a cuplike cavity and the liner acts both as adhesion layer and as oxidation barrier as illustrated in Figure 6.2. Cylinder capacitors of 550-nm height were fabricated without any collapse. Assuming that practically suitable value of aspect ratio would be less than eight, Ru electrode with LSC technology is much better than other electrode materials such as TiN, as it yields best performance of Ta_2O_5 and downscaling beyond 0.10 μm is feasible [18] for realizing a capacitance of 30 fF/cell.

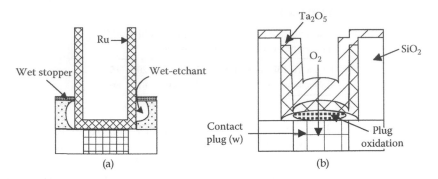

(a)   (b)

FIGURE 6.1
Difficulties in integrating Ru with cylindrical electrode. (a) Poor adhesion of Ru/SiO$_2$; wet etchant penetrates along the interface. (b) Oxygen can penetrate Ru film to oxidize plug material during Ta$_2$O$_5$ formation. (Redrawn from "Linear-Supported Cylinder (LSC) Technology to realize Ru/Ta$_2$O$_5$/Ru for Future DRAMs," Y. Fukuzumi et al., IEDM, pp. 793–796, 2000.)

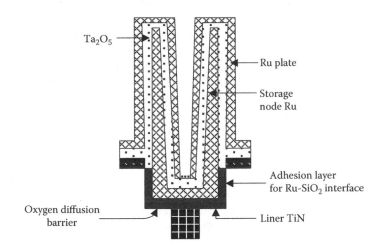

FIGURE 6.2
Schematic of a liner-supported cylinder capacitor. (Redrawn from "Linear-Supported Cylinder (LSC) Technology to Realize Ru/Ta$_2$O$_5$/Ru for Future DRAMs," Y. Fukuzumi et al., IEDM, pp. 793–796, 2000.)

A 4 Gbit DRAM fabricated in 0.11 μm CMOS technology, having cell size and the chip size of 0.1 μm^2 and 645 mm^2, respectively, was given by K.N. Kim and others in 2001 and claimed to be the first working highest density DRAM at that time [19]. The cell used 80 nm array transistor, sub-80 nm memory cell contact, and mechanically robust capacitor by using a novel capacitor structure in which concave structure to cylindrical structure is added as shown in Figure 6.3.

A multistack storage node structure having enlarged bottom size of one cylindrical storage node with much stronger mechanical stability of the

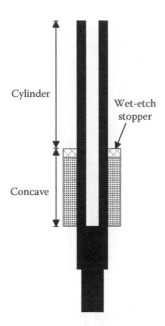

FIGURE 6.3
A capacitor fabricated using merged structure of cylinder and concave (MSCC) storage node. (Modified from "Highly Manufacturable and High Performance SDR/DDR 4 Gb DRAM," K.N. Kim et al., VLSI Tech. Dig. Tech. Papers, pp. 7–8, 2001.)

capacitor was realized in 2003. Capacitance of 22 fF/cell with ALD Al_2O_3/ HfO_2 dielectric film (tox_{eq} = 2.5 nm) was realized when the capacitor height was 18000 Å. However, 30 fF/cell capacitance could be achieved using the multistack capacitor having a total height of 22000 Å, composed of 7000 Å box-shaped (BS) capacitor stack and 15000 Å cylindrical-shaped (CS) capacitor stack [20]. The COB cell was expected to be easily fabricated in 70 nm node and beyond for longer data retention time and mechanical stability of the capacitor using the multistack structure. Figure 6.4 shows the robust multistack cell capacitor composed of BS and CS capacitor. Surface area is increased by nearly 45% in this case compared to a conventional storage node.

In 2004 two mechanically robust, COB DRAM cells for 70 nm node and beyond were given: (1) a leaning exterminated ring-type insulator (LERI) [21], and (2) mechanically enhanced storage node for virtually unlimited height (MESH) [22]; both of these used conventional MIS technology without twin bit failure. The reported LERI process realized 32 fF/cell at the design rule of 82 nm and the MESH process realized 30 fF/cell with an equivalent 2.3 nm oxide thicknesses in a MIS capacitor, with required storage node height of about 2.5 μm. For 80 nm DRAM technology leaning of storage nodes severely deteriorates yield at the height of 1.6 μm. The MESH process tried to improve the process difficulties which came up during integration of LERI with one cylindrical storage node. The cross-linking MESH supporters prevent all

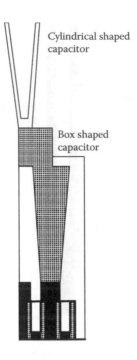

FIGURE 6.4
A robust multistack cell capacitor. (Modified from "Robust Memory Cell Capacitor Using Multistack Storage Node for High Performance in 90 nm Technology and Beyond," J. Lee et al., Symp. VLSI Techn. Dig. Tech. Papers, pp. 57–58, 2003.)

cylinders from leaning and DRAMs with high yield and extremely low level of twin bit failures were obtained. One of the key advantages of the MESH process is that it does not need additional lithography for the support structures. For further enhancement of capacitance, HSG-Si was grown only on the surface of inner cylinder and a 15–30% capacitance gain was possible with process optimization.

6.4 DRAM Cell Transistor Technology

Fabrication of transistor in a DRAM chip occupies a central position. Along with the storage capacitor, it is replicated in the core array and affects not only the area (or cost) but its performance as well. It is therefore, always fabricated in minimum size and in NMOS technology and optimized for performance and fabrication process simplicity. On rest of the chip, transistors, though not necessarily of minimum size, are to be used in every module—for example, in sense amplifiers, row and column decoders, buffers, I/O pads, and so on—and peripheral transistors are generally fabricated in CMOS technology. Because of the requirement from the transistor and the form of utilization on the chip, memory cell transistor and the peripheral transistors are different in terms of constructional details and characteristics. Before reaching technology node of nearly 100 nm, planar, single gate, and conventionally fabricated transistors were used in memory cell and peripherals. However, even at 0.25 μm, the problems of accommodating cell transistor in a small area faced difficulties which were compounded due to the loss of active width from isolation oxide and mask alignment tolerance error. Better isolation technology in the form of shallow trench isolation has given excellent results and shall be discussed in Section 6.7. A major problem faced was a near nonscalable threshold voltage (V_{th}) in cell transistor, even when drain-to-source voltage was to be scaled down with reduced channel length. Generally V_{th} of nearly 0.7 V or more was to be used to keep sub-threshold current under control. At the same time, transistor gate voltage which had to be increased by at least a threshold-voltage drop across it (in order to retain its drivability), its gate oxide thickness had to be increased. At technology node level beyond 0.25 μm, short channel effect (SCE) and narrow width effect created considerable deviations in V_{th} and the deviations increased further with decreasing feature size F. One of the commonly used techniques to increase V_{th} (or maintain) and decrease the effects of SCE was to increase substrate doping density to more than 10^{18} per cm^3 for gigabit-density DRAM. However, it increased leakage current and hence modifications were suggested in fabrication process like reduced doping density just below the source/drain, but a simple one-gate conventional structure was just not enough. Leakage current reductions have been attempted widely using circuit innovations and are described in Chapter 7. However, bottleneck minimization at around 100 nm nodes and beyond was done through the use of special kinds of transistors. First in line were recess channel array transistor (RCAT) and spherical RCAT [22,23] in which effective channel length was increased through recessing. These transistors discussed in Section 4.2 were found to be very useful up to at least 50 nm node technology. Double gate MOSFETs/FinFETs based on DELTA structure [25] and variations like body tied MOSFETs/bulk FinFETs, discussed in Sections 4.2, 4.3, and 4.4 were excellent replacements in sub 50 nm ranges with small V_{th} variation and DIBL and good SCE immunity along with process compatibility of bulk FinFETs with the conventional planar MOSFETs. Because of double gate structures these transistors were inherently providing more drive. Multichannel FET and saddle FinFET [26], discussed in Sections 4.6 and 4.8, have also shown great promise in sub-50 nm technologies with their own advantages, combining good points of RCAT as well.

In higher-density DRAMs, characteristics of MOSFETs, especially drivability decrease due to channel width decrease effect, were further improved in surrounding gate transistor (SGT) and stacked SGT in which source, drain, and gate were realized in a vertical fashion. Apart from improving transistor characteristic, it led to three-dimensional structure and was able to realize cell area per bit of less than 4 F^2 as discussed in Section 4.10. Trench capacitor cells which were overshadowed by the stacked capacitor cell with advanced transistors were given a boost in the form of BEST and VERIBEST cells [27,28] at 4 Gbit/1 Gbit level using vertical pass transistor (Section 4.11). In fact, vertical transistors, in which channel length can be chosen independent of design rules, could satisfy the requirement of I_{on} (with double gate), while keeping $I_{off} < 1$ fA/cell. Vertical transistor and its variations [29] have been discussed in Section 4.12 and these are found to be suitable at 30 nm level or beyond. Section 4.13 has included a brief discussion of few advanced recess FinFETs, like FET partially insulated RCAT (PiRCAT), partially insulated FinFET (PiFinFET) [30] and fin and recessed channel MOSFET (FiReFET) [31]. These advanced devices have further improved upon RCAT and FinFET for still small technology node.

Transistors used in the periphery, in general, have used simpler planar structure fabrication processes, with emphasis on keeping leakage low and higher breakdown voltage. Their channel length was generally larger by at least one generation compared to that for the cell transistors.

6.5 Cell Capacitor Technology

A minimum capacitance of 25 fF/cell has been selected as a guideline and has become a cornerstone for DRAM design as it controls not only the functionality but also its performance. Initially it was fabricated in a simple parallel plate form with the most compatible dielectric SiO_2 by the side of an access transistor. However, it became a challenge with increasing DRAM density and all possible ways were adopted to fabricate it in smallest possible area and with minimum of leakage. Crown-shaped capacitor and Ta_2O_5 dielectric were predominantly used as the capacitor technology up to 256 Mbit DRAM (and extendible to 1 Gbit) level which has already been discussed in Section 3.4.1. Hemispherical grained surface and CVD-Ta_2O_5 layers with rapid thermal annealing (RTA) were also used for capacitance enhancement. Other high-k dielectrics like (Ba_xSr_{1-x}) TiO_3 and $SrTiO_3$ were also under investigation at 256 Mb–1 Gbit level. Capacitors using Ta_2O_5 were based on (MIS) metal-insulator-polysilicon structure, where metal was used for the plate electrode and the polysilicon (heavily doped) was used for the storage node. Metals used for the plate electrode were either highly doped polysilicon/titanium nitride (TiN) or heavily doped polysilicon/tungsten nitride (WN), as these metals were better suited for the twin job of preventing oxygen escape from Ta_2O_5 layer and also prevented a reaction between Ta_2O_5 and the heavily doped polysilicon layer which itself was protecting TiN or WN from oxidation. Such MIS capacitors could not be used beyond

TABLE 6.2

Open Bit Line DRAM Cell Size as Function of Design Rules for Different
Value of Margins between Contact Hole and Adjacent Wirings

Design rule (μm)		0.06	0.08	0.1	0.2	0.3	0.4
DRAM density		1 Gbit			256 Mbit		
Cell size	Margin (μm)						
(μm²) requirement	0.2	0.4	0.51	0.6	0.9	1.4	2.1
	0.1	0.15	0.2	0.22	0.5	1.0	1.5
	0.05	0.08	0.1	0.12	0.42	0.7	1.25
	0.0	0.021	0.05	0.06	0.21	0.52	1.0

0.13 μm technology as sufficient capacitance could not be obtained without further thinning the Ta_2O_5 layer, which was practically not possible; otherwise it would have led to punch-through. Hence polysilicon was replaced by metal for the storage electrode and the structure was called MIM (metal-insulator-insulator). It helped in eliminating the native oxide on the polysilicon storage node surface and Ta_2O_5 could be further scaled down to improve realized capacitance. However, increase of DRAM density and variation in fabrication processes especially processing temperature, different metals for electrodes, like platinum (Pt), ruthenium (Ru), RuO_2, and Ti/Sr were used. In addition, other than Ta_2O_5, dielectrics like Al_2O_5, HfO_2, ZrO_2, BST series, and Ti-based ones were used so that combinations of new metal electrodes and high-k dielectrics were able to realize DRAM capacitance, which not only satisfied the specifications but also became manufacturable. At 0.13 μm level Ru as metal electrode was found very useful because it could obtain a capacitance of 40 fF/cell [32] and continuation of Ta_2O_5 usage restricted the enhancement of fabrication cost. However, soon BST capacitor structures in large number forced Ta_2O_5 to be replaced. All combinations of electrode metals as mentioned above were combined with BST based capacitor at gigabit DRAM level [33–40]. A comparative statement is given in Table 6.2 about a few of these realizations. Employed technologies could give even 0.5 to 0.4 nm of equivalent oxide thickness and still leakage was under control (though at reduced voltage levels). BST was found to be extremely useful, having dielectric constant between 200 to 400; however, it faced a few serious problems. For example, BST could be crystallized only on metal electrode and barrier height between BST dielectric and metal electrode was a matter of key concern. Moreover, dielectric constant of a BST film is proportional to its thickness; hence, reducing its thickness does not provide proportional increase in capacitance.

Other high-k materials based on HfO_2 and Al_2O_3 dielectrics have been used in MIM capacitors at around 70 nm node technology. Generally, TiN was used for metal electrodes and HSG was also developed for capacitor enhancement. However, HfO_2/Al_2O_3 stack could not be used for 50 nm node technology and ZrO_2 films having dielectric constants between 15 to 18 deposited on planar Si wafers have been used. Advanced fabrication techniques allowed $TiN/ZrO_2/TiN$ eDRAMs with equivalent oxide thickness of <0.8 nm.

Use of dielectrics having k >100 becomes necessary for technology node beyond 32 nm and at the same time leakage current has to be within bound as suggested by ITRS (2007/2009). Advanced MIM capacitors using TiO_3 (k 90–170) in its rutile phase has been selected, while its leakage current is controlled using Pt as top electrode instead of RuO_2. ALD $SrTiO_3$ (STO) is also a promising material, (needs precursor) with metal electrodes of Ru and Pt [41]. Obviously research and development on stable, high-k dielectrics having compatibility with a suitable metal for electrodes which keeps leakage current to minimum shall continue.

6.6 Lithography Technology

The rate of DRAM density advancement could not have been sustained without lithography technology advancements. Though the basic objective remains the same in the form of generating precise microscopic patterns in a photosensitive resist material, its practical form has continued to change over the years. A high intensity light source illuminates a mask, an image is formed on a photoresist-coated wafer through a lens, which is ultimately converted into trenches in wafer or strips of metal on silicon. Nature of light source, placement of masks, types of resists, removal of materials from wafer—all went through drastic development stages. Literature is very rich on the subject, and a good review is available in ref [42] up to 1997.

Nature of light source defines the optical resolution (R) in lithography which is related by the Rayleigh equation ($R = k_1\lambda/NA$), where k_1 is a proportionality factor, λ is the light wavelength, and NA is numerical aperture of the optics. The constant k_1 has a limiting value of 0.25 for a single exposure and as k_1 decreases, contrast is lost in the image. Memory devices like DRAM, SRAM, NAND flash, and so on, in which the layouts are highly periodic with certain fixed pitches, resolution enhancement technology (RET) is now almost always used to better the contrast in sub-micrometer technology level. Common RETs employed include off-axis illumination (OAI), phase-shift mask (PSM), and optical proximity correction (OPC), along with liquid immersion lithography and double patterning techniques.

Below 256 Mbit DRAM, i-lines and g-lines of Hg-Xe Arc lamps were used as light sources. Emission centered on 248 nm but unfortunately power available at this wavelength was very low, at the same time lithographic issues at gigabit DRAM level are different and challenging. While obtaining extremely small size patterns that are even less than the wavelength of the exposure tool, resolution, and alignment accuracy became critically important. Minimization of variations in pattern size across whole of a larger size chip became another important issue. All these requirements made it essential to use the RETs mentioned above.

In 1997, Hg i-line light sources at 365 nm were replaced with excimer lasers to satisfy volume manufacturing needs for sub 0.25 μm features in semiconductor devices. By 1995 deep ultra-violet (DUV)-grade fused silica and 248-nm resists were complete and the KrF laser became the main tool of lithography. A large number of DRAMs were manufactured not only at 0.25 μm level but at 0.15 μm [43], 0.13 μm [44,45], and even 0.11 μm level [46,47]. At 0.15 μm level, improvement of resolution and suppression of proximity effect were found to be key factors and these were obtained using *illumination optimization* and OPC correction, respectively. Suppression of mask error amplification was another important issue which was achieved through the introduction of a new mask making process such as thin resist and dry development process [43]. While fabricating a DRAM cell at 0.13 μm, importance of lithographic-centric cell has been emphasized for better results. Attributes for such a cell include a highly regular pattern layout, maximum use of one dimensional level and maximized contact/capacitor area [44]. Full CMP technology was developed in another 0.13 μm DRAM to continue its deployment. The use of newly developed PAOCS capacitor also helped its cause which restricted the aspect ratio of the metal contact to less than 10, which was not possible with full CMP [45].

Due to diffraction and/or process effects, projected images on the wafer appear with irregularities. Line widths appear to be narrower or wider than the design values and rounded corners become distorted. If these distortions are not corrected they can change the electrical characteristics. Optical proximity correction (OPC) methods are used to make up the distortions. Conventional methods include use of pre-computed tables showing spacing between different kinds of layout strips or through dynamically simulated final patterns. OPC corrections have invariably been used in all fabrications beyond 0.18 μm technology levels. Off-axis illumination (OAI) is another resolution enhancement method which is also almost invariably used. If incoming light is made to strike the photo mask at an oblique angle rather than perpendicularly, it is called OAI. This occurs because when light strikes the grating on the mask and is diffracted in different directions. For very small pitch, no pattern is created on the wafer; however, by making the illumination off-axis, all diffraction orders are tilted, which makes it easier for the higher diffraction orders to pass through the projection lamp and form image on the wafer.

In conventional photo masks, thickness of the transparent plate is the same throughout, whereas in shifted masks (PSM) it is not so and interference generated by phase difference is used to improve image resolution. In *alternating PSMs*, certain transmitting regions of the mask are made thinner or thicker, which induces a phase shift in the light passing through these regions. A suitably selected phase shift interferes with the light coming through a

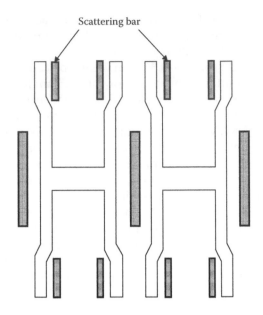

FIGURE 6.5
Layout of scattering bar pattern. (Redrawn from"Highly Manufacturable 4 Gb DRAM Using 0.11 µm DRAM Technology," H.S. Jeong et al., IEDM Tech. Dig., pp. 353–356, 2000.)

normal transparent region, improves contrast, and improves the resolution on the wafer. In *attenuated PSMs* certain opaque parts of the masks are so modified that a very small percentage of light passes through them as well. The opacity produces an interference with the light coming from transparent regions and improves the contrast. PSM has the added advantage of reducing the sensitivity to variations of feature size on the mask, which is very important for decreasing feature sizes.

A 4 Gbit DRAM at 0.11 µm technology was fabricated using KrF lithography, where maximum resolution was obtained by using OAI. However, use of hard OAI and PSM result in very small DOF margin in the peripheral area. A new scattering bar OPC pattern, shown in Figure 6.5, was developed to improve patterns and more than 0.3 µm DOF was obtained [46]. Of course full CMP was also used for DOF improvement. Another example of the continued use of the KrF lithography at 0.11 µm is the form of a 4 Gbit DRAM [47]. It uses all the performance enhancing techniques like PSM, strong OAI, OPC, and high NA exposure system. Full planarization using CMP was also done.

Beyond 100 nm, DRAM fabrication had to introduce 193 nm ArF lithography; though it could have been introduced at 0.13 µm level, economical reasons forced the continued used of KrF technology along with RETs. ArF lithography has successfully been used in a big way [48–54] down to 45 nm minimum feature size with the support of conventional RETs and few more advanced technologies like double exposure and immersion lithography [55].

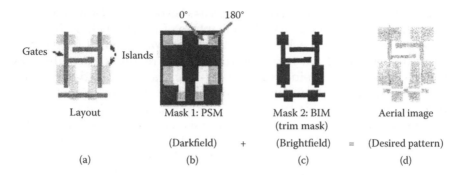

FIGURE 6.6

(a) Schematic of double-exposure phase-shift mask process. The desired pattern is decomposed into two exposure levels: (b) a dark field phase edge, and (c) bright-field trim exposure (d) the desired final image is a sum of those two exposures. ("Enhanced Resolution for Future Fabrication," M. Fritze et al., IEEE Circuits and Device Magzine, Vol. 19, pp. 43–47, 2003.)

At 90 nm technology high transmittance (HT)-PSM, scattering bar OPC and strong OAI are used with ArF lithography [48], whereas at 80 nm recessed channel of the RCAT was formed with poly mask layer and chemical dry etching process [49]. For 70 nm DRAM, extreme OAI, alternated PSM, and OPC were used [50]. Along with the RETs, layout in straight lines helped in achieving a DOF >0.3 μm for all critical layers like active, gate, self-aligned-contact pad, bit line, and storage node. For a 6 F^2 DRAM cell at 60 nm levels, double exposure technology (DET) was also used [51]. For enhancing the resolution DET is used extensively. A typical example is illustrated in Figure 6.6 with two masks and obtained pattern [52], and the final figure is a sum of two exposures.

Numerical aperture (NA) which is the product of the sine of the maximum refraction angle and the refractive index of the medium through which light passes, decides the resolution. Up to 65 nm, final lens and wafer were separated by air gap and beyond this level resolution could not be increased, resulting in blurriness. In immersion lithography (IL), instead of air gap, highly purified water is used as medium, which increases the refractive index of the UV light at 193 nm to 1.44. Enhancement in resolution is around 30–40% which depends on the medium used. In addition, DOF is also improved around two times. Example of the use of ArF IL is given in a report [53] for the critical layer of active, word line, SAC pad, bit line, and storage node (SN) of a 56 nm DRAM. A DOF of more than 0.1 μm was obtained with IL along with the use of extreme OAI, attenuated PSM, and model-based OPC. ArF IL has also been used successfully for the critical patterns such as active, fin, word line, bit line, SN contact, and storage node even at 44 nm DRAM [54].

Water-based ArF immersion lithography has successfully printed patterns down to ~45 nm with acceptable k_1 (>0.3), and NA could be extended up to 1.35. Implementation at this level did require significant OPC and lithographic friendly layouts. Further increase beyond this NA was not possible with water-based systems. The advantage of using much of the existing infrastructure helped achieve NA>1.6, which was essential for the next lithography step of 32 nm node. However, it did require new high-index final lens material, a high-index immersion liquid and high index polymer for resist formulation. Unfortunately, none of the three mentioned material developments were available even by 2008 [55]. A few other exploratory alternatives were attempted, but insufficient success forced a look for feasible alternatives: 193 nm double patterning or deep UV ($\lambda = 13.5$ nm) lithography.

There are quite a few techniques through which double patterning is applicable with the basic idea remaining same in the form of lowering k_1 while keeping NA and wavelength the same [56]. As the half pitch below 45 nm is not achievable practically in a single lithographic step, the process is split over two steps. It reduces the effective value of k_1 below the theoretical limit of 0.25 for the single step. One of the easier methods of conversion of the basic technique into practical fabrications is the one in which the first lithographic step is transferred into a hard mask layer by etch, and subsequent imaging and etching of a second photo resist layer. Two options are available in the process of litho-etch-litho-etch (LELE) either by double trench or double line patterning [55]. An important advantage lies in its development using the tools already available; hence, existing infrastructure could be used. Though the economic advantage gained is offset by the fact that those double exposures nearly double the lithography cost. As far as technology is concerned, main challenge in the process comes from aligning the first and second exposures. Critical dimensions are likely to be affected if the overlay automated solution, where dense layout is split into two sparser layouts and two masks, is used separately and the result is combined on a single wafer [57].

6.6.1 Extreme UV Lithography

Progress in lithography is mainly due to reduction in wavelength from 436 to 365 to 248 and then to 193 nm. But adoption of new wavelength as mainstream technology becomes possible only after a considerable experience in the usage of masks, tools, resists, and so on; 193 nm technology could advance further after the applicability of four generations of tools. A wavelength of 157 nm could not be successful mainly because of the unavailability of good-quality projection lenses. For various other reasons as well, wavelengths below 157 nm could not be employed. Extreme UV (EUV) wavelength at 13.5 nm can provide major relaxation in NA and k_1 and could be used for several generations in resolution extension. However, its implementation requires fulfillment of other technical and economic feasibilities as discussed in ref. [55].

Benefits of using 13.5 nm lithography are huge; if successfully implemented, benefits similar to those of optical lithography are likely to be available. For 32 nm half-pitch imaging at 0.25 NA, k_1 will be 0.59. Use of such high value of k_1 means that OPC and other corrective measures shall be minimized. Moreover, double patterning also will not be required, making the manufacturing process cost-effective. However, there are big hurdles in the practical implementation of 13.5 nm lithography. At the 2008 EUVL symposium the most important issue was the requirement of EUV light sources with sufficient power. Other important issues raised were the defect-free masks and photoresists with high sensitivity and resolution. A discussion on research and development connected with 13.5 nm lithography is documented in ref. [55].

6.7 Isolation Techniques

Considerable continuous emphasis has been given on the reduction of the size of the access transistor and the storage capacitor. Up to around 64 Mbit DRAM density at which the isolation pitch needed was 850 nm, limitations of the LOCOS (local oxidation of silicon) standard device isolation technique were not giving trouble. However, beyond that, smaller physical dimensions forced certain constraints on the device isolation technique. It required that transition between active and isolation region be abrupt, depth and width of isolation be independent of each other, and silicon surface be as planar as possible for subsequent processing steps. LOCOS isolation, with its limitations, is shown in Figure 6.7. Bird's beak formation and field implant encroachment reduced the active chip area and posed severe scaling problems. Thinning of

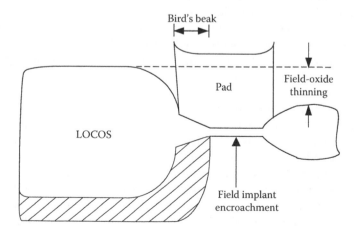

FIGURE 6.7
LOCOS bird's beak formation, narrow-space oxide thinning, and field implant encroachment. (Redrawn from "Characteristics of CMOS Device Isolation for the ULSI Age," A. Bryant, W. Häusch and T. Mii, IEDM Tech. Dig., pp. 671–674, 1994.)

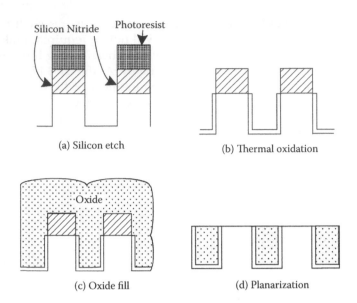

(a) Silicon etch

(b) Thermal oxidation

(c) Oxide fill

(d) Planarization

FIGURE 6.8

A typical process flow for shallow trench isolation (STI). (Redrawn from "Characteristics of CMOS Device Isolation for the ULSI Age," A. Bryant, W. Häusch and T. Mii, IEDM Tech. Dig., pp. 671–674, 1994.)

insulator reduces isolation depth and planarity becomes a casualty. Quite a few modifications were done in LOCOS [58] at 256 Mbit DRAM density level, but at considerable cost and complexity. It was observed that STI (shallow trench isolation) was a much better method in meeting the requirement of scaling at and beyond 256 Mbit. Figure 6.8 shows the STI process sequence which begins with patterning of the pad nitride and etching of the trench till required depth is reached, followed with oxide fill and planarization. Near perfect planarization is achieved with a RIE-CMP process. An abrupt active-isolation transistor becomes possible with the STI technique, hence making it highly scalable.

Another major point of comparison between LOCOS and STI is the ability of reducing junction leakage/improvement in data retention time in giga-bit DRAMs. There were two problems in LOCOS: (1) generation of many defects at the edge of volume-expanded field oxide giving rise to leakage current, and (2) high dose of channel implant, which was used to prevent isolation punch-through, increasing the electric field strength across junction. Combination of the mentioned problems enhanced trap-assisted tunneling, hence reduced DRAM cell data retention time [59]. With further scaling down, the problem becomes more acute in LOCOS because thicker field oxide and larger channel stop implantation dose were needed in this process, whereas in STI, depth of the trench is increased which suppresses isolation punch-through even with reduced channel stop implantation dose. Electric field across the junction could be reduced by a factor of three as compared to LOCOS as result of the above at the same density level [60]. This means that STI is much better in terms of reduction of leakage current/enhancement of data retention time.

It is likely that crystal defects might be generated in STI fabrication process of trench etching and densification of filled oxide, which are further aggravated in subsequent processing steps [61]. However, improvements/modifications in processes like round profile of trench, densification at elevated temperature around 1200°C [62] and optimized ion implantations have shown excellent results making STI highly suitable for high density DRAMs. The problem of crystal defects was further attended to at 0.18 μm level through the use of HDP oxide because of its better gap filling property replacing TEOS-O3, which was used at 0.25 μm level [43]. Smaller junction leakage current was obtained in HDP-oxide filled STI than with TEOS-O3 filled STI. With further reduction in device dimensions, better filling of STI gap becomes more important. Bilayers of SOG (silicon-on-glass) and HDP-oxide are used for STI gap filling at 90 nm technology for enhancing the STI properties [48]. Aspect ratio of more than 15 could be obtained for gap-fill compared to less than 4 with HDP. The bi-layer process of SOG and HDP-oxide worked satisfactorily at 70 nm as well [50]. At 60 nm, aspect ratio of isolation trench increased to 7, and the opening became smaller, but SOG and HDP bi-layer process for STI still worked well [52]. CMP planarization is the other key process of STI apart from the oxide filling. The oxide filling topology pattern is not uniform and depends on the comparative areas used by the active region and the field area. However, it is observed that CMP alone is not able to get a flat surface; and combination of RIE and CMP has shown better results [63]. In this process, RIE etch back is stopped before removing all the oxide from the nitride surface, which is then followed by CMP. A proper combination of RIE and CMP eliminates the problem of tolerance accumulation from depositing and etch back of large film thickness.

For monolithically stacked devices, it is critical to use temperature below 400°C for high quality active formation, so as to avoid degradation of Cu interconnects. Hence, to avoid thermal stress problems in Cu interconnects, laser-induced epitaxial growth (LEG) or recrystallization has been used to provide low-temperature isolation technology. Such a scheme is very useful for 3-D integration employing memory layers. A scheme for the purpose has been given for 80-nm DRAM fabrication process [64]. Crystalline quality of LEG was same as that of bulk silicon in terms of its structure and electrical characteristics. It is obvious that such techniques are going to be extremely important for 3-D integration containing memory devices that require back bias applications like DRAM and NAND flash.

6.8 Bit Line, Word Line, and Gate Technology

Sensing signal margin of a DRAM has great impact on cell design which depends on two independent parameters: the storage capacitance C_S and the bit line capacitance C_B. Realizing a minimum value of C_S (~25 fF/cell) at different levels of DRAM density is discussed in Chapter 5, whereas here a brief discussion is added about containing the value of C_B. Having smaller C_B provides larger sensing signal, which allows faster conversion of sense signal to full data value. As it is in the range of 8 to 10 times C_S, C_B has to be less than 200 fF–250 fF range for all DRAM density levels. Hence, number of cells connected to a bit lines is typically 256 irrespective of DRAM density. With scaling down the dimensions including reduction in bit line width, bit line capacitance reduces; however, at around 0.15 µm technologies bit line width became smaller than its thickness and fringe capacitance component in C_B became a dominant factor. The additional problem of providing lithographic error margin at these reduced sizes was avoided by using SAC etching technology. However, as SAC etching technology uses the different etching rate characteristic of SiO_2 and Si_3N_4, it requires fully encapsulated bit line (and word line) with Si_3N_4 film. Technology works satisfactorily, but it has the drawback of larger parasitic bit line capacitance due to higher dielectric constant of Si_3N_4. Parasitic bit line capacitance comprises inter-bit line capacitance and capacitance between bit line with substrate, storage node, word line, and plate node. In addition, interwire capacitance also increased because of reduced distance between bit lines and other conducting layers. Consequently parasitic bit line capacitance does not decrease with further reduction in technology node level beyond 100–150 nm technology unless low permittivity dielectric is used or new cell architecture is employed. Obviously, nonscalable nature of bit line capacitance beyond 100 nm has considerable impact in DRAM chip area especially in terms of number of memory cells connected to the bit while getting satisfactory level of sensing signal. As a small remedy, contact hole etching technology results in considerably reduced parasitic resistance in comparison to SAC hole etching.

Use of hierarchical bit line composed of global and segmented local bit lines has found favor over the conventional nonhierarchical bit line. The global bit line has only inter-bit line capacitance with global bit lines and not with any other conducting lines. Capacitance of local bit lines can be reduced by segmenting the bit lines. Hence, the total capacitance which is the sum of global and local bit lines is considerably reduced [65].

It is important to note that bit line capacitance has a major contributor in terms of junction capacitance, which can be up to 30–50% of the total capacitance value. A major advantage of STI is smaller junction capacitance compared to LOCOS. This reduction is mainly due to the smaller dose of channel stop implantation, and reduction in effective bit line capacitance can be up to 30% more in STI than with LOCOS at supply voltage of 2.0 V. As a result, sensing signal increases from 110 mV in LOCOS to 150 mV in STI at 2.0 V, and the percentage improves with reduction in supply voltage [61]. With the improvement in sensing signal with STI, requirement in storage cell capacitance is relaxed in gigabit DRAMs. Data access time is also improved by nearly 10% because of reduced junction capacitance. However, WSi_x could be used for bit lines up to only 0.15 μm technology even after improving its sheet resistance. Though having lower sheet resistance, $TiSi_x$ has not been found effective beyond 100 nm as its sheet resistance also increases with scaling, and it also lacks thermal instability. Hence beyond 100 nm, W and/or some new metal has to be used.

DRAMs become slow because of the RC delay of the word line (and bit lines). The delay is proportional to the square of the length of the word line, which itself multiplies by 1.5 when the DRAM density quadruples. So, the word line delay has been improved by using low-resistance metal silicides like WSi_x and $TiSi_x$ for the gate material. At early stage, a sputtered $W/Si_3N_4/$ poly-Si gate structure and H_2O_2 selective etching were used to fabricate gate structure, followed by W deposition. Therein, procedural simplification was obtained in a self-aligned W strapped transistor fabrication with a grooved gate structure, and minimization of metal contamination was also achieved using vapor HF selective etching during sidewall spacer fabrication [66]. Sufficiently low sheet resistance of 0.9 Ω/sq. and 1.8 Ω/sq. for NMOS and PMOS, respectively, was obtained for a 0.125 μm wide gate and 0.25 μm wide diffusion. Advantage of using W is that it acts as a better etch stopper than $TiSi_2$ as gate material, having a sheet resistance less than 5 Ω/sq. [67]. However, WSi_x could be used up to 0.15 μm technology even after improving its sheet resistance through optimization of its composition, morphology, and grain size. The reason behind the limitation is the increase of sheet resistance of metal silicide with decrease in the wire width. Even $TiSi_x$ having low resistivity found its utility limited to nearly 0.1 μm technology. In each case of Ti and W, a 100 nm doped polysilicon is topped with 100 nm of silicide. With a Ti polycide, a diffusion barrier in the form of 50 Å TiN has to be used between $TiSi_2$ and doped polysilicon.

Limitations of WSi_x and $TiSi_x$ meant that different material combinations were to be used for the gate. One such combination was W/WN with low sheet resistance. A W/WN/poly gate was implemented in a 135 nm vertical cell DRAM with planar support transistors. WL with sheet resistance of 7 Ω/sq. was obtained [68]. If WSi_x gate is used in sub 100 nm design, aspect ratio of the gate stack would become more than 6 so as to keep WL resistance within RC delay limits. Such an aspect ratio creates problems in patterning and gap filling and increases interwire parasitic capacitance (in the same plane). It is observed that the vertical transistor DRAM with use of W/WN gates permits same aspect ratio at 100 nm as at 135 nm node with WSi gates in planar transistor. Hence W technology was extended even for an 80 nm cell transistor with W/W_xN metal gate [69], which lowers WL resistance in 176 k cell array by a factor of three with half of the thickness of WSi_x gate.

However, gate stack was carefully optimized using selective oxidation (SO) in order to remove gate etch damage and to cure gate oxynitride. A surface treatment is required before SO to control any abnormal increase in threshold voltage and DIBL characteristics. A 50 nm PVD-W was also used in place of 100 nm CVD-WL to reduce the thickness of the BL to reduce its capacitance without increasing the resistance. Total coupling capacitance was reduced by about 30% in the process. As other alternates Ru and RuO_2 metal gates grown by MOCVD (metal-organic CVD) on rf sputtered Ta_2O_5 gate dielectric have also been used [70]. However, TiN/W metal gate is commonly used for sub-100 nm technology node, avoiding fabrication complexities, even down to 40 nm [71].

6.9 Cell Connections

For gigabit DRAM, open bit line cell size can be calculated with the help of Table 6.2. The table includes overlay margins up to 0.2 μm for contact holes for different design rules in the range of 0.06 μm to 0.4 μm. It is observed that the requirement of alignment margins becomes so dominant that cell size becomes impractically large simply because scaling down design rules reduces word and bit line pitch, but not the alignment tolerance. A self-aligned contact (SAC) pad formation has to be used for reducing wastage of chip area as well as reducing aspect ratio of the capacitor contact hole [72]. SAC pad anisotropic epitaxial growth avoids the lateral over-growth above the field oxide unlike what is happening with the isotropic epitaxial growth, hence an isotropic epi-growth becomes essential. A comparison between anisotropic and isotropic epitaxial growth and fabrication process of the selective epitaxial contact pad whose top layout view is shown in Figure 6.9 is available in ref. [72].

Si_3N_4 sidewall spacers of the gate electrode have been formed using SAC technology for a 1 Gbit DRAM [72] while forming SN contact holes (0.16 μm × 0.16 μm). Though the spacers are used to protect the gate material from the contact filling poly-Si due to any process misalignment or etching, a serious problem occurs as spacers cause drain current degradation of the MOSFET. Hence, applying Si_3N_4 spacers using SAC technology to the word lines (WLs) of DRAM requires caution. A Ge-added vertical epitaxial (GVE) Si pad technology is used to overcome the problem due to the use of Si_3N_4 spacers as well as the lateral isotropic Si growth in the two-step selective epitaxial Si pad technology [74]. In the given double structure, the first structure is a combination of SA vertical epi pad, a Si_3N_4 cap on the WL, and a contact plug (CP). As SiO_2 can be applied to the sidewalls of the gate electrode spacers, degradation of MOSFET drain current, because of the presence of Si_3N_4, is almost eliminated. For the second SAC structure, Si_3N_4 etch stopper on top

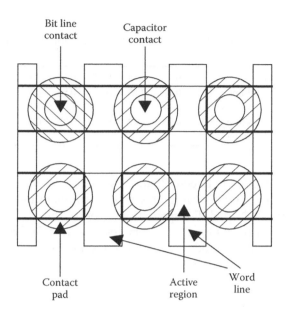

FIGURE 6.9
Schematic top view of an open-bit-line layout of cell with self-aligned contact pads. (Redrawn from H. Hada et al., "A Self-Aligned Contact Technology Using Anisotropic Selective Epitaxial Silicon For Gigabit DRAMs," IEDM Tech. Dig., pp. 665–668, 1995.)

as well as on the sides of the bit lines can be used without affecting MOSFET. The scheme could provide a 0.23 μm^2 folded bit line cell for 0.15 μm technology and alignment margin of more than 0.1 μm.

Other than the alignment tolerance, there is a severe problem of undercutting while vertically etching a thick highly doped polysilicon. A high aspect ratio (AR) of the etching sometimes causes bridge formation between adjacent SNs. A scheme in which buried contact (BC) process is merged with SN makes the SAC process much simpler and makes alignment tolerance free, and removes the problem associated with high AR etching. Such a scheme was successfully used at 0.15 μm on a 4 Gbit DRAM [75].

Instead of conventional method of SAC with contact hole printing, a reverse T-shape patterning has been used. The contact area is obtained after SAC etching and then doped poly-Si deposition and poly CMP are used to form memory cell contact pad [43]. AR of the SN contact hole is also reduced by using 100 nm thick W bit line in place of 250 nm W polycide bit line used in 0.18 μm case. Extension of SAC patterning becomes difficult at 0.13 μm technology levels because of the combination of the further reduced size of contact holes and different etching characteristics due to micro loading effect for different contact opening areas as shown in Table 6.3 [45]. To overcome the problem, landing pad technology on bar-type SAC has been developed with which wide process latitude could become available. At 0.13 μm technology level, another problem appeared in the form of the possibility

TABLE 6.3

Normalized SiO$_2$ Etch Rate as Function of Contact Opening Area in SAC Etching

Design rule (μm)		0.13	0.14	0.15	0.16	0.17	0.18	0.19
Normalized SiO$_2$ etch rate	Bit line contact (2F^2)	0.7	0.75	0.8	0.85	0.9	0.95	1.0
	Memory cell contact (F^{-2})	0.35	0.42	0.5	0.55	0.6	0.7	0.8

of unwanted connection between BL and SN. Hence, W-BL is fully encapsulated with SiN film to overcome the mentioned problem using SAC process for the self-aligned SN pad. Encapsulation of both, the WL and BL, allows the use of full SAC process for all memory cell contact [45]. Procedure of bar-type SAC patterns for SN pad and encapsulation of BL and WL with Si$_3$N4 was extended at 0.11 μm technology level as well [47].

6.10 Interconnection/Metallization Technology

As early as 0.25 μm technology level, constraints of aluminum connecting strips were becoming significant. Electromigration and stress-induced migration started to induce reliability failures in interconnections [76]. Solution was suggested in the form of a TiN/Al-Si-Cu/TiN/Al-Si-Cu/TiN/Ti layered film whose bottom TiN/Ti layer worked as diffusion barrier between the main conducting strip (Al-Si-Cu) and Si substrate, as well as stress buffer at the strip corners [77]. Contact resistance between the TiN and substrate was reduced due to the formation of a titanium layer. Layered conductors along with the sprinkling of other metals with Al improved the situation, but with continued reduction in interconnect pitch with every technology generation, interconnect delay doubled (assuming a constant metal aspect ratio). With an optimum aspect ratio of ~2, deterioration in interconnect RC delay could be restricted to a factor of 1.26, but it was not enough for further generations [78]. Use of suitable low dielectric constant material for inter-strip gap was one option, but use of higher conductivity material like copper was considered better. Along with its low conductivity of 1.7 μΩ-cm (3.0 μΩ-cm for Al-0.5% Cu alloy), Cu had 10 times better electromigration resistance than Al interconnects. However, use of copper posed a few problems like difficulty with plasma etching and hence required damascene structure [79]. Other problems included likely diffusion of Cu through the inter-strip dielectric and difficulty in its adhesion with surroundings, hence needing a barrier layer. Use of Cu and low dielectric technology did improve the performance and reduced error talk and IR drop, but it could work up to 256 Mbit DRAM generations because of the limitations of obtaining fine patterns and the global topology. At gigabit level and beyond, reduced feature size and bigger step height between the cell area and the rest of the region created problems in the form of etching and filling of tiny metal contact holes with high AR. Contact resistance also started increasing due to small area. While investigations focused on the use of Cu and low-k dielectric efficiently, development of triple-level metallization scheme [80] became a very suitable solution in overcoming the depth of focus margin problem for back-end of the line (BEOL) process. In a triple metallization scheme, first metal line was realized in W for local interconnects, and it reduced the level difference between cell array and peripherals considerably, which helped relax design rules for the next two metal lines in Al by more than 50% [67]. W was used as first metal line due to its good contact filling property, and Ti was deposited before the deposition of W to obtain reduced contact resistance. Though contact resistance reduction is not a priority as long as it is only a small percentage of total resistance, filling of contact holes with metals like W or Al requires ohmic and barrier layers between them and the substrate, for the stoppage of metal penetration [61]. Generally Ti/TiN layers are used as ohmic and barriers layers, as Ti forms $TiSi_2$, which works as ohmic layer, and TiN serves as barrier layer. Some of the early applications of triple metallization scheme include a gigabit DRAM [73] and a DRAM at 0.15 µm technology [43].

Considerable research and development concerned the use of copper and low-k dielectric integration at further reduced minimum feature size. Low-k SiOC ($k < 3.0$) integration was done with Cu at 130 nm node devices and the same technique of manufacturable Cu dual-damascene (DD) interconnect technology was extended to 90 nm node eDRAM devices [81]. Here a maximum of 11 layers, consisting of one local, four intermediate, four semiglobal, and two global layers, was used. For local layer, Cu/PE-CVD SiO_2 single damascene interconnect line was fabricated which was used for bit line. Though low-k dielectric was not used, interwire capacitance was kept low by thinning the interconnect. For intermediate layers (M2–M5) a stacked PE-CVD SiOC (k <3) film and PE-CVD SiCN (k ~4.9) film was capped with PE-CVD SiO_2 film; without SiO_2 capping, SiOC structure is damaged during etching and exhibits poor insulating properties. However, SiO_2 capping causes resist poisoning problems hence the process requires optimized deposition sequence and use of better resist material. For intermediate layers via-first DD, stacked mask process (S-MAP) technology as described in ref. [81] was developed.

Trench/via-hole DD structures present many problems beyond 100 nm node like poor step coverage of barrier metal/seed Cu, insufficient bottom fill of Cu, and poor mechanical robustness, among others. A hybrid DD structure consisting of PAE (SiLK™; k = 2.65) at wire level, second-generation SiOC (black diamond (BD2)™; $k = 2.5$) at via level, and SiC (BLOk3™; $k = 3.5$) as via etching stopper/diffusion barrier [82] was used. It gave a void-less Cu filling and good migration immunity for Cu. Use of Cu continued for 45 nm node having 70 nm wire spacing technology along with the use of porous ultra-low-k NCS (nano-clustering silica) both at wire level and via level. Such a structure was shown to have enough robustness and was called full NCS structure [83]. With full NCS structure and scaling of wire height, capacitance of intermediate-layer wires was not allowed to increase in comparison to the previous generation value of 200 fF/mm at 65 nm node with 100 nm space having NCS/SiOC hybrid structure. Care was taken to completely suppress Cu penetration into NCS, which helped in retaining its TDDB characteristics of earlier generation as well.

With full low-k/Cu interconnects becoming almost universal for high performance, attention focused on the effect of contact resistance in reduced size CMOS devices for further performance enhancement. A direct low-k/Cu DD contact line was used for a 75 nm diameter contact. It was observed that a contact resistance (CT) of 5.4 Ω only could be obtained for the Cu DD contact (Figure 6.10(a)) which was nearly one fourth compared to a conventional W plug (Figure 6.10(b)). The CT reduction ratio increased further with its size reduction as given in Table 6.4 [84].

With technology node beyond 45 nm, reliability of Cu/low-k interconnects and inter-metal-dielectric becomes more important. Problems of finer patterning, filling performance, and electromigration of Cu are also to be resolved. Metal doping has been advised to control migration of Cu atoms [85], but it results in higher resistance. CVD Co film has been investigated to replace PVD TaN/Ta in via 1/metal 2 structures for a 32 nm node [86] and has shown good results in terms of filling properties and reliability without affecting wire resistance. Use of CVD Co barrier provided void-free and

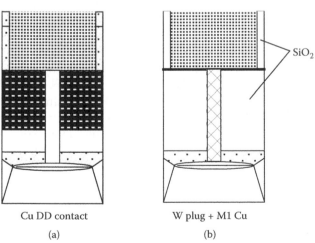

| Cu DD contact | W plug + M1 Cu |
| (a) | (b) |

FIGURE 6.10
(a) Low-k/Cu contact, and (b) conventional W plug contact. (Modified from "A New Direct Low-k/Cu Dual Damascene (DD) Contact Lines for Low Loss (LL) CMOS Device Platform," J. Kawahara et al., Symp. VLSI Tech. Dig. Papers, pp. 106–107, 2008.)

TABLE 6.4

Kelvin Contact to Node Diffusion Resistance as Function of Contact Size

Contact Size (nm)		60	80	100	120	140
Resistance (ohms)	WPlug + M1 Cu	45	20	10	5	3.6
	Cu DD Contact	9	5	3.6	2.5	2.0

better Cu migration properties and reduction of nearly 30% thickness of the barrier which provided more space for Cu. Reduction of migration in Cu is because of good adhesion with Co. Even a cursory study of the Cu/low-k interconnect underlines the requirement of continued hard work in the area.

In spite of the acceptance of low-k/Cu interconnect technology, enough reports are not available on its production level adoption for high density DRAMs. At 90 nm design rule capacitor and recessed gate DDR DRAM has been given that successfully integrated low-k and Cu wiring [87]. As the DRAM processes are sensitive to low levels of metal contamination, effect of Cu wiring needed to be studied. In a three-level metallization scheme, generally first level metal (M0) is W and second (M1) and third (M2) levels are Al/Al alloy. In the given scheme, M1, M2, and W via are replaced by Cu and FSG (fluorinated silicate glass, k ~3.5) was used as IMD (intermetal dielectric). Sheet resistance of M1 and via resistance improved by ~45% and ~60%, respectively, compared to Al-based wiring and interwire capacitance decreased by ~50% and ~43% for M1 and M2, respectively. There are additional advantages: a systematic study has also shown no degradation in yield, refresh time, or reliability takes place. Interconnects especially at BEOL at 45 nm or beyond faced additional problems while using Cu. Resistivity of the Cu increases considerably and the necessity of having liner material for preventing diffusion of Cu through the inter-layer dielectric results in reduced available width for Cu conductor [88]. One of the proposed solutions for minimization of increasing resistivity of Cu with decreasing dimensions is to increase the grain size of Cu in narrow connecting lines. However, the procedure is difficult to achieve in practice and a better option is to reduce the thickness of the metal line on Cu using improved sputtering method, ALD, or self-forming barrier layers [89].

Reduced dielectric constant was obtained for inter-level dielectric through doping of SiO_2 by fluorine or carbon [90] which formed SiCOH. Nonporous SiCOH is used up to 65 nm level whereas porous SiCOH is used at 45 nm technology node and below. Many design and process modifications became necessary while using SiCOH, especially porous SiCOH, emphasizing the need for alternatives [89]. Air gap technology [90] was again reviewed for the purpose. Although air gap technology requires costly lithography and etching, it provides several advantages over the use of SiCOH in terms of mechanical integrity and thermal conductivity maintenance throughout the die. Hence, air gap technology is also a viable option at a lower technology level of 22 nm or beyond.

6.10.1 Carbon Nanotubes and Graphene Nanoribbon-Based Interconnects

In addition to the aforementioned problems of increasing resistance of Cu at lower dimensions and difficulties faced in employing low-k dielectrics, future scaling is expected to bring more problems for the Cu interconnects including increased current density, which results in reduced electromigration reliability [91]. Hence, efforts are on to find other technologies which could replace Cu interconnects. Promising potential alternatives include graphene nanoribbons (GNRs) and carbon nanotubes (CNTs) [92]. Considerable research has been done to study the applicability of CNTs below the 45 nm technology node as better interconnects [90–94].

Despite huge interest in CNT/GNR fabrication technology, their integration into VLSI and especially with semiconductor memories is immature and much more research remains to be done. For example, to replace Cu interconnects, high density and high quality of CNT bundles are desired due to the high resistance of an isolated CNT [95]. Moreover, growth temperature of the CNT interconnect should be kept below 400°C per the requirement of CMOS BEOL process. A CNT via was shown to be integrated in ultra-low-k dielectric (k = 2.6) at a low temperature of 365°C. In fact, most of the work on CNT integration is vertical interconnects or vias because of the tendency of CNT bundle growth perpendicular to the surface [96], whereas long-length horizontal CNT interconnects along with vertical via require challenging steps. Comparatively, fabrication of GNR interconnects in horizontal direction are less challenging than CNTs. Hence, efforts have been made in this direction and in spite of difficulties, carbon films have been fabricated in DRAM trench capacitor using CVD method [97], though not with high electrical conductivity. For improving conductivity of multilayer GNRs, intercalation doping, which was done in bulk graphite, has been suggested [98]. Recently an attempt was made to combine CNTs and GNRs using CVD methods by Fujitsu [99]; however the process needed more work.

Utilization of CNT and CNT bundles as interconnects has many advantages but so far these have not been able to replace Cu mainly because of higher resistance. Methods for reducing CNT interconnect resistance are mentioned as functionalization with accepting molecules, alignment of CNTs within the interconnect, and avoidance of inter-CNT junctions [100]. A minimum sheet resistance of ~55 ohm per square could be obtained with noncovalent functionalization of CNTs. Resistance of aligned CNT interconnects remained constant even with its decreasing width, permitting it be used at lower technology level. Because of the reduced capacitance [101], it was expected that RC delay would be better compared to Cu interconnects beyond 45 nm technology. It was a fair assumption that sheet resistance of CNT interconnects could be reduced to 10Ω per square if all resistance reduction techniques are employed [100].

6.11 Advanced DRAM Technology Developments

As the DRAM density moves into gigabit levels, cell area must be decreased in inverse proportion to the density level increase; otherwise, a larger chip size has to be used. Shrinking of feature size is the main option for decreasing the cell area, but it invites a host of new problems whose solutions must be available. As the minimum feature size (F) is not reduced by the same factor as the DRAM density increases, it means chip size has to be increased, and, in combination with the reduction in feature size it shall result in yield loss, unless remedial measures are taken in terms of circuit innovations and technological advancements. As observed from Sections 6.1 to 6.9, which noted an important feature of DRAM technologies that transitions are smooth and some form of continuity is maintained even when other new processes/steps are taken. To explore the advanced DRAM technology developments, the spectrum is divided in four sections, of which three sections contain technology node descriptions of stacked technology. This bifurcation is not only to facilitate understanding but to expound upon the different features, in spite of the conventional continuity. The final section discusses deep trench technology beyond 100 nm level.

6.11.1 0.18 µm to 0.11 µm DRAM Technology Development

Capacitance density, defined as the cell capacitance divided by the projected area of the storage node has to be increased. For a consistently maintained value of 25 fF cell capacitance, a capacitance density of 145 and 330 fF/μm^2 is required for the 1 and 4 Gbit DRAMs, respectively. Hence, storage capacitor technology becomes very important. Refinements in the usage of Ta_2O_5 as capacitance dielectric, and the development of Al_2O_3 capacitors have been reported extensively for and below 1 Gbit DRAM density. With reduced feature size, and continual reduction in supply voltage/cell array voltage, charge stored in the cell capacitor becomes small, affecting the data retention time. High resolution lithography becomes a must along with suitable etching processes at this range of technology-nodes especially for the patterning of critical layers like active region, gate, and bit lines. At 0.18 µm and lower level feature size one of the additional critical factors is the variation in threshold voltage of cell transistor which greatly affects the functionality and data retention time. KrF lithography has been used in 0.18 µm to 0.13 µm technology range, mainly to save investments in a new kind of lithography. Resolution enhancement and proximity effects suppression are therefore used to decrease critical dimension variation to less than 25 nm across the 6-inch mask.

Bit line contact and storage contact hole sizes also become small with decreasing feature size and high overlay accuracy is required. Selective etching between silicon oxide and nitride has been used for making SAC to overcome the mentioned problem. For 0.20 μm design rule, SAC process has been used to get overlay accuracy within 50 nm with Si_3N_4 spacer between the gate material and the contact filling polysilicon [73]. Since silicon nitride spacers cause lowering of MOSFET current [105], silicon nitride SAC technology was not desirable for word lines of DRAM. Hence a two-step selective epitaxial Si pad has also been used which is a combination of perpendicular anisotropic and lateral isotropic silicon growth [72]. However, as the lateral expansion controllability was not satisfactory, Ge-added vertical epitaxial Si pad technology has been used for gigabit DRAMs at 0.15 μm design rules [74].

Probably first fully working 1 Gbit SDRAM fabricated in 0.18 μm CMOS technology was devised by K.N. Kim and others in 1997, overcoming major limitations at this small feature size and comparatively large size chip [106]. Critical dimension variations were improved to less than 30 nm for average critical dimensions of 220 nm by using a high-resolution technique of OPC and elimination of standing wave effect using antireflection layer. More than 25 fF/cell capacitance was obtained in a one cylinder stacked structure with hemispherical grains and Ta_2O_5 as capacitance dielectric. Another gigabit DRAM fabricated in 0.18 μm technology overcame the problems related with photolithography by employing straight line and space patterns, a fully planarized surface, and alternating PSM along with SAC technology [107]. CMP was used to completely planarize the wafer before each photolithographic step like poly-plug patterning, bit line, and first metal layer patterning. While having full planarization, deep and higher aspect ratio contact holes need to be realized, which was successfully done using dual frequency RIE equipment without hard mask. Being fully planarized, stacked capacitor necessitates capacitor height to be small for reducing aspect ratio of metal 1 contact; hence Ta_2O_5 film dielectric having oxide equivalent thickness of ~2.9 nm capacitor was used. To get 25 fF/cell, capacitor height was nearly 0.9 μm.

DRAMs based on 0.15 μm technology got greater attention as a transitory technology between 0.18 μm and 0.13 μm technology mainly on account of usage of both KrF and ArF lithography, respectively, and the increase of wafer thickness from 200 nm to 300 nm. Key technologies at 15 nm level were lithography, oxide filling in STI, polycided gate, self-aligned memory contact pad, and Ta_2O_5-based cell capacitor. Economic factors were responsible for the continued used of 248 nm KrF lithography at 0.15 μm technology level, though resolution enhancement and proximity effect suppression continued. Mask error amplification was also suppressed and this combination of improvements in lithographic process resulted in less than 25 nm variations in critical dimensions on a 6-inch mask [43]. At 0.18 μm technology

O_3-TEOS oxide filling was used in the STI process. However, at 0.15 μm node and beyond, oxide filling without voids was done using high-density plasma (HDP) oxide, resulting in lower junction leakage current than those of O_3-TEOS-filled STI.

Consistent decrease in the minimum feature size is one of the basic reasons for increasing DRAM density. However, the same rate of decrease could not be followed in terms of the alignment tolerance between contact holes and word lines or bit lines. Therefore, SAC structures were used and for 1 Gbit DRAM Si_3N_4 spacers have been used. However, presence of spacers affects drivability of transistors, hence a better option was suggested in ref. [74] in which double SAC cell in combination with a Ge-added vertical epitaxial (GVE) Si pad technology was used. First SAC structure is a combination of self-aligned vertical epitaxial pad, a Si_3N_4 cap on word line, and a contact plug. The second SAC structure is conventional using Si_3N_4 stoppers on the top as well as on sides of the bit lines. This double SAC scheme reduced the cell area from 0.39 μm² to 0.23 μm² in 0.15 μm technology with 0.1 μm alignment tolerance as shown in Figure 6.11. For 1 Gbit DRAM estimated alignment tolerance for an 8 F^2 folded bit line cell in 0.2 μm technology node is hardly 0.01 μm for a conventional pad, which improves to nearly 0.08 μm for an epitaxial pad and further increases to 0.14 μm in a GVE pad + Si_3N_4 cap. However, 4 Gbit DRAM with 0.15 μm technology node is practically not possible with a conventional pad because of the non-availability of any tolerance, whereas its corresponding tolerances are 0.06 μm and 0.09 μm, respectively, with epi pad and GVE pad + Si_3N_4 cap.

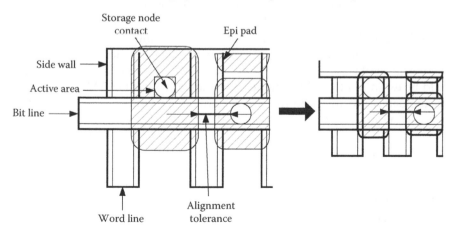

FIGURE 6.11
Memory cell layouts of 8 F^2 folded bit-line arrangement using 0.15 mm design rule with 0.1 mm alignment tolerance. The cell area is reduced from 0.39 mm² by new SAC technology using GVE Si pad with Si_3N_4 cap structure in place of former epi pad SAC technologies. (Redrawn from "A 0.23 μm2 Double Self-Aligned Contact Cell for Gigabit DRAMs With a Ge-Added Vertical Epitaxial Si Pad," H. Koga et al., IEDM Tech. Dig., pp. 589–592, 1996.)

Trench DRAM technology has also been shown to be very effective at 0.15 μm design rules [108] by improving the important parameters like cell capacitance and performance and reducing resistance and device leakage. Using isotropic silicon etching, enlargement of the bottom part technique is applied to increase the capacitance from 22 fF to 33 fF/cell. The resistance between the trench and the active area is decreased by using a self-aligned oxide hard mask STI process from 11 kΩ to 4.5 kΩ.

At 0.13 μm level most of the key technologies remained in use that were used at 0.15 μm or 0.18 μm level, for example, Ta_2O_5 MIS capacitor, self-aligned landing pads, low resistance word line and bit lines, and triple metallization. However, these high temperature processes make them suitable for stand-alone DRAMs but not for embedded DRAMs and require suitable modifications. In addition, a significant problem arises that the SAC process which was applied to form self-aligned landing pads around word lines could not be adopted due to the reduced process margin of SAC etching [45]. Around 0.13 μm technology and beyond, both word lines and bit lines need to be encapsulated with Si N film, which would allow full SAC process so that the possibility of shorting bit line to storage node is eliminated and all memory cell contacts with bit lines and word lines are formed in self-aligned form. A polyamorphous one-cylindrical capacitor (PAOCS) was realized in which capacitance was increased by using both inside and outside of the storage cylinder as capacitor plate. The outside of the storage capacitor cylinder was formed by polysilicon at less than 450°C whereas the inside was formed by amorphous polysilicon with hemispherical grain, and Al_2O_3 as dielectric which did not require high temperature annealing of Ta_2O_5. Low temperature process was suitable for embedded DRAM application, and it also aided in reducing the resistance of W bit line contact resistance. Smaller height PAOCS capacitor helped in getting the aspect ratio of metal contact within a limit of 0.1 μm with the use of full CMP, and so the KrF lithography could be extended to 0.13 μm nodes. Developed 1 Gbit DRAM cell size was approximately 0.13 μm² and the chip size nearly 267 mm².

Word line and bit lines encapsulated by Si_3N_4 for pad separation with CMP process, all pads formed by line type SAC, and low thermal budget Al_2O_3 dielectric PAOCS capacitors were some of the key technologies used at 0.13 μm level and were again used at 0.11 μm technology for the realization of a 4 Gbit DRAM, along with the continuance of KrF lithography [46,47]. Hard OAI was used in the KrF lithography for resolution enhancement and newly developed scattering bar OPC pattern was used to improve DOF margin. An important problem to be tackled at 0.11 μm level is the inter-layer dielectric gap filling having as high an aspect ratio as 10 or more. SOG was used for this purpose and after annealing it gave very satisfactory results. Metal contact which landed at bit lines reduced very high aspect ratio in conventional scheme to manageable limits and then metal contacts were formed with a borderless contact process available in ref. [46]. Some of the key technologies used at 0.11 μm technology are given in Table 6.5.

TABLE 6.5

A Few Key Process Features of 4 Gbit DRAM at 0.11 μm Level

Lithography	KrF	High NA, OAI and OPC
Isolation	STI	HDF gap fill
Transistor	GNO$_x$	4 nm
	WL	T Six
Cell structure (size .225 μm × .465 μm)	Stack pads, and pads for cell contact, storage nodes	Line-type SAC and CMP separation
Storage	Inner HSG scheme	PAOCS + HSG
Capacitance dielectric	tox_{eq} 30 angstroms (Å)	Al$_2$O$_3$ (40 Å)
Metals	Triple metal	CVD Al

6.11.2 100 nm to 50 nm DRAM Technology Development

As design rule of DRAM fabrication goes below 100 nm some new process integration technologies have been utilized along with the few technologies used in the 0.13 μm to 0.18 μm era. For DRAM cell performance improvement beyond 100 nm, parasitic resistances and capacitance in the cell structure were necessary to be reduced; otherwise, they would have offset the advantages of the *shrinkage technology* [109]. New technologies like spin-on-glass (SOG) became essential for oxide gap-filling for STI and ILD processes [48]. Dual gate of CMOS process was adopted in low voltage DRAMs because low threshold voltage PMOS could be made using conventional method of buried contact [110]. Recessed channel-array transistor (RCAT) [23] was now used to overcome the problem of short channel effect. Deep trench (DT) technology with *checkerboard* cell layout process enhancement was also employed successfully at 70 nm scale [111].

Below 100 nm, DRAM performance of multi-gigabit density is seriously affected by the memory cell characteristics and parasitic resistance and capacitance of the transistor channel, and bit lines and node resistance must be decreased as well. To reduce the parasitic resistance, the memory cell is realized having poly Si/W$_x$N/W gate, source/drain junction with Co siliciding, and CVD-W cell pad and bit line contact. Though Co saliciding on cell area increases junction leakage current [112], the limitation is overcome by using elevated source/drain selective epitaxial growth using UHV-CVD at 700°C to avoid direct formation of Co silicide on cell as well as core and peripheral region [109].

Other problems below 110 nm design rules include formation of voids while filling in STI with high-density plasma (HDP) scheme. Mechanical stress is also increased between HDP oxide and silicon due to their different thermal coefficients of expansion [113]. Moreover, in conventional STI, higher doping concentration of the channel stop increases GIDL current and deteriorates hot carrier immunity of DRAMs. To overcome these problems void-free STI was developed using polysilazane-based inorganic spin-on-glass (P-SOG). It showed void-free gap filling and lower compressive stress than HDP oxide and hence held promise for further lower feature size fabrications. As parasitic bit line capacitance is partially due to the use of Si_3N_4 bit line spacers, a top spacer storage node contact (TSC) has been developed, process flow of which is available in [49].

Use of TSC process reduces the loading capacitance by removing the additional capacitor landing pad layer that was used earlier in diamond-shaped OCS, and it also allows easy gap filling with ILD oxide. In addition, it also increases the breakdown voltage, due to lower plasma damage during TSC formation.

A highly manufacturable 90 nm DRAM technology was developed in 2002 using ArF/KrF lithography and some novel integration schemes [48] like spin coating oxide for STI and ILD processes, and diamond-shaped storage node for large capacitor area having better mechanical stability. Dual-gate oxide scheme was also used in the DRAM development for memory cell transistor and the transistors used for peripheral circuit with an objective of keeping the off current below 0.1 fA/cell. Critical patterns, such as active and gate layers, SAC pad, bit line, BC, and storage node contact, could get more than 0.3 μm DOF margin with the use of ArF lithography and hard mask etching. SOG and HDP oxide layers were used for STI gap filling and only SOG oxide filling was used for interlayer dielectric gaps. A comparison between the cell layout of a conventional OCS and newly developed OCS showed that the use of diamond-shaped structure eliminates the criticality of the layout of conventional storage node, which allowed the use of KrF lithography for it even at 90 nm technology. Mechanical stability of this structure is also better. In essence, diamond-shaped OCS provided more freedom for lithography, larger capacitance, and better mechanical stability. A MIS capacitor with AIO/HFO dielectric [114] with equivalent oxide thickness of 25 Å was used which provided 23 fF/cell capacitance.

DRAM with 80 nm technology was developed for the first time in 2003 using ArF lithography, recess channel array transistor (RCAT), low-temperature MIS capacitor, and TSC. ArF lithography is used in fabricating RCAT, which shows a distinct improvement in the performance of RCAT over planar transistor in terms of distribution of threshold voltage and the distribution of the DIBL and junction leakage current. The TSC continued to provide advantages like easy gap filling with ILD oxide, reduction in the bit line loading capacitance and higher breakdown voltage. In addition, another important feature was that the landing pad that was used in 90 nm technology [48] could be eliminated and it saved expensive ArF lithography. To obtain cell capacitance of more than 25 fF, a low-temperature MIS structure (polysilicon SN, Al_2O_3-HfO_2 film dielectric, TiN top electrode and W plate node) was realized in a low thermal budget of <500°C. Another manufacturable 6 F^2 DRAM technology at 78 nm design rule was given by Micron Technology Inc. using MIM capacitor with composite high-k Al_2O_3/HfO_2 dielectric with TiN as electrode [115], which occupied only 0.036 μm²/cell. For DDR2 DRAM design, boosted word line voltage of more than 2.5 V was used with a supply voltage of 1.5 V. Hence dual-gate oxide was used with array oxide being 5.5 nm thick and peripheral CMOS with 3 nm thick oxide. Parasitic activity and noise were reduced by using WLs and BLs covered with tungsten.

A number of DRAMs were reported at 70 nm technology level either using MIM capacitors [50,52,116], with minor variations, by adopting better part of the technologies of the previous generation for continuation sake. Technologies employed at 70 nm as well as 90 nm level which illustrate a good measure of continuity and changes in the technology are available [50]. Instead of KrF lithography, ArF lithography with simple layouts having straight lines and resolution enhancement techniques was used to get more than 0.3 μm of DOF margin for critical layers. To increase the mechanical stability of the diamond-shaped OCS, the bottom part was buried in molded oxide. The structure was further improved at manufacturable level for DDR-3 applications [116]. Important changes in the technology were made in a report given by C. Cho and others with 68 nm design rules to obtain a cell size of 0.028 μm^2 (6 F^2) and employed sphere-shaped RCAT (S-RCAT) [24] and double exposure technology (DET) for better resolution [52]. Figure 6.12 shows the layout with word lines and bit line having $2F$ and $3F$ pitches. As the word line was made of tungsten and its length was $3F$, its resistance was decreased by nearly 25%, and bit line capacitance was decreased by nearly 50% (50 fF/line with 256 cells) compared to an 8 F^2 cell. Thickness-optimized multilayer HfO-AIO dielectric was used on TiN metal electrode [117] to realize a 65 nm technology DRAM unaffected by post-thermal process. Cell capacitance of 25 fF was obtained while using equivalent oxide thickness of 11 Å, and leakage current was less than 1 fA/cell. This multilayer MIM capacitor with the dielectric used was claimed to be practicable beyond 50 nm technology.

Further reduction in feature size was reported in the form of a 6 F^2, 1 Gbit DRAM at 56 nm technology having a cell size of 0.019 μm^2, which was the smallest device with a fine performance level as well. Technologies used at

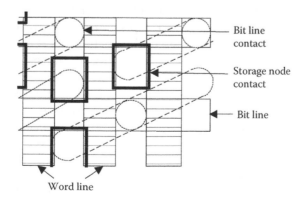

FIGURE 6.12
Part of the cell layout of a 6 F^2 DRAM. (Modified from "Integerated Device and Process Technology for Sub-70 nm Low Power DRAM," C. Cho et al., VLSI Techn. Dig. Tech. Papers, pp. 32–33, 2004.)

70 nm now needed more advancement especially for critical patterns, isolation, and inter-layer gap fill. A comparative statement is available for the
technologies used at 56 nm level with the technologies used in the previous
generations [53]. At 80 nm and 68 nm, HDP was used for isolation which is
supplemented with spin-on glass (SOG) at 56 nm, whereas the corresponding metallization changed from triple level to quadruple level. Storage node
is realized in metal for 56 nm in place of TSC or contact type poly-Si of the
previous generations. Capacitance form changed from AHO (MIS) to HAH
(MIM) and then to ZAZ (MIM) at 56 nm with SN without landing pad.

For critical layers, ArF immersion lithography was used which provided
photo lithographic process margin with DOF more than 0.1 µm. Dual poly
gate (n$^+$/p$^+$) with selective epitaxial layer were used for the elevated source
drain. SRCAT was continued to be used, however for increasing cell transistor
current, localized asymmetric channel (L-ASC) doping in which no channel
doping is done at storage node [118] was developed. Simulated 2-D junction
profiles of the different cases have been shown in ref. [53]. Junction leakage
current reduced up to one-third and the body effect also improved by ~6%
with L-ASC. One major difference over earlier DRAMs was the use of ZrO/
AlO/ZrO (ZAZ) dielectric with EOT of 75 Å for the MIM memory cell, which
provided cell capacitance of 25 fF with leakage still less than 1 fA/cell.

6.11.3 Sub-50 nm DRAM Technology Development

In two fine reports in 2005 and 2006, Kinam Kim has outlined major problems that needed to be addressed for DRAMs at 50 nm and beyond [119,120].
Shrink technology, which was the most important factor in ever-increasing
density of DRAMs, brought with it a few new issues as well. Process windows became smaller and variations in parameters due to process uncertain-
ties increased further. Many problems which have been outlined in earlier
sections like requiring sufficient cell capacitance in decreasing projected
area, leakage of current from storage node junction, short channel effect in
devices, degradation of retention time, reduction in transistor driving current, became more acute now. Along with improved lithography and etching techniques, new dielectric materials having compatibility with storage
electrode were needed. Requirement of new device structures with novel
approaches also became indispensable in order to keep pace with required
I_{on} and controlled I_{off}.

MIM structures with high-k dielectrics and metal electrodes have been
used in sub-100 nm DRAMs to provide a minimum of 25 fF/cell. Development
came in the form of the type of dielectric and the metal used for electrode.
Around 60 nm ZrO was preferred whereas for 50 nm design rules STO
or BST with TiN electrode was used. However beyond 50 nm, in place of
TiN electrode, Ru metal is preferred which ensures high value of dielectric

constant. Efforts continue to somehow increase the capacitor area/capacitor density without affecting its projected area.

To obtain the consistent value of 25 fF/cell capacitance for DRAMs at 45 nm, device needs a thin layer of oxide equivalent in the range of 6–7 Å [121]. Dielectric material layer like HfO_2/Al_2O_3 stack and $Hf_xAl_yO_z$ used up to around 70 nm, but were not able to satisfy the conditions needed at 45 nm [117]. Even the ZrO_2 film formed using atomic layer deposition (ALD) at 50 nm scale could not be used satisfactorily on account of negative biased leakage current [121]. New material like Ru has also been used for electrodes, but while the process was still to mature, dielectric film structures from ZrO_2 and Al_2O_3 have been shown to work even for mass production at 44 nm scale [54] as its equivalent oxide thickness was lowered to 6.2 Å with TiN electrode.

The problem of short channel effect which leads to enhancement of electric field at junction nodes degrades the retention time. Severity of the problem was greatly minimized through the use of RCAT and SRCAT [23,24] in which effective gate length was increased. However, beyond 50 nm other alternatives were needed for getting good performance from a cell transistor. Body tied FinFET is one such alternative which provides good progress toward short channel effect and provides excellent sub-threshold swing [122].

FinFET technology was expected to give quite satisfactory performance down to 30 nm node, along with the use of some other transistor structures like vertical pillar [123] and saddle fin [124]. However, beyond that, vertical transistors with SGT or S-SGT (slacked surround gate transistors) [2,3] are to be used.

Fully integrated, 8 F^2, 1 GB DRAM with smallest cell size of 0.15 μm² [54] was created by Hynix Semiconductor. Key process technologies used were ArF immersion lithography, peripheral transistor with WN barrier metal, and tungsten on n^+ poly-Si. Cylinder-like storage node with ZAZ dielectric ($ZrO_2/Al_2O_3/ZrO_2$) and triple-metal layer including copper were also used. To overcome the problem of nonscalability of the gate oxide thickness in line with the overall reduction in feature size at this level, saddle fin cell transistor which has the added advantage of improving short channel effect, including DIBL (drain-induced barrier lowering) and sub-threshold swing to 6 mV and 85 mV/dec was used.

With the DRAM working voltage coming down to 2 V and 1.0 V, new technologies are to be adopted. Signal margin which is further reduced due to the reduced array voltage, has to be maintained at a level that is easily picked by the sensing scheme. A buried word line (bWL) scheme has been introduced for a 6 F^2 46 nm DRAM that overcomes the problem through (including) reducing the bit line capacitance, which has size of 0.013 μm² and uses a three-dimensional EUD (extended u-shape device) array device structure [71] with a gate oxide thickness of 4 nm and a TiN/W metal gate. In fact TiN/W buried WL which is fabricated below the Si surface not only forms local interconnect but metal gate as well. In the proposed structure one bit line contact is shared by two cells, reducing chip area, and the buried WL isolates two adjacent capacitor node junctions. Simulations have shown that in the bWL structure bit-line capacitance is halved and word line capacitance became one-third that of a conventional structure. As a result, cell access time became shorter, power consumption was reduced, and the signal margin improved.

6.11.4 Sub-100 nm Trench DRAM Technology Development

As the DRAM cell minimum feature size goes below 100 nm range, short channel effect creates more and more problems. One known and workable solution was to increase substrate doping but, the remedy itself creates another problem in the form of increasing electric field [23,125]. Alternate solutions included vertical access transistors in deep trench technology [28,125] or recessed devices in stack capacitor technology [23,49]. Basically, the array transistor channel length was extended in these schemes, which avoided substrate doping increment with its resultant positive effect. However, it decreased the drive current of the array transistor and also decreased its I_{on}/I_{off} ratio; hence, the effort was to overcome this important limitation.

A planar DRAM cell device with a novel *checkerboard* cell layout has been reported and its integration at 70 nm level was achieved basically on highly asymmetric, inhomogeneous doping profiles along the channels [111]. Basics of the checkerboard (CKB) layout are shown in Figure 6.13. High symmetry in deep trench (DT) and in active area (AA) with equal lines and spaces overcomes the scalability problem faced earlier in the MINT layout [27] due to DT-DT spacing and overlaps between AA and DT. Since due to high selectivity of Si <100> to Si <110> 45° rotated substrate was used as shown in Figure 6.13; it increased capacitance by more than 25%. Hemispherical grains and the use of Al_2O_3 ALD, a high-k dielectric used first time in DT process,

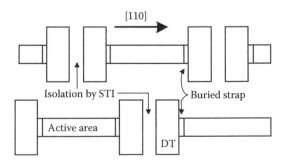

FIGURE 6.13
Schematic of MINT layout with substrate orientation. (Redrawn from "A Highly Manufacturable Deep Trench Based DRAM Cell Layout with a Planar Array Device in a 70 nm Technology," J. Amon et al., IEDM Tech. Dig., pp. 73–76, 2004.)

also increased the capacitance. However, to reduce the parasitic bit line contact to word line capacitance, oxide spacers provided good results instead of nitride spacers. A significant advantage of this technique was the objective of extending the planar DRAM cell channel length, which was achieved without compromising with the device drive current. Another DRAM cell at 70 nm generation employing CKB deep trench technology was given by C.H. Chung and others [126]. In this DRAM cell, designed buried strap was in opposite direction of each adjacent row of deep trench. Fabricated cell showed excellent refresh characteristics. Dual gate spacer technique—that is, different gate spacer thickness between cell array and core/periphery—is used to achieve the dual objective of suppressing periphery device short channel effect and ensuring sufficient process window for the array devices.

In a possible roadmap for the DRAM products, it was predicted that by the year 2010 data rate would be around 6400+ Mbit per second for DDR5 systems and supply voltage and the array voltage would have to be decreased to 1.2 V and 1.0 V, respectively, to achieve array timing of 8 ns [126]. With the expected reduced external supply voltage and the array voltage, if the storage cell capacitance is to remain nearly constant at 30 fF, bit line capacitance has to be decreased to 40 fF and 35 fF to get a cell signal of over 150 mV at 50 nm and 40 nm technology level, respectively. It may be noted that these calculations were based on the DRAM employing a trench cell array with 256 cells per twisted, folded, bit line array. To achieve mentioned shorter array timing of 8 ns and sense signal of over 150 mV at array voltage of 1.0 V (50 nm node) or 0.9 V (40 nm node), the trench resistance and the capacitance dielectric equivalent oxide thickness of the capacitor dielectric must be reduced. A hemispherical grain MIS capacitor having Al_2O_3 as dielectric (CET = 3.5 nm) and low resistance TiN liner connection of top metal capacitor plate to the poly-Si buried strap which connects to the cell transistor, reduces trench resistance by half. It results in the reduction of array timing by 1.5 ns enabling to reach the goal predicted by the DRAM product road map. For the 50 nm technology, a MIS capacitor in trench (TiN/HfSiON) has also been developed with CET = 2.5 nm [127].

To keep level of array transistor I_{off} less than standard 1 fA/cell is a huge challenge especially for technology node beyond 100 nm. Trap-assisted tunneling leakage starts taking place because of increased electrical field at any higher doping concentration area, even if localized. For a planar array device, trench DRAM was successfully scaled down by Amon and others [108] at 70 nm and RCAT [23] at 80 nm. A new structure on the lines of RCAT have been given which forms an extended U-shaped device (EUD), which meets the requirements of I_{off} and I_{on} at 60 nm node and also improved sub-threshold slope to 97 mV/dec [126]. Local damascene FinFET structure has also been developed for trench DRAM, as shown in Figure 6.14 and at 90 nm technology sub-threshold slope (SS) reduces to 77 mV/dec with I_{on} = 40 μA; for 40 nm case, predicted values were, SS = 89 mV/dec, I_{off} < 1 fA, I_{on} = 40 μA [128].

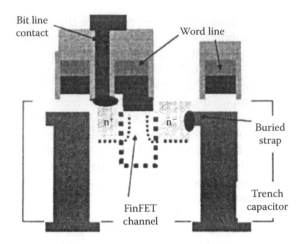

FIGURE 6.14
FinFET in trench cell integration scheme. ("Trench DRAM Technologies for the 50 nm Node and Beyond," W. Mueller et al., Int. Symp. on VLSI Tech. Sys. and Appli., pp. 1–2, 2006.)

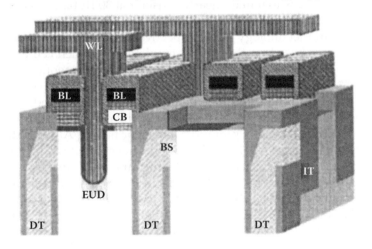

FIGURE 6.15
Schematic 3-D view of a novel word line over bit line cell. ("A Novel Cell Arrangement Enabling Trench DRAM Scaling to 40 nm and Beyond," L. Heineck et al., IEDM Tech. Dig., pp. 31–32, 2007.)

As also indicated earlier, DRAM needs to support/interface with high data rate at reduced external voltage. Schematic of a new cell arrangement called *word line over bit line* is shown in Figure 6.15, where the total bit line capacitance is reduced by nearly 30% [129]. It results in a faster cell access. In this scheme the most important innovation is that word line is realized in the first local interconnect metal layer, and the bit line is formed before the formation of array device and word line. It is also possible to realize the word line with very low resistance without affecting any other fabrication step adversely. A DRAM product at 48 nm was successfully obtained confirming the performance and shown to be scalable to 40 nm levels, which used some of the following advanced features used earlier as well. The checkerboard trench cell [111] and immersion lithography [55] helped avoidance of double patterning [52]. In addition, the EUD array device [127] helped in reducing leakage and hence satisfying data retention requirements.

6.12 Embedded DRAMs

Continued scaling, advancements in DRAM technology, and circuit innovations have made stand-alone DRAM the cheapest memory. However, DRAM's data transfer with various other digital functions created bottlenecks especially in the manufacturing domain, like input/output devices, increase in the number of pins, and so on. Around 1997, efforts started to utilize the low cost of DRAM in increasing the performance of digital function by combining the memory and logic on the same chip. Soon it was observed that further scaling was heading fast toward fabrication of complete system-on-chip (SOC) devices. Typical SOCs could contain random logic, intellectual property (IP) blocks like dedicated processors and controllers, register files, and analog components such as PLLs and D/A (A/D) converters [130].

Realization of SOC not only improved system performance, it opened possibilities of new applications because of a number of advantages. Important among these are (1) high bandwidth operation between DRAM and logic, (2) lower power dissipation, and (3) overall reduced size (hence cost) of the chip. High-speed on-chip logic in embedded DRAM reduces latency, which is much less than that of off-chip DRAM [131]. In comparison to embedded SRAM, capacity of embedded DRAM is nearly four to five times higher, although it requires special processing steps, which is not so with SRAM. Additionally, embedded DRAMs (eDRAM) consume only 30% of the SRAM power in standby mode and 10% in active mode for a general-purpose process [131], mainly because of fewer transistors per bit used. Embedded DRAM is superior in terms of soft-error-rate (SER) as well because of its larger value storage capacitor. By having DRAM and the logic on the same chip, high power consumed by the I/O modules is also avoided.

Promise of high performance has given great boost to the realization of SOCs, as SOC, which embeds DRAM array to logic, is found to be the best solution where large amounts of data are to be communicated at high speed—for example, a graphics accelerator of video games and fast Ethernet router [132]. However, merging of the DRAM and the logic requires reducing differences in manufacturing process. For an SOC, compatibility of design rule and IP module library between embedded DRAM and logic is essential

[131]. Process of reconciling compatibility for the embedded DRAM and logic is now mainly dependent on two different approaches: (1) memory circuit is incorporated in a high performance technology optimized logic, or (2) the logic circuit is incorporated in a technology-optimized high density low-performance DRAM. Obviously both the options have their advantages and limitations. When option (1) is selected additional steps are required to fabricate dense DRAM array and precautionary steps are also needed such as with deep trench (if TCC is used) dual gate oxide, and silicidation is not to be used on drain/source region. In this technology option logic circuit is not affected but the DRAM cell size becomes nearly double as SA bit contacts are now restricted. Size affects the density and architectural advantage of the commodity DRAMs.

If option (2) is selected, technology is to be tailored to improve the logic circuit/peripheral transistor performance and for the additional interconnect layers. It requires additional processing deployment of dual gate oxide which helps in optimizing transistor fabrications in DRAM cell wing I/O transistors with thicker gate oxide and random logic transistor with thinner gate oxide (with lesser gate voltage). Dual threshold voltage increases fabrication complexity but reduces leakage in DRAM cell and retains better performance in logic circuit. It is observed that process complexity of option (2) is about 15% less than option (1) [130], but the logic performance deteriorates 20 to 35%. In general, if a function needs memory exceeding 50% of the chip area, DRAM-based technology is preferred. It is important to note that minor intermediate process steps in the two options do vary with changing technology level and also depend on whether the DRAM cell uses TCC or SCC of COB or CUB type.

Huge advantages of merging the DRAM and logic on the same chip, however, do not come easy and the integration is challenging. Trade-offs are required in terms of performance/power/chip area, and depending on the variables involved, processing steps are differently selected. Since logic process and DRAM process are not compatible on many counts, options are invoked. An optimum eDRAM should combine the most advanced cell array, with minimum size and power consumption, with high performance logic at the least possible manufacturing cost. However, the process requirements of high-performance logic and the advanced DRAM are partly contradictory. Following are among the main issues in DRAM technologies compared with logic technologies:

1. Trench capacitor cell or stacked capacitor cell—High-density DRAM cells can be realized using trench capacitor cell (TCC) or stacked capacitor cell (SCC) but process steps should not be mixed with the logic realization. For examples in TCC, capacitance is formed before gate oxidation; hence, thermal process steps are not needed after the transistor module fabrication. As a result logic devices do not degrade in TCCs. However, with SCC few high temperature processes are used like storage node insulator formation, SiN deposition for SAC,

and it is always desirable to optimize these thermal operations in order to reduce their damaging effects on transistors; silicide process for the source/drain also is affected by high temperatures. Another advantage with TCC is that because of its flat topology, there is no difference between DRAM cell level and logic regions which results in having contact holes with low aspect ratio. Interconnection layers are, therefore, easier to fabricate, whereas inherent high aspect ratio creates lithography restrictions and comparatively difficult metallization in SCC.

2. Transistors in DRAM cell and logic—As requirements from the transistors are different in a stand-alone DRAM cell and for a random logic, these are fabricated differently. Some of the important features are shown in Table 6.6. Consequently in eDRAM, trade-offs are used and generally a middle of the path combination of processes is used depending on the technology used.

3. Levels of interconnects—In DRAMs, the architecture requires only two layers of metal where the second layer is generally laid with relaxed pitch; hence, metallization is rather a simple process. In the general logic, interconnect has now become very specialized with six or more layers for local, semi global, and global needs. Along with aluminum, copper with low dielectric material high aspect ratio contacts have become essential.

TABLE 6.6

Comparison of Fabrication Approach in Conventional DRAM and Logic Transistors

Parameter	DRAM	Logic
Threshold voltage, V_{th}	Higher value to reduce leakage current and off current	Comparatively lower value to improve performance (with $> I_{on}$)
Transistor gate voltage[a]	Enhanced value to improve performance, $V_{cc} = 2.5$ V	No enhancement, $V_{cc} = 1.5$ V
Gate oxide thickness[a]	Thicker gate oxide to prevent punch-through with increased gate voltage, e.g., 60 Å	No increment in gate oxide thickness, e.g., 30 Å
Silicidation of drain/source	Not used, affected at higher temperature	Silicidation is used to reduce drain/source resistance
Self-aligned bit contact	SAC is used to connect bit line to the cell transistor	Not used
Gate encapsulation to avoid shorting with gate[a]	Isolation techniques are used; gate length = 0.23 μm	Minimum gate dimensions are used which prevents use of SA contacts, gate length = 0.14 μm
Gate material	Single work-function (usually n-type) gate material is used to reduce cost	Dual work-function gates are used for better performance

[a] Numerical values of the parameters are for 0.18 μm embedded DRAM technology.

Trench cell technology was preferred for reasons mentioned above at 0.5 μm, 0.35 μm, and 0.25 μm technology levels in which commodity DRAM process was used up to trench isolation. The source/drain of the MOSFETs was silicided and W bit line and first level Al were formed using DRAM process [133,134]. A silicided-bridged substrate-plate trench (SPT) capacitor DRAM cell was fabricated using double-sacrificial-Si_3N_4-sidewall (DSS) process and silicidation at 0.25 μm design rule [135]. A closed ring of deep trench isolation (DTI) is deployed around every n-well in the logic parts, which helped in reducing chip size up to 30% in comparison to STI in a 0.17 μm design rule eDRAM [134]. As the DTI process could be completed with the deep trench capacitor realization, no additional steps were needed. Low-cost six-level Al Cu RIE metallization was used. Within a few years with aggressive scaling 65 nm SOC technology was achieved with a good yield with 8 Mbit DRAM [136]. Using tapered BF_2 implantation, cell size was 0.11 μm^2 with three layers of hybrid low-k material and Cu interconnects. Having a trench of 6 μm, storage capacitance of 26 fF was obtained. With technology enhancements, an eDRAM at 45 nm design rule could be realized having cell size of 0.069 μm^2. Bottle etching process with LOCOS collar structure and high-k node dielectric Al_2O_3 was used to maintain a storage capacitance C_s value in a scaled deep trench. Use of bottle-shaped trench and Al_2O_3 each increased C_s ~30% without any increase in leakage in comparison to nitride-oxide combination. Void-free filling of high aspect ratio STI was obtained with the application of spin on dielectric (SOD) and high-density plasma (HDP) film. Functional tests on a 256 Kbit eDRAM macro showed yield of 60% [137].

Cell size scaling of DT eDRAM has been maintained at nearly 60% for multiple generations beyond 180 nm technology, and it is expected to continue. Fabrication of 65 nm node was done on bulk substrate but beyond that SOI substrate became the choice to meet performance level and retention requirements in simplified DT process [138]. Proceeding further, SOI logic has been used in 45 nm and 32 nm node technology to obtain DT eDRAMs, which have become very useful for multicore processing [139] because of their density advantage [140]. Since concurrent refresh [141] frees eDRAM from the conventional requirement, best trade-off is possible between performance and retention. Low-k dielectric and shrunken wires help in reducing bit line capacitance by nearly 25% per generation, which relaxed the requirement of storage capacitance value. As a result, constraints on storage node/DT become less demanding. For the 45 nm, an ion implant process has been developed as required by eDRAMs for trench dimension scaling and reducing process complexity, whereas high-k SN dielectric and a metal liner have been introduced at 32 nm [140]. With reduced dimensions, controlling random variation in threshold voltage (V_{th}) of the access transistor is a major challenge. Reduction in equivalent gate oxide thickness and lowering of channel doping is a little helpful in the direction. Hence, high-k dielectrics have been used at 32 nm; however, variation in V_{th} across the chip is still a very important matter of investigation for eDRAM. Fully depleted device with low channel doping concentrations is one possible direction.

Stacked capacitor cell DRAM-based process suffered mainly from performance degradation due to high temperature processing during capacitor formation; so, initial stacked DRAM technology tried to overcome the constraint. In such an effort, DRAM-based process was integrated with 0.2 μm CMOS logic including six level metallization, which realized a cell size of 0.45 μm^2 [142]. A low-temperature metal-based cell structure with W/poly-Si gate and a W bit line were used to reduce parasitic resistance and delay. To remain in contest with high speed embedded SRAM, performance of logic in eDRAM had to be improved along with reduction in parasitic resistance of the cell. Hence DRAM cell current has been improved in a full-metal eDRAM (FMD) technology using MIM capacitor [143]. An HfO$_2$ MIM helped in reducing height of cylindrical capacitor with improvement in logic-contact resistance. Some other processes were also used to improve the logic transistor performance. Equivalent gate oxide thickness was reduced to 1.6 nm and gate poly depletion thickness was reduced by using ultra-low energy and high dose phosphorus ion implantation. As a result drive current improved by 30–40% [144]. Given eDRAM process could also be integrated with body-slightly-tied (BST) transistor with further logic performance improvement of nearly 10% compared with the bulk CMOS.

Other than DRAM-based or logic-based process, an intermediate process is *merged logic*, and as the name suggests, a middle approach is followed in which advanced CMOS and dedicated cell structure are combined so that logic performance does not deteriorate, though cell size and processing step may be relatively 20% to 50% more [145]. Proposed merged logic and FMD technology were combined to produce eDRAM having more than 8 Mbit capacity and more than 100 MHz random access read operation with an intended application in wide-band data streaming. Such an eDRAM was fabricated at 130 nm design rule with capacitance electrodes and bit lines formed with TiN/W, and source and drain of the access transistor were salicided. Care was taken for the thermal budget while forming MIM capacitor, which was kept around only 400°C. Merged DRAM technique was extended for a 90 nm eDRAM using ZrO$_2$ capacitor technology, with the formation of bottom cylindrical electrode by TiN deposition, ZrO$_2$ layer using ALD, and the top electrode TiN film through CVD. Process was shown to realize 55 nm eDRAM with the deployment of high-k dielectric, without increasing channel doping, but utilizing work-function modulation effect for improving transistor performance [145].

MIM capacitor using high-k dielectric is helpful in realizing adequate value of storage capacitance at 65 nm and 45 nm technologies. Performance was improved at these nodes and beyond by realizing higher capacitance in a technique called COLK (capacitor over low k) [146]. However, the method of realization for TiN/ZrO$_2$/TiN MIM capacitor is as before: it is placed in the first thick layer of the back-end stack. Additional advantage is the reduction in high aspect ratio contact needed for COB or CUB techniques. As a result process complexity and cost are reduced. Capacitance value of 10 fF/cell could be achieved in an area of 0.05 μm^2, effectively a density of 200 fF/μm^2. Even for a cell area of 0.025 μm^2, capacitance of 6–7 fF/cell is expected to be achievable, which was good enough for an eDRAM in 22 nm technology.

A conventional COB structure is an easy option for embedded DRAM since MIM capacitor integration is not complex. However, with decreasing feature size connecting W wire between transistor and Cu, metal 1 becomes thin and its aspect ratio increases. Resulting increased resistance and capacitance of this wire known as *bypass contact* deteriorate logic performance. A cylinder-type MIM capacitor in porous low-k film (CAPL) is given to overcome this limitation [147] as shown in Figure 6.16. However, CAPL integration faces problem of metal contamination during gas phase deposition of bottom electrode (TiN) of the capacitor on the porous low-k film. It is observed that a molecular pore stack (MPS) SiCH film ($k = 2.5$) blocks the gas phase diffusion of metal precursors like CVD TiN and hence stops the metal contamination. Use of such a technology could give eDRAM at 28 nm node and beyond.

A 128 Mbit (32 macros × 4 Mb) embedded DRAM test vehicle reported recently has been considered to be highly manufacturable at 40 nm node level. In logic-based process, only three additional critical masks are required for a low temperature MIM capacitor. Since the macro architecture could operate on a cell capacitance of 10 fF–15 fF because of reduced bit line length, height of capacitor cylinder could also be reduced with consequential advantages. The macro could operate from 25 MHz to 200 MHz which is comparable to SRAM. Using the same MIM eDRAM process, a high-speed designed macro was also shown that could operate on 500 MHz [148].

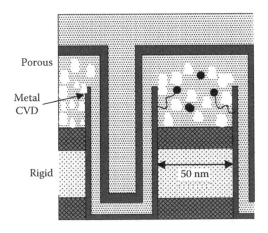

FIGURE 6.16
Structure of capacitor in porous low-k film (CAPL).

References

1. H. Akatsu et al., "A Highly Manufacturable 110 nm DRAM Technology with 8 F^2 Vertical Transistor Cell for 1 Gb and Beyond," *VLSI Tech. Dig. Tech Papers*, pp. 52–53, 2002.
2. H. Takato et al., "Impact of Surrounding Gate Transistor (SGT) for Ultra High Density LSIs," *IEEE Trans. Electr. Dev.*, Vol. 38, pp. 573–578, 1991.
3. T. Endoh et al., "2.4 F^2 Memory Cell Technology with Stacked-Surrounding Gate Transistor (S-SGT) DRAM," *IEEE Trans. Electr. Dev.*, Vol. 48, pp. 1599–1603, 2001.
4. H. Sunami et al., "A Corrugated Capacitor Cell (CCC) for Megabit Dynamic MOS Memories," IEDM Tech. Dig., pp. 806–808, 1982.
5. M. Wada, K. Heida, and S. Watanabe, "A Folded Capacitor Cell (FCC) for Future Megabit DRAMs," IEDM Tech. Dig., pp. 244–247, 1984.
6. T. Furuyama and J. Frey, "A Vertical Capacitor Cell for VLSI DRAMs," *Symp. VLSI Tech. Dig. Tech. Papers*, pp. 16–17, 1984.
7. S. Nakajima et al., "An Isolation Merged Vertical Capacitor Cell for Large Capacity DRAMs," IEDM Tech. Dig., pp. 240–243, 1984.
8. K. Nakamura et al. "Buried Isolation Capacitor (BIC) for Megabit MOS Dynamic RAM," IEDM Tech. Dig., pp. 236–239, 1984.
9. M. Nagamoto et al., "A High Density 4M dRAM Process Using Folded Bitline Adaptive Sidewall Isolated Capacitor (FASIC) Cell," IEDM Tech. Dig., pp. 144–147, 1986.
10. M. Sakamoto et al., "Buried Storage Electrode (BSE) Cell for Megabit DRAMs," IEDM Tech. Dig., pp. 710–713, 1985.
11. N.C.C. Lu et al., "The SPT Cell—A New Substrate-Plate Trench Cell for DRAMs," IEDM Tech. Dig., pp. 771–772, 1985.
12. W. Wakamiyia et al., "Novel Stacked Cell for 64 Mb DRAM," *Symp. VLSI Techn. Dig. Tech. Papers*, p. 31, 1989.
13. W.F. Richardson et al., "A Trench Transistor Cross Point DRAM Cell," IEDM Tech. Dig., pp. 714–717, 1985.
14. M. Sakao et al., "A Straight-Line Trench Isolation and Trench-Gate Transistor (SLIT) Cell for Gigabit DRAMs," *Symp. VLSI Techn. Dig. Tech. Papers*, pp. 19–20, 1993.
15. Y. Kawamoto et al., "A 1.28 µm² Bit-line Shielded Memory Cell Technology for 64 Mb DRAM," *Symp. VLSI Tech. Dig. Tech. Papers*, pp. 13–14, 1990.
16. W.D. Kim et al., "Development of CVD-Ru/Ta_2O_5/CVD-TiN Capacitor for Multi Gigabit-Scale DRAM Generation," *Symp. VLSI Technol. Dig. Tech. Papers*, pp. 100–101, 2000.
17. C.M. Chu et al., "Cylindrical Ru/$SrTiO_3$/Ru Capacitor Technology for 0.11 µm Generation DRAM," *Symp. VLSI Technol. Dig. Tech. Papers*, pp. 43–44, 2001.
18. Y. Fukuzumi et al., "Linear-Supported Cylinder (LSC) Technology to Realize Ru/Ta_2O_5/Ru for Future DRAMs," *IEDM*, pp. 793–796, 2000.
19. K.N. Kim et al., "Highly Manufacturable and High Performance SDR/DDR 4 Gb DRAM," *VLSI Tech. Dig. Tech. Papers*, pp. 7–8, 2001.
20. J. Lee et al., "Robust Memory Cell Capacitor using Multi-Stack Storage Node for High Performance in 90 nm Technology and Beyond," *Symp. VLSI Techn. Dig. Tech. Papers*, pp. 57–58, 2003.

21. D.H. Kim et al., "A Mechanically Enhanced Storage Node for Virtually Unlimited Height (MESH) Capacitor Aiming at Sub 70 nm DRAMs," *IEDM*, pp. 69–72, 2004.

22. J.M. Park et al., "Novel Robust Cell Capacitor (Leaning Exterminated Ring type Insulator) and New Storage Node Contact (Top Spacer Contact) for 70 nm DRAM Technology and Beyond," *Symp. VLSI Tech. Dig. Tech. Papers*, pp. 34–35, 2004.

23. J.Y. Kim et al., "The Breakthrough in Data Retention Time of DRAM Using Recessed Channel-Array Transistor (RCAT) for 88 nm Feature Size and Beyond," *Symp. VLSI Tech. Dig. Tech. Papers*, pp. 11–12, 2003.

24. J.Y. Kim et al., "S-RCAT (Sphere-shaped-Recess-Channel-Array Transistor) Technology for 70 nm DRAM Feature Size and Beyond," *Symp. VLSI Tech. Dig. Tech. Papers*, pp. 34–35, 2005.

25. D. Hisamoto, T. Kaga, and E. Takeda, "Impact of Vertical SOI 'DELTA' Structure on Planar Device Technology," *IEEE Trans. Electr. Dev.*, Vol. 38, pp. 1419–1424, 1991.

26. S-W. Chung et al., "Highly Scalable Saddle-Fin (S-Fin) Transistor for Sub-50 nm DRAM Technology," *Symp. VLSI Tech. Dig. Tech. Papers*, pp. 32–33, 2006.

27. L. Nesbit et al., "A 0.6 μm^2 256 Mb Trench DRAM Cell with Self-Aligned Buried Strap (BEST)," IEDM Tech. Dig., pp. 627–630, 1993.

28. U. Gruening et al., "A Novel Trench DRAM Cell with a VERtical Access Transistor and BuriED Strap (VERY BEST) for 4 Gb/1 Gb," IEDM Tech. Dig., pp. 25–28, 2000.

29. R. Weis et al., "A Highly Cost Efficient 8 F^2 DRAM Cell with a Double Gate Vertical Transistor Device for 100 nm and Beyond," IEDM Tech. Dig., pp. 415–418, 2001.

30. S.Y. Han et al., "A Novel DRAM Cell Transistor Featuring a Partially Insulated Bulk FinFET (Pi-FinFET) with a Pad-Polysilicon Side Contact (PSC)," *Symp. VLSI Techn. Dig. Tech. Papers*, pp. 166–167, 2007.

31. J.Y. Song et al., "Fin and Recess Channel MOSFET (FiReFET) for Performance Enhancement of Sub-50 nm DRAM Cell," *ISDRS*, 2007.

32. W.D. Kim et al., "Development of CVD-Ru/Ta_2O_5/CVD-TiN Capacitor for Multigigabit-Scale DRAM Generation," *Symp. VLSI Techn. Dig. Tech. Papers*, pp. 100–101, 2000.

33. K. Koyama et al., "A Stacked Capacitor with (Ba_xSr_{i-x}) TiO_3 for 256 M DRAM," IEDM Tech. Dig., pp. 823–826, 1991.

34. E. Fujii et al., "ULSI DRAM Technology with $Ba_{0.7}Sr_{0.3}TiO_3$ Film of 1.3 nm Equivalent SiO_2 Thickness and $10^{-9}A/Cm^2$ Leakage Current," IEDM Tech. Dig., pp. 267–269, 1992.

35. Y. Ohno et al., "A Memory Cell Capacitor with $Ba_xSr_{1-x}TiO_3$ (BST) Film for Advanced DRAMs," *Symp. VLSI Tech. Dig. Tech. Papers*, pp. 149–150, 1994.

36. S. Yamamichi et al., "An ECR MOCVD (BaSr) TiO_3 Based Stacked Capacitor Technology with RuO_2/Ru/TiN/TiSix Storage Nodes for Gbit-Scale DRAMs," IEDM Tech. Dig., pp. 119–122, 1995.

37. Y. Nishioka et al., "Giga-bit Scale DRAM Cell with New Simple Ru/(Ba,Sr) TiO_3/Ru Stacked Capacitor Using X-Ray Lithography," IEDM Tech. Dig., pp. 903–906, 1995.

38. A. Yüki et al., "Novel Stacked Capacitor Technology for 1 Gbit DRAMs with CVD-(Ba,Sr) TiO₃ Thin Films on a Thick Storage Node of Ru," IEDM Tech. Dig., pp. 115–118, 1995.
39. H. Yamaguchi et al., "A Stacked Capacitor with an MOCVD-(Ba,Sr) TiO₃ Film and a RuO₂/Ru Storage Node on a TiN-Capped Plug for 4 Gbit DRAMs and Beyond," IEDM Tech. Dig., pp. 675–678, 1996.
40. R.B. Khamankar et al., "A Novel BST Storage Capacitor Node Technology Using Platinum Electrode for Gigabit DRAMs," IEDM Tech. Dig., pp. 245–248, 1997.
41. N. Menou et al., "0.5 nm EOT Leakage ALD SrTiO₃ on TiN MIM Capacitor for DRAM Applications," IEDM Tech. Dig., pp. 1–4, 2008.
42. J.H. Bruning, "Optical Lithography—Thirty Years and Three Orders of Magnitude: The Evolution of Optical Lithography Tools," Proc. SPIE, Vol. 3051, pp. 14–17, 1997.
43. K.N. Kim et al., "A 0.15 μm DRAM Technology Node for 4 Gb DRAM," Symp. VLSI Tech. Papers, pp. 16–17, 1998.
44. S. Bukofsky et al., "Extending KrF Lithography to 0.13 μm² Sub-8F² DRAM Technology: The Importance of Lithography-Centric Design," Proc. European Sol. St. Dev., Res. Conf., 2000.
45. K.N. Kim et al., "A 0.13 μm DRAM Technology for Giga Bit Density Stand-Alone and Embedded DRAMs," Symp. VLSI Tech. Dig. Tech. Papers, pp. 10–11, 2000.
46. H.S. Jeong et al., "Highly Manufacturable 4 Gb DRAM Using 0.11 μm DRAM Technology," IEDM Tech. Dig., pp. 353–356, 2000.
47. K. Kim and J-S. Park, "A 0.11 μm DRAM Technology for 4 Gb DRAM and Beyond," 6th Int. Conf. Solid State and Integ. Cir. Tech., pp. 178–182, 2001.
48. Y.K. Park et al., "Highly Manufacturable 90 nm DRAM Technology," IEDM Tech. Dig., pp. 819–822, 2002.
49. H.S. Kim et al. "An Outstanding and Highly Manufacturable 80 nm DRAM Technology," IEDM Tech. Dig., pp. 17.2.1–17.2.4, 2003.
50. C. Cho et al., "Integrated Device and Process Technology for Sub-70 nm Low Power DRAM," VLSI Techn. Dig. Tech. Papers, pp. 32–33, 2004.
51. C. Cho et al., "A 6 F² DRAM Technology in 60 nm Era Gigabit Densities," Symp. VLSI Techn. Dig. Tech. Papers, pp. 36–37, 2005.
52. M. Fritze et al., "Enhanced Resolution for Future Fabrication," IEEE Circuits and Devices Magazine, Vol. 19, pp. 43–49, 2003.
53. Y.K. Park et al., "Fully Integrated 56 nm DRAM Technology for 1 Gb DRAM," Symp. VLSI Techn. Dig. Tech. Papers, pp. 190–191, 2007.
54. H. Lee et al., "Fully Integrated and Functioned 44 nm DRAM Technology for 1 Gb DRAM," Symp. VLSI Techn. Dig. Tech. Papers, pp. 86–87, 2008.
55. K. Ronse et al., "Lithography Options for the 32 nm Half Pitch Node and Beyond," IEEE Trans. Circuit and Systems, Vol. 56, pp. 1883–1890, 2009.
56. W.H. Arnold, "Challenges for Lithography Scaling to 32 nm and Below," Proc. VLSI Techn. Systems and Appl. (VLSI-TSA), 2007.
57. V. Wiaux et al., "A Methodology for Double Patterning Compliant Split and Design," Proc. SPIE 7140, pp. 71401X–71401X, 2008.
58. A. Bryant, W. Häusch, and T. Mii, "Characteristics of CMOS Device Isolation for the ULSI Age," IEDM Tech. Dig., pp. 671–674, 1994.

59. T. Hamamoto, S. Sugiura, and S. Swada, "Well Concentration: A Novel Scaling Limitation Factor Derived from DRAM Retention Time and Its Modeling," IEDM Tech. Dig., pp. 915–918, 1995.

60. B.H. Roh et al., "Shallow Trench Isolation for Enhancement of Data Retention Times in Gigabit DRAM," *SSDM Tech. Dig.*, pp. 830–832, 1996.

61. K. Kim, C-G. Hwang, and J.G. Lee, "DRAM Technology Perspective for Gigabit Era," *IEEE Trans. Electr. Dev.*, Vol. 45, pp. 598–608, 1998.

62. K. Ishimura et al., "Mechanical Stress-Induced MOSFET Punch-Through and Process Optimization for Deep Submicron TEOS-03 Filled STI Device," *VLSI Tech. Dig. Tech. Papers*, pp. 123–124, 1997.

63. B. Davari et al., "A New Planarization Technique, Using a Combination of RIE and Chemical Mechanical Polish (CMP)," IEDM Tech. Dig., pp. 61–64, 1989.

64. Y.H. Son, et al., "Highly Manufacturable Device Isolation Technology Using LASER-Induced Epitaxial Growth for Monolithic Stacked Devices," *IEEE Trans. Electr. Dev.*, Vol. 58, pp. 3863–3868, 2011.

65. K. Kim and M-Y. Jeong, "The COB Stack DRAM Cell at Technology Node below 100 nm-Scaling Issues and Directions," *IEEE Trans. Semiconductor Manufacturing*, Vol. 15, pp. 137–143, 2002.

66. M. Sekine et al., "Self-Aligned Tungsten Strapped Source/Drain and Gate Technology Realizing the Lowest Sheet Resistance for Sub-quarter Micron CMOS," IEDM Tech. Dig., pp. 493–496, 1994.

67. G-T. Jeong et al., "A High Performance 16 Mb DRAM Using Giga-Bit Technology," *IEEE Trans. Electr. Dev.*, Vol. 44, pp. 2064–2069, 1997.

68. R. Mallick et al., "W/WN/Poly Gate Implementation for Sub-130 nm Vertical Cell DRAM," *Symp. VLSI Tech. Papers*, pp. 31–32, 2001.

69. S. Choi et al., "Highly Manufacturable Sub-100 nm DRAM Integrated with Full Functionality," *Symp. VLSI Tech. Dig. Tech. Papers*, pp. 54–55, 2002.

70. M. Tapajna et al., "Electrical Characteristics of Ru and RuO_2/Ta_2O_5 Gate Stacks for Nanoscale DRAM Technology," *7th Int. Conf. Adv. Semiconductor Dev. μ-Systems*, pp. 267–270, 2008.

71. T. Schloesser et al., "A 6 F^2 Buried Word Line DRAM Cell for 40 nm and Beyond," IEDM Tech. Dig., pp. 1–4, 2008.

72. H. Hada et al., "A Self-Aligned Contact Technology Using Anisotropic Selective Epitaxial Silicon For Gigabit DRAMs," IEDM Tech. Dig., pp. 665–668, 1995.

73. K.P. Lee et al., "A Process Technology for 1 Giga-Bit DRAM," IEDM Tech. Dig., pp. 907–910, 1995.

74. H. Koga et al., "A 0.23 μm² Double Self-Aligned Contact Cell for Gigabit DRAMs With a Ge-Added Vertical Epitaxial Si Pad," IEDM Tech. Dig., pp. 589–592, 1996.

75. Y-S. Chun, "A New DRAM Cell Technology Using Merged Process with Storage Node and Memory Cell Contact for 4 Gb DRAM and Beyond," IEDM Tech. Dig., pp. 351–354, 1998.

76. K. Hinode, I. Asano, and Y. Homma, "Void Formation Mechanism in VLSI Aluminium Metallization," *IEEE Trans. Elect. Dev.*, Vol. ED-36, pp. 1050–1055, 1989.

77. T. Kikkawa et al., "A Quarter-Micron Interconnection Technology Using Al-Si-Cu/TiN Alternated Layers," IEDM Tech. Dig., pp. 281–284, 1991.

78. M.T. Bohr, "Interconnect Scaling—The Real Limiter to High Performance ULSI," IEDM Tech. Dig., pp. 241–244, 1995.

79. A.V. Gelalos et al., "CVD of Copper from a Cu+1 Precursor and Water Vapour and Formation of TiN Encapsulated Submicron Copper Interconnects by Chemical Mechanical Polishing," *Int. Elect. Dev. Tech. Dig.*, pp. 123–124, 1994.

80. J.S. Kim et al., "A Triple Metallization Technique for Gigabit Scaled DRAM's," *Proc. VMIC 95*, pp. 28–33, 1996.

81. K. Higashi et al., "A Manufacturable Copper/Low-k SiOC/SiCN Process Technology for 90 nm-Node High Performance eDRAM," *IEEE Int. Interconnect Tech. Conf.*, pp. 15–17, 2002.

82. A. Kajita et al., "Highly Reliable Cu/Low-k Dual-Damascene Interconnect Technology with Hybrid (PAE/SiOC) Dielectrics for 65 nm-Node High Performance eDRAM," *Proc. IEEE Int. Interconnect Tech. Conf.*, pp. 9–11, 2003.

83. T. Sugiura et al., "45 nm-Node BEOL Integration Featuring Porous-Ultra-Low-k/ Cu Multilevel Interconnects," *Proc. Inter. Interconnect Tech. Conf.*, pp. 15–17, 2005.

84. J. Kawahara et al., A New Direct Low-k/Cu Dual Damascene (DD) Contact Lines for Low Loss (LL) CMOS Device Platform," *Symp. VLSI Tech. Dig. Papers*, pp. 106–107, 2008.

85. S. Yokogawa et al., "Analysis of Al Doping Effects on Resistivity and Electromigration of Copper Interconnects," *IEEE Trans. Dev. Mat. Reliability*, Vol. 8, pp. 216–221, 2008.

86. H.K. Jung et al., "Formation of Highly Reliable Cu/Low-k Interconnects by Using CVD Co Barrier in Dual Damascene Structures," *IRPS*, pp. 307–311, 2011.

87. N. Kwak et al., "BEOL Process Integration with Cu/FSG Wiring at 90 nm Design-Rule DDR DRAM and Their Effect on Field, Refresh Time and Wafer-Level Reliability," *Proc. Inter. Interconnect Tech. Conf.*, pp. 150–152, 2007.

88. S.M. Rossnagel et al., "Interconnect Issues in Post 45 nm," *IEDM Tech. Dig.*, pp. 89–91, 2005.

89. J.P. Gambino, "Copper Interconnect Technology for the 22 nm Node," *Int. Symp.*, *VLSI-TSA*, pp. 1–2, 2001.

90. S. Nitta et al., "Performance and Reliability of Air Gaps for Advanced BEOL Interconnects," *Interconnect Tech. Conf.*, pp. 191–192, 2008.

91. K. Ueno, "Material and Process Challenges for Interconnects in Nanoelectronics Era," *Int. Symp. VLSI-TSA*, pp. 64–65, 2010.

92. International Technology Roadmap for Semiconductors, 2009, Edition: Interconnects.

93. A. Nieuwoudt and Y. Massoud, "Evaluating the Impact of Resistance in Carbon Nanotubes Bundles for VLSI Interconnect Using Diameter-Dependent Modeling Techniques," *IEEE Trans. Elect. Dev.*, Vol.53, pp. 2460–2466, 2006.

94. H. Li et al., "Carbon Nanomaterial for Next-Generation Interconnects and Passives: Physics, Status, and Prospects," *IEEE Trans. Elect. Dev.*, Vol. 56, pp. 1799–1821, 2009.

95. A. Neemi, R. Sarvari, and J.D. Meindl, "Performance Comparison between Carbon Nanotube and Copper Interconnects for Gigascale Integration (GSI)," *IEEE Elect. Dev. Lett.*, Vol. 26, pp. 84–86, 2005.

96. P.G. Collins et al., "Current Saturation and Electrical Breakdown in Multiwalled Carbon Nanotubes," *Phys. Ress. Lett.*, Vol. 86, no. 14, pp. 3128–3131, 2001.

97. D.K. Brock et al., "Carbon Nanotube Memories and Fabric in a Radiation Hard Semiconductor Foundry," *IEEE Aerospace conf. paper*, pp. 1–9, 2005.

98. H. Li et al., "Circuit Modeling and Performance Analysis of Multi-Walled Carbon Nanotube Interconnects," *IEEE Trans. Electr. Dev.*, Vol. 55, pp. 1328–1339, 2008.
99. B.Q. Wei et al., "Organized Assembly of Carbon Nanotubes," *Nature*, Vol. 416, no. 6880, pp. 495–496, 2002.
100. G. Aichmayr et al., "Carbon/High-k Trench Capacitor for the 40 nm DRAM Generation," *Proc. IEEE Symp. VLSI Technol.*, pp. 186–187, 2007.
101. C. Xu et al., "Graphene Nano-Ribbon (GNR) Interconnects: A Genuine Contender or a Delusive Dream?," IEDM Tech. Dig., pp. 201–204, 2008.
102. D. Kondo, S. Sato, and Y. Awano, "Self-Organization of Novel Carbon Composite Structure: Graphene Multi-Layers Combined Perpendicularly with Aligned Carbon Nanotubes," *Appl. Phys. Express*, Vol. 1, no. 7, p. 074003, 2008.
103. J.W. Ward et al., "Reduction of Interconnect Resistance for the Replacement of Cu for Future Technology Nodes," *IEEE Trans. Nanoelectronics*, Vol. 11, pp. 56–62, 2012.
104. S.-N. Pu et al., "Crosstalk Prediction of Single- and Double-Walled Carbon Nanotubes (SWCNT/DWCNT) Bundle Interconnects," *IEEE Trans. Elect. Dev.*, Vol. 56, pp. 26–37, 2007.
105. T. Mizuno et al., "Si_3N_4/SiO_2 Spacer Induced High Reliability in LDD MOSFET and Simple Degradation MODEL," IEDM Tech. Dig., p. 234, 1988.
106. K.N. Kim et al., "Highly Manufacturable 1 Gb SDRAM," *Symp. VLSI Tech. Dig. of Tech. Papers*, pp. 9–10, 1997.
107. K. Itabashi et al., "Fully Planarized Stacked Capacitor Cell," *Symp. VLSI Tech. Dig. of Tech. Papers*, pp. 21–22, 1997.
108. T. Rupp et al., "Extending Trench DRAM Technology to 0.15 μm Groundrule and Beyond," IEDM Tech. Dig., pp. 33–36, 1999.
109. B-C. Lee et al., "High Performance Cell Technology Featuring Sub-100 nm DRAM with Multi-Gigabit Density," IEDM Tech. Dig., pp. 835–838, 2002.
110. Y. Hiura et al., "Integration Technology of Polymetal (W/WSiN/Poly-Si) Dual-Gate CMOS for 1 Gbit DRAMs and Beyond," IEDM Tech. Dig., pp. 389–392, 1998.
111. J. Amon et al., "A Highly Manufacturable Deep Trench-Based DRAM Cell Layout with a Planar Array Device in a 70 nm Technology," IEDM Tech. Dig., pp. 73–76, 2004.
112. M. Hamada et al., "A High Performance 0.18 μm Merged DRAM/Logic Technology Featuring 0.45 $μm^2$ Stacked Capacitor Cell," IEDM Tech. Dig., pp. 45–48, 1999.
113. J-H. Heo et al., "Void Free and Low Stress Shallow Trench Isolation Technology Using P-SOG for Sub 0.1 μm Devices," *Symp. VLSI Tech. Dig. Tech. Papers*, pp. 132–133, 2002.
114. J.H. Lee et al., "Practical Next Generation Solution for Stand-Alone and Embedded DRAM Capacitor," *Symp. VLSI Tech. Dig. Tech. Papers*, pp. 114–115, 2002.
115. F. Fishburn et al., "A 78 nm DRAM Technology for Multigigabit Densities," *Symp. VLSI Tech. Dig. Tech. Papers*, pp. 28–29, 2004.
116. H. Kim et al., "70 nm DRAM Technology for DDR-3 Application," *Int. Symp. on VSLI Technology (VLSI-TSA)*, pp. 29–30, 2005.
117. D-S. Kil et al., "Development of Highly Robust Nano Mixed $Hf_xAl_yO_z$ Dielectric for $TiN/Hf_xAl_yO_z/TiN$ Capacitor Applicable to 65 nm Generation DRAMs," *Symp. on VLSI Tech. Dig. Tech. Papers*, pp. 126–127, 2004.

118. J.W. Lee et al., "Improvement of Data Retention Time in DRAM Using Recessed Channel Array Transistors with Asymmetric Channel Doping for 80 nm Feature Size and Beyond," *E.S.S.D. RC*, pp. 449–452, 2004.

119. Kinam Kin, "Technology for Sub-50 nm DRAM and NAND Flash Manufacturing," *VLSI-TSA*, pp. 88–94, 2005.

120. Kinam Kim, "Memory Technologies for 50 nm and Beyond," *Solid State and Int. Cir. Tech. (ICS ICT)*, pp. 685–688, 2006.

121. D.S. Kil et al., "Development of New $TiN/ZrO_2/Al_2O_3/ZrO_2/TiN$ Capacitors Extendable to 45 nm Generation DRAMs Replacing HfO_2 Based Dielectrics," *Symp. on VLSI Techn. Dig. of Tech. Papers*, pp. 38–39, 2006.

122. C.H. Lee et al., "Novel Body Tied FinFET Cell Array Transistor DRAM with Negative Word Line Operation for Sub 60 nm Technology and Beyond," *Symp. VLSI Techn. Dig. Tech. Papers*, pp. 130–131, 2004.

123. J-M. Yoon et al., "A Novel Low Leakage Current VPT (Vertical Pillar Transistor) Integration for 4 F^2 DRAM Cell Array with Sub 40 nm Technology," *Device Research Conference*, pp. 259–260, 2006.

124. S-W. Chung et al., "Highly Scalable Saddle-Fin (S-Fin) Transistor for Sub-50 nm DRAM Technology," *VLSI Techn. Dig. Tech. Papers*, pp. 32–33, 2006.

125. R. Divakaruni et al., "Technologies for Vertical Transistor DRAM Cells to 70 nm," *Symp. on VLSI Tech. Dig. Tech. Papers*, pp. 59–60, 2003.

126. C-H. Chung et al., "A Novel DRAM Cell Design and Process for 70 nm Generation," *Int. Symp. on VLSI Tech. Sys. and Appl.*, pp. 1–2, 2006.

127. W. Mueller et al., "Challenges for the DRAM Cell Scaling to 40 nm," IEDM Tech. Dig., pp. 226–230, 2005.

128. W. Mueller et al., "Trench DRAM Technologies for the 50 nm Node and Beyond," *Int. Symp. on VLSI Tech. Sys. and Appli.*, pp. 1–2, 2006.

129. L. Heineck et al., "A Novel Cell Arrangement Enabling Trench DRAM Scaling to 40 nm and Beyond," IEDM Tech. Dig., pp. 31–32, 2007.

130. D.K. Schulz and N. Wehn, "Embedded DRAM Development: Technology Physical Design, and Application Issues," *IEEE Design and Test of Computer*, pp. 7–15, 2001.

131. S. Natrajan et al., "Searching for the Dream Embedded Memory," *IEEE S.S. Circuit Magazine*, pp. 34–44, 2009.

132. T. Yoshida et al., "A Fabrication Method for High Performance Embedded DRAM of 0.18 μm Generation and Beyond," *IEEE Custom Integ. Cir. Conf.*, pp. 61–64, 2000.

133. H. Ishiuchi et al., "Embedded DRAM Technologies," IEDM Tech. Dig., pp. 33–36, 1997.

134. H. Takato et al., "Embedded DRAM Technology: Past, Present, and Future," *Int. Symp. VLSI Techn., System & Appl.*, pp. 239–242, 1999.

135. M. Togo et al., "A Salicide-Bridged Trench Capacitor with a Double-Sacrificial-Si_3N_4-Sidewall (DSS) for High-Performance Logic-Embedded DRAMs," IEDM Tech. Dig., pp. 37–40, 1997.

136. M. Matsubara et al., "Fully Compatible Integration of High Density Embedded DRAM with 65 nm CMOS Technology (CMOS5)," IEDM Tech. Dig., pp. 423–426, 2003.

137. T. Sanuki et al., "High Density and Fully Compatible Embedded DRAM Cell with 45 nm CMOS Technology (CMOS6)," *Symp. VLSI Techn. Dig. Tech. Papers*, pp. 14–15, 2005.

138. G. Wang et al., "A 0.127 μm² High Performance 65 nm SOI Based Embedded DRAM for On-Processor Applications," IEDM Tech. Dig., pp. 1–4, 2006.
139. R. Kalla et al., "Power 7™: IBM's Next Generation POWER Microprocessor," *Hot Chips*, 2009.
140. G. Wang et al., "Scaling Deep Trench Based eDRAM on SOI to 32 nm and Beyond," IEDM Tech. Dig., pp. 259–262, 2009.
141. J. Barth et al., "A 500 MHz Random Cycle 1.5 ns Latency, SOI Embedded DRAM Macro Featuring a 3T Micro Sense Amplifier," *I.SSCC Dig. Tech. Papers*, pp. 486–487, 2007.
142. M. Yoshida et al., "An Embedded 0.405 μm² Stacked DRAM Technology Integrated with High Performance 0.2 μm CMOS Logic and 6-Level Metallization," IEDM Tech. Dig., pp. 41–44, 1999.
143. S. Arai et al., "A 0.13 μm Full Metal Embedded DRAM Technology Targeting on 1.2V, 450 MHz Operation," IEDM Tech. Dig., pp. 403–406, 2001.
144. Y. Aoki et al., "Ultra-High Performance 0.13 μm Embedded DRAM Technology Using TiN/HfO₂/TiN/W Capacitor and Body-Slightly-Tied SOI," IEDM Tech. Dig., pp. 831–834, 2002.
145. Y. Yamagata et al., "Device Technology for Embedded DRAM Utilization Stacked MIM (Metal-Insulator-Metal) Capacitor," *IEEE Custom Integ. Cir. Conf. (CICC)*, pp. 421–427, 2006.
146. S. Cremer et al., "COLK Cell: A New Embedded DRAM Architecture for Advanced CMOS Nodes," *Proc. European Solid-State DRC (ESSDRC)*, pp. 158–161, 2010.
147. K. Hijioka et al., "A Novel Cylinder-Type MIM Capacitor in Porous Low-k Film (CAPL) for Embedded DRAM with Advanced CMOS Logic," IEDM Tech. Dig., pp. 756–759, 2010.
148. C.Y. Chen et al., "A High-Performance Low-Power Highly Manufacturable Embedded DRAM Technology Using Backend Hi-K MIM Capacitor at 40 nm Node and Beyond," *Symp. VLSI Techn. System & Appl.*, pp. 58–59, 2011.

7

Leakages in DRAMs

7.1 Introduction

Continued reduction of minimum feature size and the advancements in the fabrication processes were the main reasons for the DRAM density enhancement. As soon as multi-megabit DRAM appeared on the horizon, it opened completely new application areas for the DRAMs. The use of megabit DRAMs in portable equipment, which required low-voltage operation, such that they could operate on battery for longer duration, became highly desirable. At one point in time, a 5 V supply was standard but was not suitable for long-term battery-based operation. In the low-voltage operating range a 1.5 V DRAM given by M. Aoki and others [1] was one of the first such DRAMs. It also reduced power consumption as bit line swing was reduced only up to the sum of the threshold voltages for NMOS and PMOS transistors in the sense amplifier. A 2 Kbit test DRAM worked successfully at the reduced voltage, and calculated value of the operating current was shown to be reduced from 15 mA at 5 V to less than a third of its value and the DRAM was expected to operate for 500 hours on eight 2-ampere-hour dry batteries. At this stage, reduction in supply voltage created two major problems: (1) decrease in signal-to-noise (S/N) ratio, and (2) degradation in operating speed. Soon after, a major breakthrough came in the form of another landmark 1.5 V DRAM, at 64 Mbit level with increased S/N ratio, given by Nakagome and others [2], using 0.3 μm triple-well CMOS process with no back-gate bias. The new DRAM used three new circuit modifications for making it operative at low voltage with improved performance. It used an I/O sense amplifier circuit with complementary sensing scheme having PMOS-driven current sense amplifier and NMOS read-out gates. From word line swing to bit line swing, a boost ratio of more than 1.75 was needed for low-voltage operation of 1.5 V; hence, a new feedback charge pump circuit was used. To improve S/N ratio, an accurate and highly regulated half-V_{cc} voltage level was also used. All these features resulted in an RAS access time of 50 ns and power dissipation was only 44 mW with active current of 29 mA and standby current of 1 mA at room temperature.

In stand-alone commercial DRAMs, standard supply voltage V_{DD} was reduced to 1.8 V by 2004 and trend is to continuously decrease it [3]. Table 7.1

TABLE 7.1

Predominantly Used Supply Voltage and Adopted Technology Node
and (Expected) Progress with DRAM Density

Year	1980	1984	1986/87	1990	1993	2002	2003
DRAM density	256 kb	1 Mb	4 Mb/16 Mb	64 Mb	256 Mb	512 Mb	1 Gb
V_{DD} (V)	5.0	5.0	5.0/3.3	3.3	3.3/2.5	2.5	2.5/1.8
Technology node (μm)	—	—	—	0.35	0.18	0.13	0.1

Year	2004	2005	2006	2008	2010	2011	2012
DRAM density	2 Gb	2 Gb	2 Gb/4 Gb	4 Gb/8 Gb		16 Gb	
V_{DD} (V)	1.8/1.5	1.5/1.3	1.5/1.3	1.3	1.1	1.1	1.0
Technology node (nm)	80/70	65	65	55	44	40	36

Source: Y. Nakagome, et al., "Review and Future Prospects of Low-Voltage RAM Circuits,"
IBM J. Res. and Dev., Vol. 47, no. 5/6, pp. 525–552, 2003; International Technology
Roadmap for Semiconductors, http://public.itrs.net/Files/2001. ITRS/Home.
hlm, 2001 Edition; and ITRS 2009.

shows trend in supply voltage with DRAM density over the years. For
eDRAMs the voltage was to be lowered even more as it had to be compatible
with surrounding peripherals and logic and its suitability with the use of
single NiCd cell having a minimum level of supply voltage of 0.9 V only [4].
At the same time advances in technology forced DRAM cell area along with
its storage capacitor size to continuously decrease through using innovative
cells and reduction in the technology node, and resulting in increased DRAM
density accompanied with decrease in gate oxide (t_{ox}) thickness bringing a
host of challenges to be taken up. Table 7.2 shows the trend of decreasing
t_{ox} with DRAM density. Increased DRAM density meant, more number of
cells on a single bit line, which increased bit line capacitance C_{BL}, resulting
in reduced charge transfer ratio ($C_S/(C_S + C_{BL}) \approx C_S/C_{BL}$), which would result
in unreliable performance in presence of soft errors and other noises unless
measures were taken. Another cause of unreliable operation and variable
response was the variation in physical dimensions and resultant variations
in the characteristics of DRAM cell transistors and capacitors and transistors used elsewhere on the chip. As with other parameters, it caused varied

TABLE 7.2

Change in Gate Oxide Thickness with Increase in DRAM Density

DRAM density	64 Mb	128 Mb	256 Mb	512 Mb	1 Gb	2 Gb
t_{ox} (nm)	10	7.5	6.0	4.5	4.0	3.0
F (μm)	0.35	0.25	0.18	0.13	0.10	0.07

values of threshold V_{th} on the same chip. It affected the performance in different ways depending on the use of transistors. However, a more serious problem was the increase in power dissipation due to the increase in leakage currents, both in the active and inactive condition of a cell, module, or block. This chapter will discuss the ways in which the challenges that arose in designing low-voltage/low-power DRAMs were faced and the problems resolved especially with regard to power dissipation.

A closely related issue is reducing the data retention power in DRAMs. Because of inherent leakages and destructive nature of READ operation, continuous refresh also requires power. This requirement of data retention power is necessary to be reduced for battery backup-based operations. Continuous efforts have been made in this direction. Before discussing the techniques/circuit solutions or technological advancements for reducing leakage current and improving data retention time, different kinds of leakages for DRAM cells and peripherals are discussed.

7.2 Leakage Currents in DRAMs

A DRAM chip can broadly comprise (1) the memory cell array, (2) row and column decoders, and (3) other peripheral circuits. Each component has undergone some changes from its basic form over the years. However, there are several reasons for leakages due to which the stored signal changes and power is dissipated irrespective of the modifications in the DRAM chip components: Some of the significant ones are as follows [5]:

1. Reverse-biased junction leakage current from the storage node
2. Sub-threshold leakage current of the access transistors
3. Capacitor dielectric leakage current
4. Gate-induced drain leakage current (GIDL) at the storage node
5. Gate dielectric leakage current
6. Leakage current between the adjacent cells

While the total leakage current is the sum of all leakage currents, each of the above mentioned components has different weight and their weight has varied considerably with the changes in the technologies, fabrication processes, minimum feature size and the supply voltage. Moreover, all the peripheral circuits, like row and column decoders, sense amplifiers, voltage up/down converters, refreshing circuit, etc. do have similar types of leakages, and power is needed to recover the leaked charge as well as for the normal functioning of the DRAM cell and peripherals.

7.2.1 Junction Leakage and Sub-Threshold Currents

Leakage current flows through the reverse-biased p-n junctions wherever these are located on the chip. For the sub-micron feature size, leakage current is in the range of few $pA/\mu m^2$ at room temperature and the total contribution may not be very high. However, the junction leakage currents increase exponentially with junction temperature, due to the thermally generated carriers. Hence, keeping the operating temperature low is highly desirable to keep this kind of leakage under control. Another important source of leakage is sub-threshold current of the transistors on the chip. Drain-to-source current does flow in MOSFETs when gate-to-source voltage V_{GS} is less than its threshold voltage V_{th} and even when $V_{GS} = 0$. The closer is the threshold voltage to zero volts, the larger is the leakage current, and the larger is the static power consumption. Standard practice to keep sub-threshold leakage small is to keep V_{th} not below 0.5 V–0.6 V or even higher (~0.75–1.0 V); though, keeping V_{th} high becomes a problem with reduced supply voltage V_{DD}. Transistors fabricated through such technologies which produce sharper turn-off characteristics ($V_{DS} \sim I_D$) are preferable, to reduce sub-threshold current.

The junction leakage current from the storage node is a dominant component among all the possible leakages when boron concentration of p-well is increased and the working temperature rises. However, this increased leakage could not be accounted for with normal reasons, that is, diffusion current and generation–recombination current, as both of these are inversely proportional to the boron concentration of the p-well [6]. The anomalous situation is well characterized by thermionic field emission (TFE) current, which has an exponential relationship to the activation energy at the deep level. Thermionic emission from a deep level is enhanced by the tunneling effect due to the strong field in the depletion region [5]. As this increased electric field is due to the enhancements of substrate doping, junction leakage current can be reduced by lowering the substrate doping to an optimum level, keeping in mind that it will affect the sub-threshold characteristics of the transistor.

Gate tunnel leakage current is the sum of the leakage currents through the dielectrics of the gates of all transistors on the chip. This kind of leakage becomes a problem around 3 nm gate oxide thickness (t_{ox}) and increases after that at a rate of one order with t_{ox} decrement of only 0.2 nm [7]. As the DRAM cells needed a high operating voltage (because of larger V_{th}) for stable memory operation, t_{ox} of standard DRAMs has not been reduced as fast as the rate for static RAMs. A thinner t_{ox} could be used for peripheral circuits as these could operate on low voltage, though normally same gate oxide thickness transistors were preferred and used on the whole chip. However, for eDRAMs, a dual-V_{DD} and dual-t_{ox} device approach has also been adopted for getting higher speeds, and it is expected that even stand-alone DRAMs would use dual-t_{ox} approach [8]. Different solutions have been proposed to reduce the leakage, such as shutting off the supply path by inserting a power switch [9]; however, these are applicable for standby mode and shall be discussed later.

7.3 Power Dissipation in DRAMs

Irrespective of the nature of leakage currents (charge) and their weights, power has to be supplied to the DRAM, which, along with the normal chip-functioning requirement, shall also include the power dissipated due to the leakage currents. For discussing different components of power dissipation, Figure 7.1 shows a simplified chip architecture displaying three major blocks of cell array, row/column decoders, and peripherals. Here the core cell array is comprised of m cells in a row with n number of rows. In the simplified form one word line having m cells is activated and the remaining $m(n - 1)$ cells remain unselected. For a CMOS DRAM total power consumption shall be $P = V_{DD} I_{DD}$, where I_{DD}, the total chip current, shall be the sum of the following components [9]:

$$\text{For selected memory cells} = m \times i_{act} \tag{7.1}$$

$$\text{For inactive memory cells} = m\,(n - 1) \times i_{hld} \tag{7.2}$$

$$\text{For row and column decoders} = (n + m)\,C_{DE}\,V_{int} \times f \tag{7.3}$$

$$\text{For total capacitance } C_{PT} \text{ of CMOS peripherals} = C_{PT} \times V_{int} \times f \tag{7.4}$$

$$\text{For refresh-related circuits and on chip voltage converters} = I_{DCP} \tag{7.5}$$

In Equations (7.1) to (7.5), i_{act} and i_{hld} are the effective currents in active/selected cells and effective data retention current of inactive/nonselected cells, respectively. C_{DE} is the output node capacitance of each decoder, V_{int} is the internal supply voltage, I_{DCP} is the total static or quasi-static current of periphery, and f is the operating frequency being equal to $(1/t_{RC})$ with t_{RC} being the cycle time. The working frequency (or t_{RC}) has considerable impact in changing the weights of different components in the power requirement. At the same time, levels of V_{int} and threshold voltage of transistors V_{th} do affect the power component requirement ratio. For example, for Equation (7.3), the decoder charging current may become negligible when a CMOS NAND decoder is used because only one each of the column and row decoders is operative, hence $(n + m)$ is replaced by 2 only [10]. As far as Equation (7.5) is concerned,

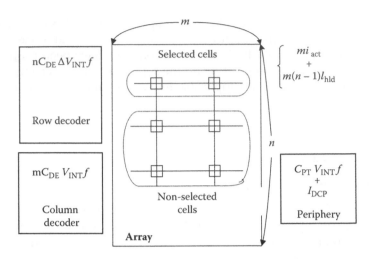

FIGURE 7.1
Important current components for power dissipation in a DRAM chip. (Redrawn from "Trends in Low-Power RAM Circuit Technologies", K. Itoh, K. Sasaki and Y. Nakagome, Proc. IEEE, Vol. 83, No,4, pp. 524–543, 1995.)

I_{DCP} is mainly due to the on-chip voltage converters of different types. With efficient design, its contribution can become small and shall remain constant even with increasing value of f. Ordinarily, the component could have been neglected below gigabit range DRAMs; however, for low-voltage, low-power DRAMs, special efforts are needed to control this component. A discussion of on-chip voltage converters is included in Section 7.7. Even for megabit-density DRAMs, the data retention current given by Equation (7.2) was small at high frequencies in comparison to other components, but not for gigabit DRAMs. Considerable effort is now being made to contain the data retention current; the issue shall be discussed in latter parts of the chapter.

A sense amplifier is a necessity for a DRAM cell for the refreshing process, wherein bit line is charged and discharged with a relatively large swing ΔV_{BL}, with charging current of $(C_{BL} \Delta V_{BL}* f)$, where C_{BL} is the bit line capacitance. At higher operating frequency f, power requirement components depending on it become large and while neglecting smaller components, can be combined and approximated as

$$V_{DD} I_{DD} \approx [m\, C_{BL}\Delta V_{BL} + C_{PT}\, V_{int}]f * V_{DD} \qquad (7.6)$$

Expressions for the DRAM power requirement from Equations (7.1 to 7.6) show obvious ways of reducing it. Reduction in charging capacitances $m\, C_{BL}$ and C_{PT} and lowering of external and internal voltages V_{DD}, V_{int}, and ΔV_{BL} need careful consideration; however, reduction of bit line dissipation charge $(mC_{BL}\, \Delta V_{BL})$ needs special attention as it dominates the total active power.

7.4 Cell Signal Charge

Sufficient signal-to-noise (S/N) ratio must be maintained for reliable DRAM operation as the cell signal is small in magnitude and resides on the floating bit line, which itself is susceptible to noise. Charge transfer ratio between a cell and a bit line is equal to $C_S/(C_S + C_{BL})$, where C_S is the storage cell capacitance. Hence generated signal v_S for the half-V_{DD} precharging scheme is expressed as

$$v_S = (C_S/(C_S + C_{BL}))^* (V_{DD}/2) \approx (C_S/C_{BL}) \Delta V_{BL} = Q_S/C_{BL} \qquad (7.7)$$

where $Q_S = C_S. \Delta V_{BL}$ is the cell signal charge. Hence reduction in C_{BL} is advantageous in two ways; it reduces I_{DD} while increasing signal v_S. Whereas reducing ΔV_{BL} reduces I_{DD} but degrades the signal charge [4].

Magnitude of the signal charge has been reduced considerably with the increase in DRAM capacity, mainly due to reduction in the minimum feature size (F) and supply voltage reduction, and the type of cell, as shown in Table 7.3. It is important to keep Q_S above a critical value for reliable DRAM operations that should make ΔV_{BL} distinct from the memory array noise and soft errors. Value of Q_S is also affected due to higher V_{th}, V_{th} variation, and V_{th} mismatch among devices. Increasing the value of V_{th} is a necessity when the memory capacity is increased, even with lower V_{DD}, because the maximum refresh time, t_{REFmax}, of DRAM must increase with the memory capacity as mentioned in Section 7.6.1.1. Variations in V_{th} change the half-V_{DD} sensing in DRAM and V_{th} mismatch between cross-coupled MOSFETs present in large number of DRAM sense amplifiers also increases with increased memory capacity and decreased F; that further degrades the sensing signal of the DRAMs. Calculated maximum V_{th} mismatch in NMOS used in the DRAM sense amplifiers is shown in Table 7.4; it also shows improvement while using redundancy [8]. The mismatch is doubled with F changing from 0.35 μm to 0.07 μm. Though not a very serious problem for DRAM around 0.1 μm technology, it becomes serious with further small values of F, due to high process

TABLE 7.3

Cell Signal Charge Range Variation with DRAM Density and Cell Type

DRAM density	4 Kb	16 Kb	64 Kb	256 Kb	1 Mb	4 Mb	16 Mb	64 Mb
Cell type	←	Planar		→	←	Trench		→
Cell signal charge (fC) range	900–270	650–140	400–80	150–65	180–75, 120	110	100	80

DRAM density	1 Mb	4 Mb	16 Mb	64 Mb	256 Mb	1 Gb
Cell type	←		Stack			→
Cell signal charge (fC) range	150–80	130–70	75–50	70–32	35–20	24–12

TABLE 7.4

Maximum V_{th} Mismatch without and with Redundancy

DRAM density	64 Mb	128 Mb	256 Mb	512 Mb	1 Gb	2 Gb
F (μm)	0.35	0.25	0.18	0.13	0.1	.07
$\lvert \delta V_{th}\rvert$ Max (mV)						
Without redundancy	17	20	24	28	33	38
With redundancy	12	13	15	17	19	22

sensitivity. One of the main problems created by transistor mismatch (along with the bit lines deviating from design characteristics) is the development of offset noise. An offset-compensating bit line sensing was proposed in [11] but could not be applied to commercial DRAMs because the proposed compensating circuit was too large to fit in the cell pitch. Another offset cancellation sense amplifier, which could shrink to fit in the cell pitch, was also proposed [12]. However, it uses extra chip area and consumes extra time for offset cancellation before the word line activation can cause significant reduction in sensing speed. The large current drawn by differential amplifiers is an additional problem. Direct current sensing technique improves sensing performance by removing timing constraints of column-select line signals, but the low-voltage operation requires a multistage amplification, because of small value of ΔV_{BL} [13]. An offset compensated pre-sensing scheme was employed along with the direct sensing scheme, which effectively reduces the total time for the read operation [14]. However, it requires additional charge pumping for possible leakage of the boost voltage source for equalization and needs at least 3% chip size overhead.

Use of column-redundancy technique seems a better option for overcoming the V_{th} mismatch/offset noise problem as it can replace a certain percentage of sense amplifiers with excessive variation in V_{th} (δV_{th}). For example, if the ratio of spare columns to normal columns is 1/256 (0.4% of area), memory capacity limitation is extended by at least three generations [8]; Table 7.4 shows advantage of using redundancy for reducing V_{th} mismatch. An efficient test method to detect and replace defective sense amplifiers or excessive δV_{th} sense amplifier is also needed. On chip error-checking and correcting (ECC) schemes are almost essential and shall be taken up separately.

7.5 Power Dissipation for Data Retention

While the DRAM is in data retention mode, the refresh operation retains the data. In normal course, *m* cells of a word line are read simultaneously and restored. The process is done for the *n* word lines. Obviously *n* also becomes the number of refresh cycles and current given by Equation (7.6) flows during every cycle. The frequency f at which the current flows is (n/t_{REF}), where t_{REF} is the refresh time of cells in the retention mode and reduces with increasing junction temperature. So, from Equation (7.6), data retention current is given as [4]

$$I_{DD} \sim [m\, C_{BL}\, \Delta V_{BL} + C_{PT}\, V_{int}]\, (n/t_{REF}) + I_{DCP} \tag{7.8}$$

When the DRAM is in the active mode, cell leakage current and junction temperature become maximum. Hence refreshing of the cell is required at a very high rate and t_{REF} becomes $t_{REF\,max}$, which makes it much smaller than t_{REF}. Thus, I_{DCP} becomes relatively small in active mode but becomes larger than AC current component because of small (n/t_{REF}) during refresh-only duration and it also needs minimization.

7.6 Low-Power Schemes in DRAM

The necessity for reduction in power consumption on a DRAM chip cannot be overemphasized. Continuous attempts have been made to decrease it on the face of increasing memory density and capacity. At each generation low-power circuits have been developed and combinations of technology developments have resulted in a downward trend in power consumption [4]. Continued reduction in power dissipation could become possible mainly with the applicability of the following:

1. Partial activation of multi-divided bit line and shared input/output
2. Use of CMOS technology in place of NMOS technology, including half-V_{DD} bit line precharging of the bit line
3. Reduction in the supply voltage (V_{DD}) and use of on-chip voltage converters

7.6.1 Bit Line Capacitance and Its Reduction

Figure 7.2 shows basic architecture of a ($m \times n$) DRAM array comprised of m columns and n rows with k subarrays. The rows have been divided such that each subsection has (n/k) rows, thereby dividing the bit-line capacitance to a manageable value. All the m cells connected to a word line are refreshed simultaneously, and the process is repeated sequentially for the remaining ($n - 1$) rows one at a time, without selecting a bit line. For proper functioning (i.e., to avoid conflict with refreshing process), normal READ/WRITE operation is done during the rest period. However, a successful refresh operation needs to be performed for each row within the maximum allowable refresh

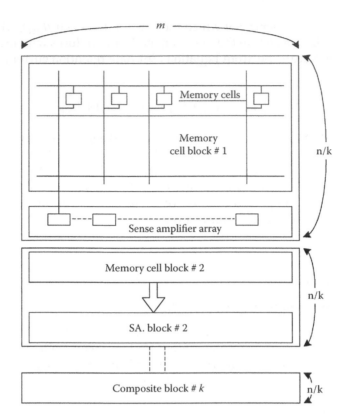

FIGURE 7.2
Basic architecture of an m * n DRAM core with subdivided *k* arrays. (Redrawn from "Trends in Megabit DRAM Circuit Design", K. Itoh, IEEE J. Solid-State Circuits, Vol. 25, No. 3, pp. 778–788, 1990.)

time, $t_{REF\ max}$, which depends on the maximum leakage current of the memory cell; the maximum refresh cycle time is ($t_{REF\ max}/n$).

As the memory capacity ($m*n$) increases, the rows are further divided, that is, k is increased to limit (n/k), or in other words to keep the bit line capacitance C_{BL} within safe bounds. It results in a sharp rise in the number of amplifiers used on the chip, $m*k$; a 1 Gbit chip may require even a million amplifiers. The reason behind the increase in the number of sense amplifiers was that in the initial scheme one sense amplifier at each column decoder division of sub-block was used as shown in Figure 7.2. In another approach, called *shared sense amplifier* scheme, which was in practice at 256 Kbit–4 Mbit DRAM capacity level, two sub-data lines were allowed to share one sense amplifier [15,16]. In this scheme cell signals become double as the bit line capacitance becomes half the conventional scheme. There was reduction in chip size because of reduced number of sense amplifiers and associated circuitry and hence power consumed in sense amplifiers was also reduced. However,

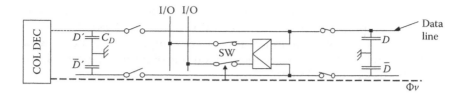

FIGURE 7.3
Memory array organization for showing shared column decoder. (Modified from "An Experimental 1 Mb DRAM with On-Chip Voltage Limiter", K. Itoh et al., ISSCC, pp. 282–283, 1984.)

drawback of power dissipation due to bit line capacitance still remained equal to the conventional case because for the purpose of (dis) charging the bit lines, there was no change. This line of action exhausted its usefulness at 16 Mbit DRAM level with combined use of shared sense amplifier, shared I/O, and shared column decoders; an arrangement is shown in Figure 7.3. Advantage of using shared I/O is that it further halves the multi-divided bit line [17]. Only one sense amplifier is activated along the bit line to achieve its partial activation, which lowers the power dissipation. More reduction in power is achieved by increasing the value of n, as discussed later.

In addition to the division of bit line, a multi-divided word line structure is also available. In hierarchical word line structure, a word line is divided into a few sub-word lines; hence partial activation of word lines becomes possible. Though the architecture has some speed performance drawback, it has great potential for power reduction [18]. The architecture can easily reduce chip power dissipation to half.

7.6.1.1 Refresh Time

Use of partial activation of multi-divided data line reduces its effective charging capacitance, but word line length m is not reduced without increasing t_{REFmax} of the cell. To know the reason for it, the refresh-busy rate γ, expressed in Equation (7.9), needs a bit of consideration [16]:

$$\gamma = t_{RC\ min}/(t_{REF\ max}/n) = (M^* t_{RC\ min})/(m^* t_{REF\ max}) \tag{7.9}$$

Here M and t_{RCmin} are the memory capacity and the minimum cycle time, respectively, and it is better to have smaller γ for having larger active time. Hence, for a given DRAM capacity M and fixed $t_{RC\ min}$, it is necessary to maintain the product of m and $t_{REF\ max}$, which will keep γ constant. For increased capacity of DRAM, $m * t_{REF\ max}$ shall have to be increased proportionally. Normally m and $t_{REF\ max}$ are increased in the same proportion, to have a compromise between cell array power consumption and the cell leakage current. Value of $t_{REF\ max}$ has been doubled at each generation though it is difficult to maintain in practice because of its dependence on the cell leakage current [19]. Hence one of the solutions is to use a new refreshing scheme that employs multi-divided word line [18]. With divided word line, given that m is reduced, maximum power consumption is reduced, which can allow increased $t_{REF\,max}$. More recent methods of improving refresh time shall be discussed in Section 7.8.

7.6.2 CMOS Technology

CMOS technology has shown conclusively that its application reduces power dissipation. For DRAMs, there are different reasons which combine together, like use of half-V_{CC} (array voltage) bit line precharging, and use of CMOS peripheral circuits. For example, as in Equation (7.7), generated cell signal v_S has been derived assuming half-V_{DD} (supply voltage) precharged bit line, has become a standard practice now. Charging and discharging of bit lines need reduced power as the precharge voltage is not through the charging of bit lines to V_{DD} but due to the charge sharing between two halves of the bit line. In addition, peak currents are nearly halved while sensing and precharging. It results in narrow supply lines decreasing parasitic capacitance of the wiring, which not only makes the DRAM cell fast but also reduces the generated noise.

For the NMOS case, all row decoders except the one selected were discharged from the precharge voltage, whereas for the CMOS case only one selected decoder is discharged and all the rest of the decoders remain at the precharge level. This results in a sharp decrease in power dissipation for CMOS decoders and could reach 4% of that in NMOS in normal course [10]. Peak current is also reduced to nearly half that of NMOS case, with consequential advantage.

Clock generators are essential components on DRAM chips. CMOS clock generators have been shown to consume about half the power (of NMOS clock generators) including the dissipation for (dis) charging the load capacitance. CMOS clock generators have smaller node capacitances and occupy nearly half the chip area compared to usual NMOS circuits, on account of simpler circuitry. It all results in lower power consumption [10].

7.6.3 On-Chip Voltage Reduction/Conversion

For NMOS DRAMs normal supply voltage was +12 V in the early 1970s, which soon changed to 5 V for CMOS DRAMs for better reliability and the minimization of hot electron injections through gate oxide. It was essential to standardize it from the point of view of IC manufacturers as well as users. Around 64 Mbit DRAM density levels it became obvious to the manufacturers that continuation of single supply voltage, V_{DD} of 5 V was not feasible on twin counts of excessive power dissipation and the scaling down of the transistors on the chip. Although practically it was also not possible to reduce supply voltage with each generation of reduced minimum feature size of the transistor, it was predicted very early that beyond 1 Mbit, it would be very difficult to reduce power dissipation without further reducing supply voltage. From the early 1990s, with 1.5 V DRAMs design for mobile device applications, sharp reduction in the voltage was foreseen. At the same time at low operating voltage of bit line (which was further reduced because of $V_{CC}/2$ level precharging), it becomes necessary to increase the word line voltage even above V_{CC} for overcoming the threshold offset voltage. A back-bias voltage (changing with the DRAM generations) was also needed on the chip. Conversion to up and down fluctuations of the standard power supply becomes essential for reducing power consumption and reliable operation. A large number of circuits/techniques are now available for this purpose and shall be discussed in Section 7.7.

7.6.4 Signal-to-Noise Ratio

At 64 Kbit DRAM density level even when supply voltage was reduced to 5 V, disturbance due to noise was not prohibitive. Architecture was simple NMOS, full V_{DD} precharge and open bit line with enough spacing between bit lines. With DRAM density moving into Mbit range spacing between bit lines decreased, giving rise to large coupling capacitance. Introduction of CMOS technology led to lower power consumption DRAMs and then to low voltage battery-operated systems, which reduced the amount of cell signal, which was already reduced with ½ V_{CC} precharge scheme. With the generation of larger noises and reduction of cell signal, signal-to-noise ratio started to decrease rapidly. At 4–16 Mbit stage, it became clear that unless measures were taken, differentiation of noise from signal would become difficult. The second half of the 1980s saw a lot of activity in the study of different noises on the DRAM chip and efforts in improving SNR; either cell signal was increased or noises were decreased.

For signal improvement, charge transfer $(\sim C_s/C_{BL})$, was maintained by increasing the storage capacitance C_s, through the use of three-dimensional capacitors like trench or stacked capacitors and later on other advanced versions and techniques like hemispherical grains and use of high permittivity dielectric. However, a limit of 25–30 fF was considered practically safe and the bit line capacitance C_{BL} was tried to be reduced. Half-V_{CC} precharging was also an important scheme for enabling doubled storage capacitance for a fixed electric field across the capacitor insulator [19]. Multidivision of data lines in blocks, as discussed in Section 7.6.1, was found essential in this regard. Reduction in bit line wire width reduced C_{BL}; however, interbit capacitance increased when the DRAM density increased as spacing between bit lines decreased.

With storage capacitance C_s remaining almost constant, and C_{BL} also nearly same within practically reducible limits, only choice was to reduce noises of all types, which continued to increase with rise in DRAM density.

Around 16 Mbit DRAM density, coupling or interference to noise became a major problem in realizing high-speed, high-density DRAMs. There were mainly two types of noises, due to interbit line (BL) coupling capacitance and BL-word line (WL) feedback noise in the access transistor. Inter-BL coupling noise was originating because of signals in adjacent BL pairs and intrapair coupling noise was between true and complementary BLs. A twisted bit line (TBL) scheme in which BLs are divided in four sections with each BL pair twisted two times and a modified twisted bit-line (MTBL) scheme in which BLs are not placed adjacent to their pair BL, in addition to the twisting like that in TBL, eliminated inter-BL coupling noise and claimed to have eliminated intra-BL coupling noise as well [20,21]. The TBL and MTBL schemes were applicable to both the open and folded BL approaches, where BLs are equally divided into four sections and each BL pair is twisted at two of the four boundary points of the sections.

The TBL scheme has been used in some DRAM chips [21,22], but it could not suppress intra-BL coupling noise and required additional chip area for dummy cells and BL twisting. In addition, the signal loss in a 1 Gbit DRAM is estimated to be >30% of total signal. To overcome these limitations multiple TBL has been proposed [23]. It eliminated both the inter-and intra-BL coupling noise but imposed extreme constraints on chip design and required special process and design technologies, and it could not be applied to the folded-bit line arrangement, commonly used in DRAMs. In a different technique, a data-line shielded stacked capacitor (STC) cell was employed in which position of the data line was changed with respect to the storage and plate layer, compared to earlier STC cells; the data-line is shielded by either the storage or the plate layer [24]. Interference noise was reduced below 7% at 2.8 μm data-line pitch, without transposing the pair of data lines. Extra chip area was not needed.

The capacitor-over-bit-line (COB) cell structure suppresses the inter- and intra-BL interferences, but its fabrication process was considered difficult. Storage node contact formation needed small and deep contact node etching along with a large topology difference between the cell array and the peripherals, which was expected to be more pronounced in gigabit range [25]. Another modified twisted bit-line scheme (MTBL) structure was proposed to overcome the problems of earlier TBL schemes and the COB structure. Figure 7.4(a) shows one of the structures proposed [26,27] in which twisting of BLs is done in a different way. Hence, the inter-BL interference is converted into common mode noise and the intra-BL noise is suppressed by a shielding effect. The twisting can be applied to any BL precharging scheme, and compared to conventional folded bit line proposed twisting reduced 50% of inter- and intra-BL noise. The scheme does not need any special layout and/or sense amplifier and can be applied to open BL structure with same benefits. Further improvement in noise reduction is possible by twisting sets of six or eight BLs as shown in Figure 7.4(b) and (c). Resultant reduction in noise by approximately 66% and 75%, respectively, is possible by twisting the sets of six or eight BLs.

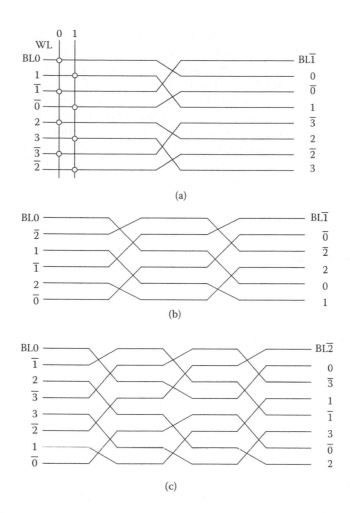

FIGURE 7.4
MTBL configuration having (a) four BLs with single twisting, (b) six BLs with double twisting, and (c) eight BLs with triple twisting. ("Multiple Twisted Data Line Techniques for Multigigabit DRAMs", D.-S. Min and D.W. Langer, IEEE J. Solid State Circ., Vol. 34, pp. 856–865, 1999.)

In high density DRAMs, word line coupling noise also becomes trouble-some mainly because of higher WL voltage than supply voltage and metal WL strapping. Even for the DRAMs with sub-WL driver schemes, with scaled down threshold voltage and supply voltage, WL coupling can become more problematic than BL-coupling noise. Similar to MTBL, a multiple twisted word-line (MTWL) scheme has been effectively used. Four WLs are twisted at the center of the WLs in the same way as it was done for BLs shown in Figure 7.4. It reduces the effective coupling capacitance of adjacent WLs. While twisting, two WLs are separated by two WL pitches, considerably reducing capacitance coupling [27]. A 256 Mbit DRAM was fabricated, with transistor threshold V_{th} = 1.0 V, sheet resistance of 25 and 0.07 Ω/sq. for poly-cide and metal WLs, respectively, and noise-to-signal ratio (NSR) was simu-lated and measured for the cases using (1) conventional WL, (2) proposed MTWL, and (3) combined MTWL and MTBL. Value of NSR increases rapidly as WL pitch is scaled down. For WL pitch of 0.37 μm in case of 1 Gbit DRAM level, NSR is 23% for the convention WL scheme, which when combined with BL coupling noise becomes unacceptably high for a functional DRAM. With MTWL scheme, 35% reduction in NSR is achieved in comparison to the con-ventional WL. For a 256 Mbit DRAM case combinations of MTWL and MTBL schemes achieved a 64% reduction in NSR compared to the conventional WL and TBL implementation.

7.7 On-Chip Voltage Converter Circuits

Operation of DRAMs at low voltage became an extremely important require-ment to save power consumption. It was also essential for maintaining reli-ability of the DRAMs. With continued reduction in oxide film thickness (t_{ox}), excess electrical stress had to be avoided and generation of hot electrons was to be minimized. With successive technology generation, internal chip volt-ages were different and their ratio with supply voltage continued to change. At the same time standardization of supply voltage could not be done too frequently. Only the on-chip voltage converters reconciled the conflicting requirements; hence became essential. For example, supply voltage was con-verted on chip to nearly 2.5 V for 64 Mbit DRAM having t_{ox} = 10 nm, which was further reduced to 2–2.5 V and 1.5–1.8 V for 256 Mbit and 1 Gbit level DRAMs [28]. Another extremely important reason for further reduction in operating voltage was the DRAM demand for battery operated hand-held mobiles and digital devices. On the other hand, voltage boosting was needed for the word line, and other on-chip voltage levels were needed for half-V_{CC} operation of bit line, back-bias voltage, and so on.

Probably the first circuit of an on-chip supply voltage converter for DRAMs was given by Mano and others in 1983 for immunizing MOS transistors from hot carrier injections [29]. A 5 V supply was converted to 3 V for cell array circuitry and the rest of the peripheral and interface worked on 5 V. Soon a voltage converter was given by Itoh and others as shown in Figure 7.5, in which limiter block converted V_{DD} to V_L (nominal value 3.7 V) and the word line voltage was ($V_L + V_{th}$). Bit line was precharged to V_L (not $V_{CC}/2$) and precharge clock also received ($V_L + V_{th}$). A charge pump was used to gener-ate $V_{CC} + 2V_{th}$ for the voltage limiter. For on-chip voltage down converter (VDC), it is extremely important to provide regulated and accurate voltage when large DRAM array current changes from zero to peak and vice-versa. The VDC should also be suitable against changes in external power supply

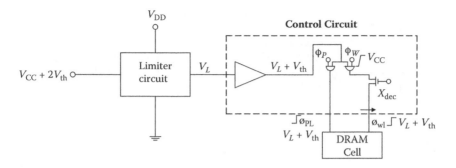

FIGURE 7.5
A voltage limiter arrangement for a DRAM cell. (Modified from "An Experimental 1 Mb DRAM with On-Chip Voltage Limiter", K. Itoh et al., ISSCC, pp. 282–283, 1984.)

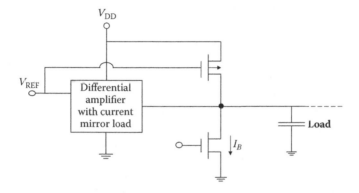

FIGURE 7.6
A typical voltage down converter. (Redrawn from "Trends in Low-Power RAM Circuit Technologies", K. Itoh, K. Sasaki and Y. Nakagome, Proc. IEEE, Vol. 83, pp. 524–543, 1995.)

and temperature. At the same time, it is also important to have a provision of on-chip burn-in capability. Figure 7.6 shows an early VDC, which basically consisted of a current mirror differential amplifier and common source output transistor [30] and converts 5 volts V_{DD} to a stable 3.3 V. Gate width of the output transistor has to be very large and the gate voltage has to respond quickly when the output goes low; VDC needs larger value of amplifier current. The bias current is needed to minimize output voltage deviation when the load current becomes nearly zero. An important requirement of the on-chip voltage converter circuits is that the reference voltage V_{REF} must be accurate and stable with supply voltage operating temperature variation, and if there is variation in transistor parameter due to the limitation of fabrication processes. Band-gap V_{REF} generator is considered to be a good choice and in earlier reports bipolar transistors were deployed for it [31]; later on, CMOS V_{REF} generators making use of threshold voltage differences were proposed [32].

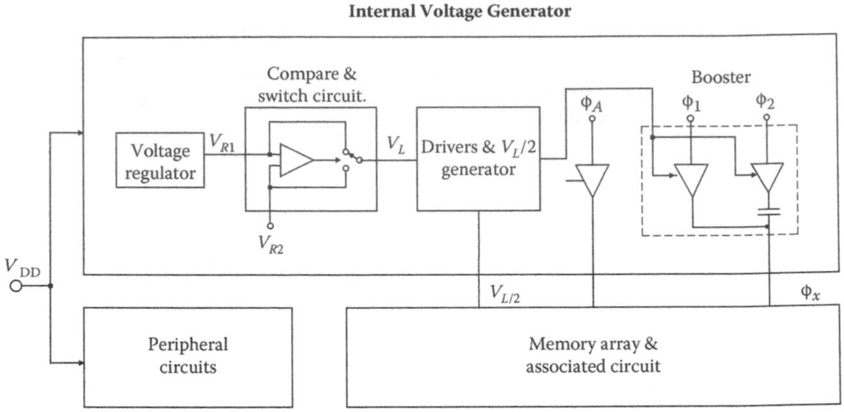

Internal Voltage Generator

FIGURE 7.7
An internal voltage generating scheme. (Redrawn from "Dual-Operating-Voltage Scheme for a Single 5V 16 Mbit DRAM", M. Horiguchi et al., IEEE J. Solid State Circ., Vol. 23, pp. 612–617, 1988.)

At 16 Mbit DRAM level a dual-operating-voltage scheme for a single 5 V supply was suggested in which memory array operating voltage was chosen to be 3.3 V since up to that stage of density level internal voltage converters were providing satisfactory performance. Moreover, a lower voltage <3.3 V at that stage would have required a larger storage capacitance for obtaining reliable cell signal and >3.3 V would have made excessive electric field across the memory cell capacitor [33]. Voltage converters used large standby current and dual operating-voltage was shown to be the best choice in terms of speed and reliability of the devices used; and also reduced power consumption and kept cell signal charge at a suitable level. Proposed internal voltage generator is shown in block form in Figure 7.7 along with different voltage levels. In the dual operating voltage scheme, there is an inherent problem of racing due to mismatch between memory array and peripherals. Compare-and-switch circuit is designed to overcome this problem by raising memory array operating voltage when supply voltage is considerably higher than 5 V. To achieve sufficient drivability and voltage accuracy, a driver using a simple differential amplifier with a PMOS load is found to be a suitable choice.

Because of the importance of on-chip voltage limiters, several designs, including the just discussed dual-operating voltage scheme, were reported. However, all of them had some common deficiencies like imprecise voltage regulation and effect of threshold voltage variation; both were due to variations in fabrication processes. In addition, effects of voltage bounce and feedback stability were also not attended to [31,33–36], and these internal voltage generators depended on the threshold voltage of MOSFETs (with inherent variations) [34,35], which had large dependence on the external supply voltage, operating temperature, or else depended on a band-gap reference [31,36], which used bipolar transistor on a CMOS chip, with consequential limitations including fabrication complexities. To overcome the deficiencies

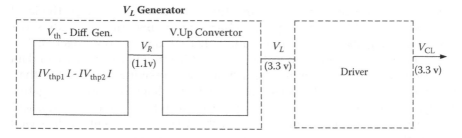

FIGURE 7.8
A voltage limiter in block schematic form. (Redrawn from "A Tunable CMOS-DRAM Voltage Limiter with Stabilized Feedback Amplifier", M. Horiguchi et al., IEEE J. Solid State Circ., Vol. 25, no. 5, pp. 1129–1135, 1990.)

and some process limitations, a CMOS DRAM voltage limiter comprising a precise internal voltage generator and a stabilized driver, as shown in block form in Figure 7.8 was given by Horiguchi and others [32]. In this scheme the internal voltage generator comprising a PMOS-V_{th} difference generator with a voltage-up converter and fuse trimmings, was preferred. Generated voltage ($|V_{thp1}| - |V_{thp2}|$) maintains stability, and it is not affected because of bulk bias voltage (V_{BB}) fluctuations which (dis) charges large bit line capacitance and is not affected by the noise of the main chip supply voltage.

Boosted sense grounded (BSG) and negative word line (NWL) schemes are used to extend the data retention time of the DRAMs. In the BSG scheme, *low* level of the bit line is slightly boosted to suppress sub-threshold current of the unselected memory cell word line transistor in the active memory array. In the BSG scheme suggested by Akasura and others [37], reference voltage V_{REF} was set at 0.5 V as shown in Figure 7.9. Large NMOS M_2 turns on at the beginning of sensing, which suppresses unwanted rise in boost sense ground voltage (VBSG) due to sensing current in bit line. In the standby

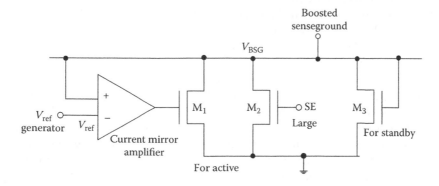

FIGURE 7.9
Circuit diagram for boosted sense-ground scheme; SE enables the beginning of sensing. (Adapted from "A 34 ns 256 Mb DRAM with Boosted Sense-Ground Scheme," M. Akasura et al., Proc. ISSCC, pp. 140–141, 1994.)

mode, current mirror amplifier is inactive and NMOS M_3 clamps the level of VBSG. Ground line for supplying charge to the BSG line is totally separate from the global ground lines, and other circuits are not affected. The BSG scheme reduces junction leakage current helping in the increase of data retention time. However, in the BSG scheme, bit line swing becomes less than V_{CC} which makes it a little unsuitable for low voltage operation. Negative voltage word line is another technique for the refresh time improvement [28] through the suppression of sub-threshold leakage current. As it is well known that junction leakage current under the storage node and the sub-threshold leakage current affect the refresh characteristics of the DRAMs, a small back-bias voltage V_{BB} is used to reduce electric field between the storage node and the p-well under the memory cell. However, small V_{BB} level reduces the threshold voltage V_{th} of the access transistor resulting in increased sub-threshold leakage current. Negative-voltage word line (NWL) technique suppresses the just mentioned increase of sub-threshold current. In a conventional scheme WL is controlled with $V_{BB} = -2$ V; however, an improvement of 2.5- to 3-fold in the refresh time was made possible by using both V_{BB} and low WL voltage level of -0.5 V. Since low values of V_{th} can be used, another advantage of the NWL technique is that is allows one to use a lower level than the conventional high level of the word lines. This suggests that even at 1.2 V, cell voltage level can be obtained in full using a boosted voltage, V_{pp}, $(= 2V_{CC})$, and it can be easily generated by a usual V_{PP} generator. Figures 7.10(a) and (b) illustrate some voltage levels in conventional, BSG, and NWL schemes.

The NWL scheme requires highly regulated and correct *high* levels of the word line and back-bias voltage V_{BB}; otherwise presence of noise and/or variation in chip power supply affect access transistor threshold voltage causing serious signal loss due to increased sub-threshold current. Hence, a precise on-chip voltage generator is a necessity especially for gigabit level DRAMs. Even a 0.1 V decrease in gate source voltage of the cell access transistor increases the sub-threshold leakage current by an order. Charge pump generators shown in block form in Figure 7.11(a) are used in conventional procedure for the provision of *high* and *low* WL voltages in a DRAM. The output generally has a ripple of nearly 0.2 V, which is reduced by combining series pass regulator with charge pump regulator as shown in Figure 7.11(b). The series pass regulators need accurate WL offset voltages, which are made available by combining band gap reference with a differential amplifier and current mirror offset voltage generator, where accurate offset voltages were produced using mirror current as illustrated in ref. [39]. Band gap reference voltage generator can be selected among many available circuits; however, one MOSFET is preferred over a BiCMOS for high-density DRAMs.

It is well established that on-chip supply voltage conversion has become essential and a number of schemes are available. However these conventional methods used as much as half to two-thirds of total chip power at 1 Gbit/4 Gbit DRAMs. As an alternative, two internal circuits were connected

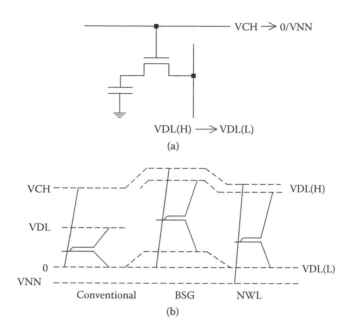

FIGURE 7.10
(a) Basic cell, and (b) voltage levels, in conventional, BSG, and NWL schemes. (Adapted from "A Precise On-Chip Voltage Generator for a Giga-Scale DRAM with a Negative Word-Line Scheme", H. Tanaka et al., Symp. VLSI Circ. Dig. Tech. Papers, pp. 94–95, 1998.)

in series between supply rails [38]. Both the internal circuits had identical DRAMs with same core and peripherals except input/output buffers and both circuits operated with the same clock. As a result, the AC current waveform in both circuits was the same and more importantly voltage across each DRAM was fixed at $V_{DD}/2$ without using any conversion process. The technique was successfully tested using two 4 Mbit DRAMs.

7.7.1 Back-Bias Generator

Substrate requires enough negative bias when chip is active. Obviously power dissipation of the bias generator should be low, and it should be able to provide adequate voltage level at greater efficiency. As the working voltage of the DRAM chip decreases, back-bias voltage level also has to go down; its value (range) is decided based on the following. Back-bias voltage (V_{BB}) level depends on the value of the threshold voltage of the access transistor in the DRAM whose upper and lower limits are decided by the boosted word line voltage and suppression of the sub-threshold leakage current, respectively. For the upper limit, the word line voltage needs to be ($V_{CC} + 1.2\ V_{th}$), where V_{th} is the increased threshold voltage of the access transistor with its source at V_{CC} level. [40]; effective back-bias voltage becomes ($V_{CC} + |V_{BB}|$). For a practically available word line voltage of ~1.7 V_{CC}, an upper limit of

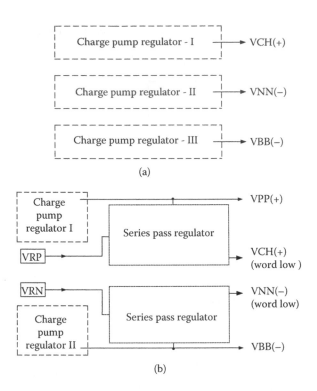

(a)

(b)

FIGURE 7.11
Schematic of a voltage regulator for negative word line scheme. (a) Charge pump regulator (conventional), and (b) hybrid regulator. (Adapted from "A Precise On-Chip Voltage Generator for a Giga-Scale DRAM with a Negative Word-Line Scheme", H. Tanaka et al., Symp. VLSI Circ. Dig. Tech. Papers, pp. 94–95, 1998.)

V_{th} is 0.88 V at V_{CC} = 1.5 V. As mentioned, the sub-threshold leakage current decides the minimum required value of the threshold voltage. The accepted value of memory cell capacitance of 30 fF (~25 fF has also been considered safe by a large number of reports) and the data hold time of 100 ms, V_{tho} (with source connected to ground) should be larger than 0.6 V for keeping sub-threshold leakage current less than 11 fA per cell. It requires that V_{BB} of lower than –1.0 V is essential in a 1.5 V DRAM [40].

Few circuits are available for on-chip back-bias generation. Figures 7.12(a) and (b) show a conventional pumping circuit (CPC) and hybrid pumping circuit (HPC), respectively [40]. In CPC, which uses two PMOSs, the V_{BB} could not be pumped lower than $|V_{thp}| - V_{CC}$, whereas in HPC it could reach $(-V_{CC})$. Working of the HPC can be understood by following the clock when it is low and then high. When CLK is *low*, node voltage N5 is clamped to ground level. When CLK changes to *high*, node N4 rises to $|V_{thp}|$ and by capacitive coupling node N5 voltage level and V_{BB} become $-V_{CC}$. In HPC no threshold voltage is lost while generating V_{BB}. An important precaution for the HPC is that the NMOS used in the pumping circuit needs to be fabricated in a

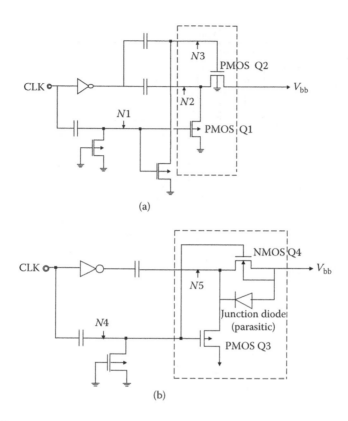

FIGURE 7.12
Back-bias generator: (a) conventional circuit, and (b) hybrid pumping circuit (HPC). ("An Efficient Back-Bias Generator with Hybrid Pumping Circuit for 1.5V DRAMs", Y. Tsukikawa, el.al., Symp. VLSI Circ. Dig. Tech. Papers, pp. 85–86, 1993.)

triple-well structure to avoid the minority carrier injection, which could destroy stored data.

7.7.2 Voltage Limiting Schemes

On-chip voltage limiters are extremely important for reducing DRAM power dissipation and the enhancement of device reliability. Utility of the voltage limiter is considerably improved when it also generates precise voltage during burn-in and stress conditions are applied automatically. A DRAM voltage limiter with a burn-in test mode was given by Horiguchi and others as shown in Figure 7.13(a) [41]. It is based on the simple arrangement where the DRAM core (circuit L_1) operates on the internally generated voltage V_L, and the peripherals (circuit L_2) and the voltage limiter operate on the external supply voltage V_{DD}. In a burn-in test V_L and V_{DD} are raised and a number of schemes are available; however, these have focused mainly on voltage stability under normal operation [41]. A dual-regulator dual-trimmer scheme

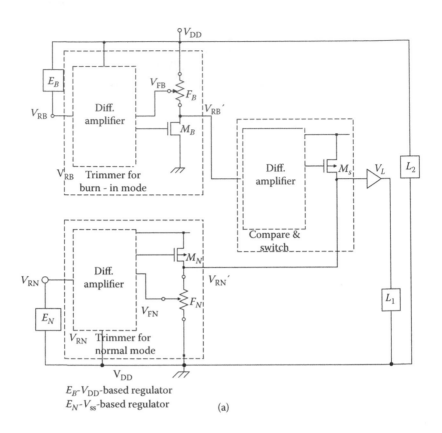

FIGURE 7.13
Voltage limiter with symmetrical dual trimmers using decoding scheme having (a) trimmers and compare-switch circuit, and (b) details of negative feedback circuit. (Redrawn from "Dual-Regilator Dual-Decoding-Trimmer DRAM Voltage Limiter for Burn-in Test", M. Horiguchi et al., IEEE J. Solid State Circ., Vol. 26, pp. 1544–1549, 1991.)

shown in Figure 7.13(a) is one such practical scheme in which not only a precise high voltage for burn-in test is available, but it also maintains a constant limited voltage under normal condition. Here the compare and switch outputs exceed the two regulated and trimmed voltages V_{RN} and V_{RS}. The regulator E_B keeps ($V_{DD} - V_{RB}$) constant, independent of temperature—an important condition for proper circuit operation. Accurate burn-in voltage is obtained by simply raising V_{DD}. In addition, the two sets of trimmers T_B and T_N reduce any deviations in the generated voltage due to the process variations, especially the change in threshold voltage V_{th}. Circuit realization of V_{RB} regulator and V_{RN} regulators used biasing circuit which employed PMOS threshold voltage difference scheme; otherwise any other suitable circuit can also be used. Each trimmer block comprises a differential amplifier, an output transistor, and a variable negative feedback circuit (F_B or F_N), some details of which are shown in Figure 7.13(b). Deviations in burn-in

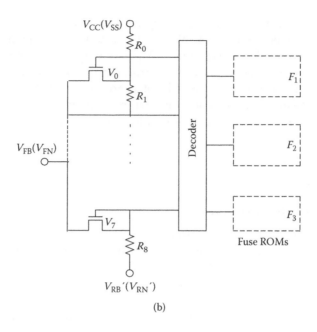

(b)

FIGURE 7.13 (continued)

voltage as well as in the normal operating voltage were reduced to ± 0.13 V while using only six fuse ROMs. Main limitation of the reference generator is the use of 100–1000 kΩ resistors and large amount of standby current because of the presence of quite a few differential amplifiers and DC current paths in the voltage divider, making it unsuitable for low power battery-operated DRAMs. A dynamic reference voltage generator that consumes considerably less current has been proposed, as shown in Figure 7.14 [42]. Threshold voltage difference ΔV_{th} and resistance R_R determine the reference

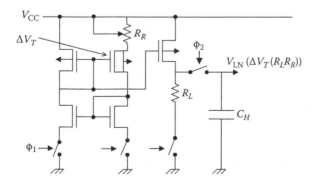

FIGURE 7.14
A dynamic reference voltage generator. (Modified from H. Tanaka et al., "Sub-1-µA Dynamic Reference Voltage Generator for Battery-Operated DRAM", Symp. VLSI Circ. Dig. Tech. Papers, pp. 87–88, 1993.)

current I_R, which is mirrored to the output node and flows through R_L, and output voltage depends only on ΔV_{th} and resistance ratio. Hence, accurate output voltage is available even when ΔV_{th} may vary due to fabrication process limitations, as polysilicon fuses accurately trim the resistance R_R. In the experimental verification, pulse widths of φ_1 and φ_2 were taken as 200 ns and 100 ns, respectively, and the generator current could be reduced to less than 1 μA, making it suitable for battery-operated DRAMs [42].

7.8 Refresh Time Extension

Improvement of DRAM cell retention time is a critical factor in realizing high-density DRAMs, since it needs to be doubled with every generation and chances of failure of weak cells rise. At the same time the main source of trouble is that leakage currents continue to grow with reduction in minimum feature size. Several approaches have been followed for successful extension in refresh time. Reduction in leakage currents is very important in this direction and is a matter of discussion in Section 7.7. Another kind of scheme depended on the knowledge that data retention capability of the cell depended on operating temperature and voltage, and random variations in parameters which are inherent in fabrication processes. Quite a few schemes were given which tried to set optimum internal refresh periods by using temperature detectors [43–45] and internal voltage converters [45], or through measuring voltage degradation of weak cells [46]; however, DRAM design had to continue to improve their speed, which resulted in increased power dissipation as well. At the same time, the JEDEC standards adopted an extended temperature range of (85 to 95°C) from the earlier temperature range of 0 to 85°C, and it is now operating in servers [47,48]. DRAM data retention time became half of that in this extended temperature range compared to the standard temperature range, which was 64 ms. Corresponding to 64 ms refresh interval (t_{REFI}), the interval at which refresh command must be sent to each DRAM (from an internal counter to the next part of the chip as per JEDEC standard) was 7.8 μs (64 ms divided by number of rows) for 256 Mbit DDR2 DRAM. DRAM density requires doubling of refresh time (t_{REF}) with every generation using conventional methods. The problem was circumvented in some cases by doing simultaneous refreshing of a number of rows with single command [49], but it requires larger charging current and hence delay in terms of refresh cycle time (t_{RFC}), the amount of time that each refresh takes. For the current in-use DDR3 4 Gbit DRAM, it requires t_{RFC} of 300 ns, and it may go to 350 ns for 8 Gbit [50].

Error control codes have also been used to minimize errors along with conventional techniques for refresh time extension. The combination of longer refresh time along with shorter refresh time for a few rows has also been used [51,52]. The major drawback of all such approaches was that multiple refreshing was done on a row basis and with cumbersome measurement methods. Probably, the main reason was the failure to identify the weak cell efficiently and develop a systematic approach. A multiple refresh scheme in terms of an algorithm, which depended on error correction, was also proposed for optimal selection of multiple refreshing periods [53]. Boosted sense ground (BSG) schemes discussed in Section 7.7 have been effectively used to double the refresh period [37].

Instead of a row-based refreshing method, a novel method of refreshing on a block basis was given in which a large number of multiple refresh periods match the required refresh period of the blocks as closely as possible [54]. Consider a 16-cell memory as shown in Figure 7.15(a) where data retention time of each cell is written in each square (cell). In conventional refreshing t_{REF} must not exceed the minimum retention time of all cells, that is, 2. If the array of Figure 7.15(a) is broken into a 4 cell block as shown in Figure 7.15(b) then four different refreshing periods can be used; 8 for row 1, 2 for rows 2 and 4, and 6 for row 3. If the array is further divided in blocks of two cells as shown in Figure 7.15(c), blocks may be refreshed with t_{REF} of 2, 4, 6, 7, 8, 9, 12, and 15. DRAM architecture of the block-based multiple period is also made available for the purpose of generation of refresh signal, bit selection, and multiple period refreshing. A polynomial time algorithm is provided that computes set of optimal refresh periods for the selected blocks. As the selected blocks comprise cells having closely valued t_{REF}, these refresh periods are obtained during post-fabrication testing of memory array. Moreover,

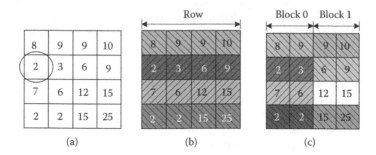

(a)   (b)   (c)

FIGURE 7.15
Memory with 16 cells: (a) data retention times, (b) necessary refresh period using blocks with 4-cell, and (c) 2-cells. (Redrawn from "Block-Based Multiperiod Dynamic Memory Design for Low Data-Retention Power", J. Kim and M. Papaefthymiou, IEEE Trans. on VLSI Systems, pp. 1006–1018, 2003.)

the method needs nearly 6% chip area overhead and was not tested at higher operating temperature.

The JEDEC DDR$_x$ standards have allowed for spacing in refresh operations. If a small number of t_{REF} periods are delayed, data are not lost provided that all cells have been refreshed satisfactorily. This *deferral* of refresh count is eight for the DDR3 standard [55]. A memory controller receives read and write instructions from the CPU and places in the input queue and then moves to an available space. The memory controller also has to execute refresh operation. The selection between various operations including refreshing is done by the memory scheduler. A significant report has investigated for optimum priority scheduling read/write and refresh [55]. Scheduling of refresh command has not been used often. Usually the refresh scheduling algorithm asked for refresh operation as soon as t_{REF} period expired as t_{RFC} was not affected much, and at the same time it required simple hardware logic control. In a much better scheme, which has been given the name *defer until empty* (DUE), refresh operations are not selected over read/write until refresh deferral count reaches seven refresh operations. However, even these kinds of designs are not good enough in containing refresh penalties. For example, take the case of low-multilevel parallelism (MLP) workloads. There are many time slots when the memory controller bank queues are empty but refresh scheduler will continue to provide refreshing when t_{REF} counter expires; along with long t_{RFC}, it results in large penalties by the memory controller. Even with high DRAM bus utilization refresh penalties accrue due to scheduler inefficiencies [55]. The *elastic refresh* algorithm has been proposed, which betters the other schemes including the DUE approach, in which an additional period of time is added for waiting for the rank to remain idle before the refresh command is given. A detailed study was undertaken for evaluating the elastic refresh scheme and was compared with best-known algorithm DUE.

The temperature and supply voltage dependence of t_{REF} and the identification of weaker cells are helpful in devising a mechanism of extending refreshing duration. An important study has shown another cause on which t_{REF} depends. For analysis purposes, three chips were fabricated in 54 nm technology with different types of cell structures like recessed and buried word line, and variation of t_{REF} was studied for different data patterns like all cell *high* and all cells *low* with only one cell *high*. It was observed that t_{REF} is determined not only by the cell leakage but also by the bit line sense amplifier (BL SA) offset. As data patterns determine the interference between bit line and word line, this also affects t_{REF}. Dependence of t_{REF} on data patterns is found by studying relation between cell leakage characteristics and its own offset variation [56]. The chip having smallest variation in t_{REF} shows the best offset variation and the worst leakage, whereas the chip having largest variation in t_{REF} shows the best cell leakage characteristics. The study concluded that it is only important to improve the BL SA offset present due to data patterns, and it will reduce t_{REF} variation.

7.9 Sub-Threshold Current Reduction

Sub-threshold current in an NMOS is given as

$$I_{sub} = I_s \, e^{\dfrac{(V_{GS} - V_{tho} - \gamma\left(\sqrt{|(-2)\varphi_F + V_{SB}|} - \sqrt{|(2)\varphi_F|}\right) + \lambda V_{DS})}{nkT/q}} \left(1 - e^{\dfrac{-qV_{DS}}{kT}}\right) \quad (7.9)$$

where V_{tho} is the threshold voltage when source and bulk substrate are at same potential, and γ is the body effect coefficient. V_{GS}, V_{BS}, and V_{DS} are the respective device terminal voltages, λ is the drain-induced barrier lowering (DIBL) factor, k is the Boltzmann constant, T is the absolute temperature, and q is the electron charge. Here, I_s is the drain current coefficient, and n is the sub-threshold swing parameter, which is related with slope factor S, a quality metric of sub-threshold region as to how much reduction in V_{GS} produces an order of magnitude reduction in the sub-threshold current. It is given as

$$S = \ln 10 \, dV_{GS}/d\,l_n\,(I_{sub})$$

$$= \frac{nkT}{q} \Bigg| \, n10 \sim \frac{kT}{q} l_n 10 * (1 + C_D/C_i) \qquad (7.10)$$

where C_D is the depletion layer capacitance, and C_i is the insulator capacitance of the MOSFET.

In low-voltage DRAMs, if threshold voltage V_{th} is also scaled down (though not in proportion to the supply voltage), it will result in increased I_{sub}. The most effective way of overcoming this problem of increased I_{sub} is to keep V_{th} high as before in the DRAM cells and peripheral circuits, both in the active mode and in the standby mode. Depending upon the DRAM topology and its circuit operation, the importance of reducing I_{sub} in different modules varies and shall be discussed accordingly. During fabrication, value of V_{th} is increased mostly by increasing the doping level of the MOSFET substrate, but application of reverse bias(es) is the most effective way even after fabrication and it can easily be applied to only the selected low V_{th} MOSFET circuits. Idea of affecting I_{sub} through device terminal voltage reverse biasing is obvious from Equation (7.9) and almost all suggested techniques fall in this broad category. A fine description of the classification of reverse biasing techniques and the respective leakage reduction efficiencies is available in ref. [8].

As mentioned earlier, I_{sub} has to be reduced in the DRAM core, and in the peripheral circuits in the active and sleep modes. Different techniques are available, some of which are specific for active or for sleep mode and some are applicable in general. For better understanding, Section 7.10 describes only those techniques which are better suited for peripheral circuits.

7.10 Multithreshold Voltage CMOS Schemes

Multithreshold voltage CMOS technology scheme (MTCMOS) proposed by Mutoh and others not only reduces standby current, but it also obtains high-speed performance at low supply voltage using low threshold MOSFETs [57]. Figure 7.16 shows basic MTCMOS scheme in which all logic gates use low V_{th} (0.2–0.3 V) MOSFETs, and its terminals are not directly connected to supply rails but to virtual supply rails. High threshold (0.5–0.6 V) MOSFETs M_1 and M_2 link the actual and virtual power lines, and they act as sleep mode control transistors through select signals SL and $\overline{SL}$. Sleep control transistors are relatively wider with low on-resistance value; hence, when asserted, virtual supply lines function as real. However, larger I_{sub} of low-V_{th} logic is almost completely suppressed by M_1 and M_2 having large V_{th}. Performance of the MTCMOS circuit depends on the size of the control transistors and the capacitances of the virtual power supply lines. Voltage deviations in supply lines due to the switching of the logic gates are suppressed by having larger values of supply line capacitances. It is claimed that MTCMOS operates almost as fast as low V_{th} logic and at supply voltage of 1.0 V its delay time is nearly 70% less than that for the conventional logic gate with normally high V_{th}.

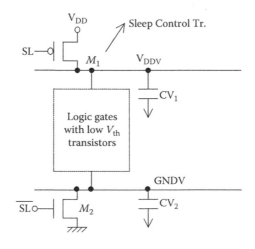

FIGURE 7.16
Basics of an MTCMOS scheme. ("1-V Power Supply High-Speed Digital Circuit Technology wth Multithreshold-Voltage CMOS", S. Mutoh et al., IEEE J. Solid-State Circuits, Vol. 30, pp. 847–854, 1995.)

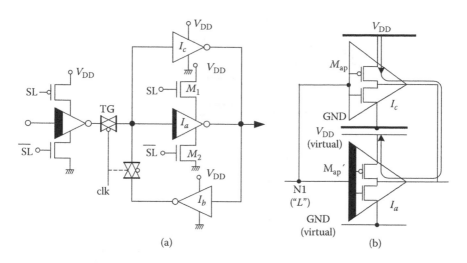

FIGURE 7.17
(a) MTCMOS latch circuit, and (b) problematic leakage current path.

Since latches and flip-flops are memory elements and should retain their data even in sleep mode they need special attention in MTCMOS. An MTCMOS latch circuit is shown in Figure 7.17(a); it has conventional inverters I_b and I_c, which have high V_{th} and are connected directly to the actual supply rails. Data is retained by the latch even in sleep mode because the latch path consisting of inverters I_b and I_c continues to receives power. Inverter I_a and the transmission gate TG (both with low V_{th} MOSFETs) form the forward path and provide high-speed operation. In this path, control transistor M_1 and M_2 (high V_{th}) are also included for maintaining proper operation which is better understood through Figure 7.17(b). Since node N_1 shall be in *low* state during sleep mode, real and virtual rails would be connected through on PMOS transistors Map and Map', thereby increasing current in the sleep mode. Therefore, inclusion of the control transistors M_1 and M_2 becomes essential.

The MTCMOS circuit technique is an excellent method so that logic can work at low voltage, however, use of high V_{th} MOSFETs in the critical path for holding data creates bottlenecks. So, special circuits called *balloon circuits* have been used which help in preserving the state of nodes even during the sleep mode, without having high V_{th} MOSFETs in the critical paths [58]. The balloon circuit, as shown in Figure 7.18(a), is connected to the real supply directly, but it does not operate in the active period; hence, there is no requirement on its speed. Balloon circuits are therefore realized with normal high V_{th}, minimum size MOSFETs with low standby power, and chip area penalty. Signals B_1 and B_2, as shown in Figure 7.18(b), control the transmission gates (TGs), and their states decide the active mode or sleep mode period. In the active mode, node A does not float because leakage current of the low V_{th} TG flows and the balloon is not in the critical path, which avoids the bottleneck of the earlier MTCMOS technique.

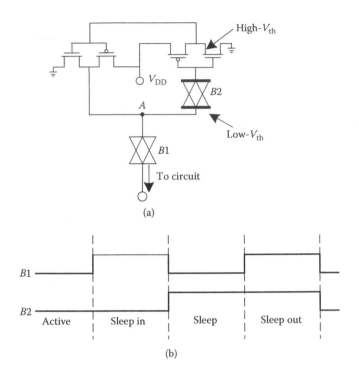

FIGURE 7.18
(a) Typical balloon circuit and (b) sleep operation of the balloon circuit. (Modified from "A 1-V high-speed MTCMOS circuit scheme for power-down applications", S. Shigematsu et al., Symp. VLSI Circuit, Dig. Tech. Papers, pp. 125–126, 1995.)

Use of balloon circuits improves the operation of MTCMOS circuits; however, it costs chip area and needs a timing control scheme for switching of operating modes. Another data holding circuit is proposed that uses intermittent power supply (IPS) scheme with low V_{th} transistors without any increase of leakage currents [59]. In this scheme, virtual power lines are cut off from the mains and connected only intermittently during sleep mode; hence, no extra state holding circuit is required. In an experimental latch, fabricated in 0.35 μm CMOS technology, saving of 30% chip area, 10% reduction in delay, and 10% reduction in active power consumption were obtained in comparison to the conventional MTCMOS case.

For battery-based application, along with performance (speed), energy consumption is critical. Hence while developing a dual V_{th} MOSFET process at 0.18 μm level, energy-delay product (EDP) was extracted from measured data. Investigations done with V_{DD} and V_{th} variations gave following important inferences: (1) for minimum EDP, optimum $(V_{th}/V_{DD}) \sim (120\,\text{mV}/300\,\text{mV})$, (2) optimum V_{th} is a logarithmic function of the activity factor of the application, and (3) dual V_{th} process gives good results with high sleep mode durations operation cases [60].

In general, leakage-sensitive circuits like dynamic NOR gates, which require level keeper for compensating loss of charge, are not preferred in DRAMs [61]. However decoders of modern CMOS DRAMs do use dynamic NAND gates for rows. With technology at 0.1 μm and beyond, use of dynamic circuit was bound to increase, even when dynamic circuits are significantly worse in terms of I_{sub} in comparison to the static circuits. Consequently, low V_{th} devices, as normally used in static MTCMOS circuits, cannot be used for reducing delay in critical paths. In spite of the limitation, dynamic circuits became a necessity in high-speed submicron process technologies as they were faster by at least 25% in comparison to the static counterparts [62]. A study at 0.25 μm technology optimized for 1.8 V V_{DD} observed that in dynamic circuits with $V_{th} > 300$ mV, I_{sub} can be maintained at an acceptable level of nearly 1 nA/μm (at 30°C), but for $V_{th} < 300$ mV, I_{sub} is substantially higher than 1 nA/μm, and it is independent of channel length. Moreover, I_{sub} multiplies steeply with rise in working temperature and a functionality problem is created in dynamic circuits. A solution was proposed that dynamic substrate bias be used to raise the threshold voltage in standby mode, which would reduce I_{sub} several orders of magnitude. Performance is not degraded since source-to-well reverse bias is not applied in active mode, retaining low values for V_{th} [62].

A high V_{th} significantly reduces I_{sub} of a MOSFET; however, it also results in its higher equivalent on-resistance deteriorating propagation delay. Empirically, V_{th} is kept around 20% of V_{DD} supply voltage for proper maintenance of balance in I_{sub} and propagation delay [63], but for low voltage DRAMs, V_{th} (20% of V_{DD}) becomes too small. The problem has been minimized by using varieties of circuits with dual-threshold MOSFETs. In one such scheme, low V_{th} devices are used in critical path(s) for achieving high performance, with the lower limiting value of V_{th} depending on noise margin, whereas high V_{th} devices are used in noncritical paths and the value of the V_{th} may range between low V_{th} to 0.5 V_{DD}. However, a major concern is in the distinction and selection of paths that would use higher V_{th} devices, as it may convert some of the noncritical paths to critical ones; it also depends heavily on the value of higher V_{th} used. If the value of higher V_{th} is slightly more than low V_{th}, a large number of MOSFETs can be assigned this value without turning the path into critical one but improvement in I_{sub} shall be small. On the contrary, with too large a V_{th}, only few paths can use such devices, though with improved I_{sub}. Therefore, it needs a solution for optimum value of V_{th}, and a levelization back-tracing algorithm is one such attempt which selects and assigns optimal high-V_{th} [64]. The algorithm begins with initializing a circuit with one value of low V_{th}, then high V_{th} is assigned to some devices lying in noncritical paths (within the constraints of certain performance limits). This assignment is done by "back-tracing of the slack of each node level by level." Here slack means possible slowdown of a gate without affecting the overall performance of the circuit. The value of V_{th} is increased till such time that slack becomes zero. Use of the algorithm on certain ISCAS benchmark circuits has shown reduction in active and standby leakage power even up to 80% [64].

7.10.1 Stacking Effect and Leakage Reduction

Insertion of an extra transistor between the supply line and the pull-up transistor of driver circuit was shown to create a reverse-bias between gate and source of the driver transistor when both transistors were off [58]. Result was a substantial decrease in I_{sub} and this phenomenon is referred to as the *stacking effect*. For a more general case a transistor stack of arbitrary height was studied, taking into account the body effect as well as drain-induced barrier lowering factor (DIBL), which also becomes significant for submicron devices [65]. It was observed that the leakage power also depended on primary gate-input combination. In another model for the stack effect, a *stack effect factor*, ratio of the leakage current in one off device to that for a stack of two off-devices, was found to be dependent on the process parameters, sub-threshold swing S, and the power supply voltage V_{DD} [66]. The stack effect reduction factor was shown to increase with downward technology scaling, which was basically due to expected increase in DIBL factor λ_d and possible reduction in supply voltage. Stacking of devices does reduce I_{sub}, but drive current of a forced stack is lowered, resulting in increased propagation delay. It suggests that stack forcing would be used in noncritical paths, similar to the dual V_{th} scheme. Of course there can be a delay leakage trade-off and paths which are faster than as per local requirement can be slowed down with a condition that such slow down does not affect the overall performance. The forced stack technique, which effectively reduces I_{sub} of noncritical paths, can be used with and without the dual V_{th} process. Common gates like NAND, NOR, or more complex ones do have stacked gates in their original form. During standby, if numbers of stack devices are off, I_{sub} can be reduced. However, it is not practical to have all stacked devices off throughout the time duration of sleep mode. In case, if stacking is forced in both n- and p-networks of a gate, leakage will surely be reduced, irrespective of the input logic levels [66].

Analysis of the scaling of stack effect and improvement in gate leakage also showed possibility of performance degradation and required sleep mode input vectors to take full advantage of stacking. An enhanced MTCMOS scheme is also available, as shown in Figure 7.19 with stack transistors M_1 and M_2 placed between low V_{th} logic and the ground for leakage control in standby mode [67]. This scheme has an important characteristic of eliminating the need of input vector set for minimizing leakage current and works on a single sleep-mode signal S for turning off both the transistors in the sleep mode. It has been shown that optimum stack height is only two, but at the same time, the size of the two MOSFETs is to be optimized, which will minimize performance degradation and leakage power consumption [67]. Since the method of optimization given here took into account only sub-threshold

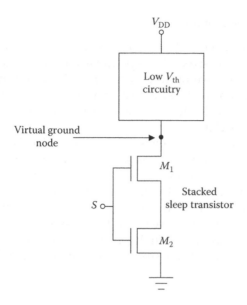

FIGURE 7.19

Enhanced MTCMOS scheme with stacked sleep devices for leakage reduction. (Modified from "Analysis and Optimization of Enhanced MTCMOS Scheme", R.M. Rao, J.F. Burns and R.B. Brown, Proc. 17th International Conf. VLSI Design (VLSID'04), 2004.)

leakage in standby mode and not the gate leakage, a better proposition takes into account the increasing gate leakage, which is expected to be a dominant component in total leakage [68,69]. At the same time it is also important to consider leakage power drawn in the active mode, during such optimization.

For the forced stacking scheme, it was shown in ref. [70] that to ensure same performance as that of a conventional MTCMOS scheme with one MOSFET of width W inserted between logic and ground, sizing of the MOSFETs of Figure 7.19 (W_1 and W_2) must be done according to the following relation and constraint:

$$W_1 = \frac{W.W_2}{W_2 - W} \text{ and } W_1 > W \tag{7.11}$$

Though the device sizes optimize sub-threshold leakage, there is always an increase in the gate leakage current if identical performance is a constraint as mentioned above, and for optimizing total gate leakage, including I_{sub}, size of the upper device M_2 must be smaller than that of M_1; unfortunately, it is a contradictory condition for I_{sub} optimization. For a circuit with smaller active duration than sleep-mode, input occurrence probability is also small and if the circuit is optimized in terms of total leakage saving, it will give more benefit than a case when it is optimized in terms of sub-threshold leakage only. Hence, it is always better to optimize for total leakage [55].

7.10.2 Sleepy Stack Concept

The sleepy stack leakage reduction technique suggests a structure which combines previously discussed sleep transistor technique and the forced stack technique to achieve up to two orders of leakage power reduction in comparison to the forced stack. It also retains the original state unlike the sleep transistor approach, though the advantages come at a small price of some chip area and delay cost.

Figure 7.20 shows a sleep transistor inverter which isolates the existing logic network whereas in the forced stack transistor inverter, existing transistor is broken into two to take advantage of stacking. In the sleep stack mode existing transistors are divided into two with each one being half the size of the original transistor width [71]. During active mode, as shown in the figure, all sleep transistors are turned on to switch faster than the forced stack structure. In addition high V_{th} transistor and its parallel transistor may be used for the sleep transistor. In the sleep mode both the sleep transistors are turned off, keeping the sleepy stack structure its original logic. Leakage current is reduced by high V_{th} sleep transistor/parallel transistors and in addition stacked and turned-off transistors produce stack effect with further reduction in leakage current. Combined effect of the above achieves extremely low leakage power consumption while retaining the original logic.

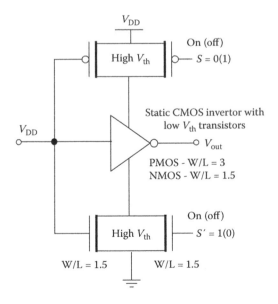

FIGURE 7.20
Sleepy stack inverter showing W/L of each transistor in active mode assertion, and sleep mode assertion. (Redrawn from "Sleepy Stack Leakage Reduction", J.C. Park and V.J. Mooney, IEEE Trans. VLSI Systems, Vol. 14, pp. 1250–1263, 2006.)

7.10.3 Importance of Transistor Sizing

Circuit designers would like to size the NMOS sleep transistor large enough for achieving a good performance. Virtual ground is preferred to be as close to the actual ground as possible, which would force drain-to-source voltage of the high-V_{th} sleep transistor to a small value so that it is biased in the linear region, enabling it to be represented by a linear resistor. However, if the sleep transistor is sized too large, not only the chip area shall be wasted, energy overheads during sleep-active mode switching would increase; whereas the circuit becomes slow during high-to-low transition if the NMOS transistor size is too small because of the increase of its resistance. Hence, deciding the optimum size of the sleep transistor needs further analysis. At first, let us consider the effect of its on-resistance $R_{s,on}$ with the assumption that parasitic capacitance $C_{s,on}$ as shown in Figure 7.21 is negligible. A voltage drop developed across $R_{s,on}$ during any charge flow from low-V_{th} logic will affect the working of the MTCMOS scheme. It will reduce the drive of the logic transistor from V_{DD} to $V_{DD} - V_{VG}$, and because of the raised potential of the source of pull-down logic NMOS transistors, their threshold voltage would increase; both the effects increase high-to-low transition time t_{pHL}. With the continued downscaling of supply voltage and almost constant V_{th}, effective $R_{s,on}$ of the sleep transistor is bound to increase and to keep it at the same level (if not able to reduce it), another large size device is to be used. In addition for increasing its drive, overdriving of the sleep transistor gate can also be used.

The parasitic capacitance shown in Figure 7.21, caused by wiring and junction capacitances serves as local charge sink (and source) and reduces

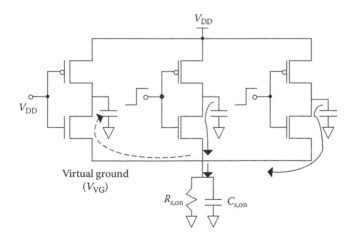

FIGURE 7.21
MTCMOS block illustrating equivalent resistance, capacitance, and reverse conduction effects. ("Dual-Threshold Voltage Techniques for Low-Power Digital Circuits", J.T. Kao and A.P. Chandrakasan, IEEE J. Solid-State Cir., Vol. 35, pp. 1009–1018, July 2000.)

transient spikes in the circuit while switching between sleep and active modes. But $C_{s,on}$ is not able to reduce the effect of the IR drop of $R_{s,on}$ unless it becomes too large, which is not desirable; hence, proper sizing of the sleep transistor is very important [72]. As shown in Figure 7.21, there are some other problems as well. For example, when the current flows backward through the low V_{th} NMOS from the virtual ground and charges up the output capacitance for PMOS sleep transistor, its output is raised from zero to V_{VG}, and a PMOS with *low* input might experience a reverse conduction [72]. This charging current gets accumulated while other gates discharge from *high* to *low* and only a small portion of this current flows through the sleep transistor. Consequently, MTCMOS structure becomes a bit faster as voltage drop across the sleep transistor shall be a little less than if all the current would have flown through it. Another effect of the reverse conduction is to precharge the output to V_{VG} instead of its original value of 0 V and required charging effort from *low* to *high* becomes a little less, thus lowering t_{pLH}. On the flip side, noise margins are reducing, which in worst cases may lead to malfunctioning.

Optimum sizing of sleep transistor becomes increasingly difficult in complex MTCMOS circuits because critical paths, depending upon the input-vectors in a basic CMOS circuit, do not translate into critical circuits in MTCMOS. Exhaustive simulation of the complete circuit for all combination of inputs needs to be done with the sleep transistor's size as a variable parameter. To avoid this cumbersome process, a suggested alternative for optimal transistor sizing is based on *mutual exclusive discharge pattern* [73], which ensures the performance of a complex MTCMOS circuits within pre-scribed performance limits for all possible input patterns. In this procedure, at first it is ensured that individual gates are allowed to be degraded by only a fixed percentage, which guarantees that a complex MTCMOS constructed from these gates shall not degrade by more than the same limit from its original CMOS version. While implementing this method, it is not necessary to determine worst case input vector for the whole circuit, which makes it an easier alternative. An individual gate is exhaustively simulated to determine its own high-V_{th} sleep transistor, then these sleep transistors are merged as they can be shared among mutual exclusive gates [72]; the gates do not discharge at the same time. Through merging of sleep transistors, important chip area is saved. An algorithm has been developed and shown to work on LSI logic for transistor sizing and merging for fixed percentage of performance degradation [74].

In MTCMOS circuits, high V_{th} series sleep transistor always degrades performance while reducing the leakage power consumption. An alternate to it is dual-V_{th} domino logic where already existing devices in the logic are assigned high and low V_{th}; hence, an extra sleep transistor is not needed. Dual V_{th} domino style has the sleeping mode leakage consumption of a purely high-V_{th} logic, but performance is decided by the low-V_{th} devices. It happens because of the fixed transition directions in domino logic, which allows using the dual-V_{th} domino gate with low leakage, and high V_{th} devices are used in noncritical paths without affecting performance [75]. Dual-V_{th} domino logic does not have to do the device sizing and at the same time performance degradation due to series sleep transistor is also avoided.

7.11 V_{GS} Reverse Biasing

Peripherals of DRAM contain a large number of logic circuits, like decoders, word drivers, and column drivers, and these circuits are in multiple numbers. I_{sub} of these drivers/circuits becomes substantial as their total width becomes more than widths of transistors on the rest of the chip and threshold voltage of the driver transistors is also necessarily low for obtaining high speed. Therefore, reduction of I_{sub} for these circuits assumes significant importance.

In a conventional CMOS driver, either NMOS or PMOS is off during steady-state condition. However, even for $V_{GS} = 0$, small threshold current flows in the off MOSFET with its gate under weak inversion. Expression of the I_{sub} under this condition is given as

$$I_{sub} = I_o \cdot \frac{W}{W_o} \cdot \exp \frac{|V_{th}|}{S/l_n 10} \tag{7.12}$$

When the threshold voltage is reduced from $-V_{th1}$ to $-V_{th2}$, leakage subthreshold current at $V_{GS} = 0$ increases from I_{sub1} to I_{sub2} for a PMOS as shown in Figure 7.22, which results in larger power consumption in the standby mode of the MOSFETs. One of the most effective ways to minimize the increase in I_{sub} is to have a circuit arrangement in which V_{GS} is increased automatically either by lowering the source voltage or increasing the gate voltages and I_{sub2} can be reduced to I_{sub1} if V_{GS} reversed by ΔV_{GS} as shown in the figure.

Basic scheme for self-reverse biasing is shown in Figure 7.23 wherein a low-V_{th} PMOS, M_{SP} is inserted between the common source of PMOS transistors M_{P1} to M_{Pn} of the drivers and the power supply V_{DD}. The inserted PMOS turns on during active mode and switches off in the standby mode [76,77]. Since, in the decoded drivers, the same kind of circuits are arranged repeatedly, but only a few (or even one) operate, the width of the transistor M_{SP} is barely sufficient to provide *on current* to only a few active drivers. In the standby mode, gate NMOSs are turned on and PMOSs off, and the switching PMOS-M_{SP} is also turned off. Therefore, the large sub-threshold current of all the drivers is reduced to a small value, which is the sub-threshold current of the M_{SP} only, because of its automatic reverse bias to M_P. This reverse biasing is due to the stacking effect of M_{SP} working like a power switch and

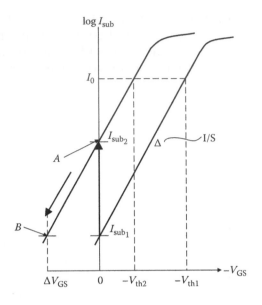

FIGURE 7.22
Transfer from state A to state B due to lowering of the threshold voltage from V_{th1} to V_{th2}. (Redrawn from "Sub-threshold Current Reduction for Decoded-Driver by Self-Reverse Blasing", T. Kawahara et al., IEEE J. Solid State Circuits, Vol. 28, pp. 1136–1143, 1993.)

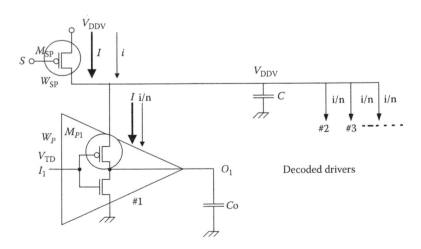

FIGURE 7.23
Sub-threshold-current reduction by self-reverse biasing through decoded-driver. (Redrawn from "Sub-threshold Current Reduction for Decoded-Driver by Self-Reverse Blasing", T. Kawahara et al., IEEE J. Solid State Circuits, Vol. 28, pp. 1136–1143, 1993.)

providing source impedance to M_P. It is important to note that reverse biasing and hence I_{sub} are controllable with the threshold voltage V_{ths} and channel width W_{SP} of the inserted PMOS. Larger reverse biasing can completely cut off the switch but recovery time from sleep mode to active mode becomes large and creates spike noise, whereas smaller reverse biasing allows some leakage flow, but problems are reduced [77]. Generally, a low V_{ths} MOS is preferred as it also improves its trans-conductance and drivability with smaller channel width.

7.11.1 Offset Gate Driving

Basics of offset gate driving indicate where the input voltage is overdriven. Since the logic swing of the output is smaller than the input, the technique is slightly difficult to be applied to random logic, but it has been applied for reducing leakage current in bus drivers, as shown in Figure 7.24 [78], in power switches having low actual V_{th} [79] and in RAM cells [80,28]. The bus driver in Figure 7.24 is a conventional CMOS inverter with low V_{th}, but supply lines are V_{DL} and V_{SL}; hence, its output voltage swing is reduced, making it faster than a conventional inverter with normal supply V_{DD}. For proper operation of the scheme, a bus receiver is also required which converts reduced voltage swing of bus to full voltage swing for the logic. Power dissipation can be reduced up to two thirds compared to conventional architecture.

Logic circuits can be operated in 0.5 V–0.8 V V_{DD} range, provided V_{th} is in the range of 0.1–0.2 V. Super cutoff CMOS (SCCMOS) scheme has been proposed which overcomes the problem of high leakage in standby mode with small value of $V_{th;}$ [79]. Figure 7.25 shows the basic scheme of SCCMOS in which a low-V_{th} (0.1–0.2 V) cutoff PMOS is inserted in series with the logic circuits consisting of low-V_{th} MOSFETs. Gate voltage (V_G) of M_1 is at ground level when logic is in the active mode and in the standby mode V_G is overdriven to $V_{DD} + 0.4$ V for completely cutting off the leakage current. In a test chip fabricated with 0.3 μm triple-metal CMOS technology, working on 0.5 V V_{DD}, standby current per logic gate could be reduced to 1 pA. However,

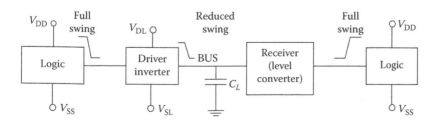

FIGURE 7.24
Concept of offset driving applied to a BUS, with $V_{DL} < V_{DD}$ and $V_{SL} > V_{SS}$. (Modified from "Sub-1-V Swing Bus Architecture for Future Low Power VLSIs", Y. Nakagome et al., Symp. VLSI Circuits, Dig. Tech. Papers, pp. 82–83, 1992.)

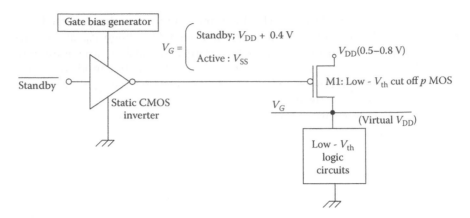

FIGURE 7.25
Concept of super cutoff CMOS through pMOS insertion. (Redrawn from "A Super Cut-Off CMOS (SCCMOS) Scheme for 0.5V Supply Voltage with Picoampere Stand-by Current", H. Kawaguchi, K. Nose and T. Sakurai, IEEE J. Solid-State Circuits, Vol. 35, pp. 1498–1501, 2000.)

problems mentioned in connection with a perfect switch remain present. Negative word line technique described in Section 7.7 to suppress the increase in I_{sub} due to shallow V_{BB} is another example of offset gate driving.

7.11.2 Substrate Driving

For high-speed and low-power operation, both supply voltage V_{DD} and threshold voltage V_{th} can be lowered. The problem of higher leakage current can also be solved by applying a substrate bias during standby mode, which increases the threshold voltage. However, in the active mode, the substrate bias is not applied, retaining low V_{th} and higher performance.

In a test DRAM, fabricated in 0.3 µm technology, V_{th} could be increased to 0.7 V from 0.3 V by applying a substrate voltage of –2 V. Basic scheme of such a standby power reduction is shown in Figure 7.26, consisting of a level-shifter and a voltage switch. During active mode n-well bias mode V_{NWELL} = V_{DD} = 2 V and p-well bias V_{PWELL} = V_{SS}, whereas during standby mode respective values are set at 4.0 V and 2 V. Standby to active mode transition and vice versa takes about 50 ns [80]. Substrate biasing technology has successfully been used in the fabrication of processors [81,82] and for the reduction of leakage in power switches [83].

7.11.3 Offset Source Driving

Offset source driving is similar to the V_{GS} reverse biasing scheme to the extent that it also uses source switches, but its behavior is considerably different as the input gate voltage and output drain voltage are not up to full swing and the source terminal voltage is also not at V_{DD} or V_{SS}. This makes a huge

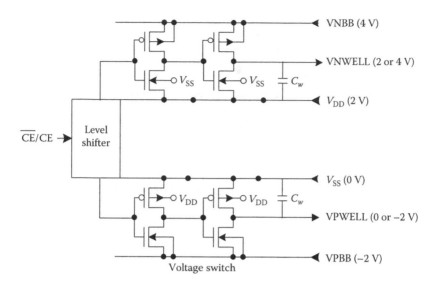

FIGURE 7.26
Standby power reduction (SPR) circuit. Well capacitance (C_w) is supposed to be 1000 pF. (Modified from "50% Active-Power Saving without Speed Degradation Using Standby Power Reduction (SPR) Circuits", K. Seta et al., ISSCC, pp. 318–319, 1995.)

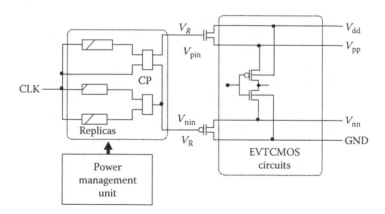

FIGURE 7.27
EVTCMOS circuit design. (From M-Mizuno et al., "Elastic-VT CMOS Circuit for Multiple On-Chip Power Control", ISSCC, pp. 300–301, 462, 1996.)

difference with the leakage current reduction efficiency. Control of the MOS transistor source voltage is shown in Figure 7.27 in which threshold voltage is changed to a high value during sleep mode and to a low voltage during active mode [84]. Additional advantage of the arrangement is that fluctuation in the device parameters due to the process can also be controlled. Through the power management circuit, the virtual V_{DD} (V_{pp}) and ground (V_{nn}) can be

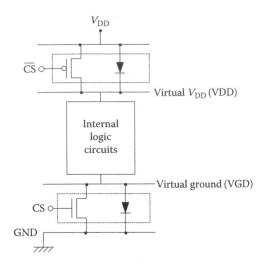

FIGURE 7.28
A virtual rail clamp scheme. (Modified from K. Kumagi et al., "A Novel Powering-down Scheme for Low Vt CMOS Circuits," Symp. VLSI Circuits, Dig. Tech. Papers, pp. 44–45, 1998.)

varied while switching from active to sleep mode or vice versa. A deviation-compensated loop (DCL) sets the threshold voltage to the required value and making the leak current small in the sleep mode. The DCL contains three types of replica circuits. A delay line for speed adjustment, a phase detector (PD), and a charge pump (CD) for controlling the voltage levels of V_{pin} and V_{nin}. Another similar scheme of offset source biasing is shown in Figure 7.28 providing a voltage lesser than V_{DD} (VVD) and higher than ground (VGD) using only two extra MOS transistors and two diodes. Levels of voltages VVD and VGD are clamped by the diode. In this scheme speed degradation happens as in multithreshold CMOS circuits, but it also does not need timing design as there is only one control signal (CS) for switching between active and sleep mode. The V_{th} of the switching transistors is the same as that for the transistors used in the controlled circuit leading to no enhancement in fabrication cost [85].

Discussed methods for reducing I_{sub} can be compared on the basis of their performance and circuit requirements, limitations, and advantages. Switching time between active and sleep mode is an important factor. In substrate driving for V_{th} control, the requirement of voltage larger than V_{DD} creates limitations especially for scaled-down technology. A good comparison among these techniques is available in [8].

7.11.4 Simultaneous Negative Word Line and Reverse Body Bias Control

Applicability of negative word line (NWL) for extending data retention time of DRAMs was discussed in Section 7.7. Though the scheme has been extensively used for low voltage DRAMs, it has a disadvantage in terms of GIDL [86]. A selective negative word line scheme (SNWL) combines the advantages of conventional NWL and the ground word line scheme [87], in which NWL scheme is applied only to the active cell block. The scheme was applied in a 54 nm DRAM chip and provided lower GIDL and improvement in dynamic refresh time with lower off-state current.

The NWL scheme permits the use of lower threshold access transistor with consequential advantages related with large on-current, while reverse body bias voltage scheme reduces sub-threshold current [4,88] and increases reliability of cell operation. Unfortunately, most of the schemes are effective for one or two leakage components but are not effective in reducing overall leakage current at nanometer level DRAMs. Recently, a scheme has been given in which the body and the WL bias voltage are controlled simultaneously [89]. Architecture of the biasing scheme consists of a leakage monitoring circuit which classifies the amount of each leakage component and decides the biasing levels. Other components in the scheme are V_{BB} and VNWL chains (for generating regulated bias voltages), a control signal generator, which generates a number of signals during monitoring of leakage, and a counter. The counter generates a multi-bit signal which determines the self-refresh period and the self-refresh is updated dynamically on a real-time basis. Experimental result at 46 nm CMOS technology level has shown an improvement of ~60% data retention time compared to a fixed biasing scheme.

7.12 Leakage Current Reduction Techniques in DRAMs

Weightage of leakage current in the active mode is different from DRAM in sleep mode. Moreover, even in active mode, only a few selected components are active for a short duration. In practice, decoders of the CMOS DRAMs mostly consist of dynamic gates for the rows and static NAND gates for the columns and the NAND decoders discharge only one selected node. It makes the leakage reduction in active mode extremely difficult. However, a large number of row and column decoders with wide size transistors are iterative in nature, which makes control of leakage current easier and effective. There is another important feature in DRAMs. Most of the modules are composed of input-predictable circuits; hence, all node voltages can be predicted and effective leakage current reduction schemes can be applied. Those nodes that are not input-predictable can be made so by using level-fixing input buffers [90]; arrangement is shown in Figure 7.29. Here each address signal A_i is gated with standby at high level so that internal address signals including a_i and $\overline{a_i}$ are at low voltage level, irrespective of A_i. In a similar approach leakage power of CMOS circuits is reduced during logic design. It is based on the principle that CMOS logic gate leakage in steady state depends on the

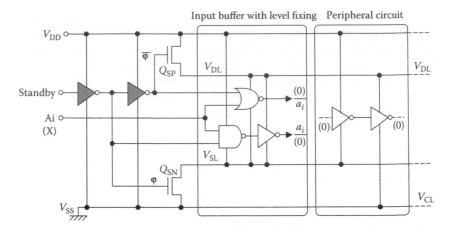

FIGURE 7.29
Switched-source-impedance scheme applied to memory LSI to make internal nodes of RAMs predictable. Enhanced-V_{th} transistors are used for Q_{SP} and Q_{SN} and the shaded inverters. ("Switched-Source-Impedance CMOS Circuit For Low Standby Sub-threshold Current Giga-Scale LSI'S", M. Horiguchi et al., IEEE J. Solid-State Circuits, Vol. 28, pp. 1131–1135, 1993.)

gate input state; hence, the original logic design can be modified into a low-leakage state during an idle period. To find a proper input vector, which finds a low leakage state, an algorithm has also been developed [91]. Obviously, the circuitry/gate which modifies the original design should be such that effects on speed and chip area consumed are minimum; some of the suitable options are pass-gate multiplexers and CMOS NAND and NOR gates. Moreover, the latches are also to be modified to either force 0 or 1 during sleep mode. With the technique in use, leakage power reduction up to 54% could be obtained with minimum overheads. Standard static "jam" modified to force a value at the output during sleep mode and modified dynamic C^2 MOS latches, respectively [91], can be viewed as representative applications.

7.13 Analysis of Sub-Threshold Leakage Reduction

Different techniques for reducing sub-threshold leakage in CMOS digital circuits have been proposed and are briefly discussed in previous sections. These techniques have broadly been classified as (1) leakage control in standby mode, and (2) leakage control in active mode. Fairly good analysis of all these techniques has been made in ref. [92] while using a 28-transistor standard 1-bit full adder with the TSMC 180 nm technology circuit. The analysis is based on the following performance criteria: leakage power dissipation, dynamic power dissipation, and propagation delay.

One of the options for reducing leakage is to cut off supply from the circuit during sleep mode. Important techniques in this category are (1) power gating, and (2) super cutoff CMOS (SCCMOS). Power gating is done using either NMOS or PMOS only or both NMOS and PMOS having high V_{th}. With only PMOS gating, 9× reduction was achieved in leakage power, but delay was 1.046× and there was a slight increase in dynamic power. For only NMOS gating leakage power was reduced up to 12× with delay increase of 1.084× and again a small increase in dynamic power consumption. With both NMOS and PMOS gating leakage power was reduced by two orders of magnitude but with increase in delay of 1.14×. In the SCCMOS reduction in leakage power comparative to two sleep transistor circuits is less; at the same time, the delay introduced also becomes less. The technique has further advantage of easier fabrication, as all transistors are realized with standard V_{th}.

Important techniques for reducing leakage in active mode analyzed using 1 bit full adder were forced stacking, input vector control (IVC), and sleepy stack. In the forced stacking technique leakage saving was up to 1.45× but delay was increased by 1.025×. The IVC technique provides higher saving in leakage power but an exhaustive simulation is required for determining critical input vector. Use of sleepy stack gave lesser reduction in leakage power, but propagation delay improves, and it also requires a little complicated signal circuitry. The analysis confirms a strong correlation among the three important factors of leakage power, dynamic power, and propagation delay.

Different leakage reduction techniques for the active mode of operation like reverse body biasing (RBB), forced stacking, input vector control (IVC), and use of high V_{th} MOSFET have been analyzed from the point of view of finding the effect of technology scaling. For the purpose of analysis, inverter, two and three input NAND and NOR gates were simulated using predictive technology models at 65 nm, 45 nm and 32 nm and its important observations are as follows [93]. Use of high V_{th} MOSFETs scales well with technology in both active as well as standby modes of operation and reduces the leakage with only small performance degradation. The RBB scheme has been found to be less effective with technology downscaling because of increased band-to-band tunneling (BTBT) current and the scaling of body effect coefficient. After an optimum RBB value, its value decreases with scaling [94]. With technology scaling, pin ordering was found to be attractive at low voltage levels as it could be combined with any other technique for leakage reduction, with minimal adverse impact. Another point of consideration is the comparative bigger size of PMOSs in pull up blocks, thus having bigger leakage component. Hence transistor sizing assumes significance. In addition, logic manipulation can also result in an implementation with low leakage logic gates.

7.14 Sub-Threshold Leakage Reduction for Low-Voltage Applications

It was expected that supply voltage would be scaled down to below 0.5 V and the leakage power would become a dominant component even in the active mode of operation, unless the threshold voltage is more than 0.1 V [95]. In a slightly different technique, dynamic threshold voltage hopping (V_{th}-hopping) is used in which dynamic adjustment of clock frequency and V_{th} is done through back-gate bias variation based on the workload of the system. Figure 7.30 shows the schematic of the V_{th}-hopping method. Here the power control block generates signals as shown which in turn controls substrate bias of the system. Control signal is controlled by software through a *software feedback loop* scheme [70]. The clock frequency has only the values of $f_{CLK}, f_{CLK}/2$, and so on, to avoid any synchronization problem at the interface. Frequency controller generates f_{CLK} and $f_{CLK}/2$, with V_{th} low-enable and V_{th} high-enable assertion, respectively. Determination of V_{thlow} and V_{thhigh} is based on the maximum achievable performance of the processor at the clock frequency f_{CLK} and $f_{CLK}/2$, respectively. An important consideration in the scheme is that the back-gate bias is possible for its positive and negative values for the V_{th}-hopping. Such a combination is suitable for reduced technology node. Value of V_{thlow} and V_{thhigh} is to be determined for maximum performance in such a way that system works at the possible discrete frequencies as mentioned above. Hence, an algorithm to dynamically change the V_{th} depending on the variation of the workload is important. The applied algorithm is based on the *run-time voltage hopping* scheme [96]. It was observed that the V_{th}-hopping scheme can achieve up to 82% power reduction compared to

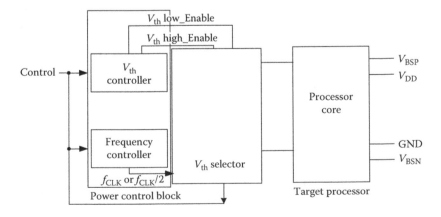

FIGURE 7.30
Schematic diagram of V_{TH}-hopping. (Redrawn from "VTH-Hopping Scheme to Reduce Subthreshold Leakage for Low-Power Processors", K. Noise et al., IEEE J. Solid-State, Vol. 37, pp. 413–419, 2002.)

the fixed low-V_{th} circuits when 0.5 V power supply is used. A similar range of power saving was obtained in a RISC processor where zero back-gate bias was applied in comparison to a fixed positive back-bias.

SRAMs have been very successful in embedded memories. However, at low voltage level of 0.5 V V_{DD}, DRAMs are favorable to replace SRAMs mainly because of the non-scalability of its V_{th} which stays at about 1 V [97]. DRAMs are more immune to the V_{th} variation at V_{DD} of 0.5 V, and active power consumption is heavily reduced, making it a preferred memory for mobile applications. A low voltage DRAM at 0.5 V V_{DD} needs suitable sense amplifier (SA) and word line boosters. High performance, SAs are available and discussed in Chapter 8 [98,99]. A number of circuits for another key element-word line booster are available, however conventional boosters suffered from either V_{th} loss or inefficient charge transfer mechanism. An efficient booster overcoming the problem of V_{th} loss is given by Tanakamaru and Takeuchi, which generates word-line voltage V_{PP} of 1.4 V for 0.5 V DRAM operations [100]. This booster raises V_{PP} from 0.5 V to 1.4 V within three clock cycles compared to other boosters which take eight clock cycles to raise V_{PP} to 1.4 V. Another major advantage of the booster is its reduced power consumption of 60 pJ which is ~68% of the conventional circuits.

7.15 Data Retention Time and Its Improvement

In Section 7.8, it was shown that as the DRAM size M increases, say quadrupled, then both, the number of cells in a word line m and the maximum refresh time $t_{REF\ max}$ are doubled. Increase in m and $t_{REF\ max}$ is necessary to keep power consumption due to cell leakage in control and maintain constant refresh interval independent of DRAM density increase. Because of this continued rise in the value of $t_{REF\ max}$, study of retention time, defined as the time duration during which stored cell signal can be read reliably, became extremely important. The retention time has to be increased with every successive generation; otherwise, reliability and functionality of the DRAM shall be at stake.

For the purpose of enhancing retention time, its distribution was plotted as a function of boron concentration of a p-well for a specific memory cell; studies revealed a very significant aspect. It was observed that the distribution of the retention time is clearly divided in two parts: (1) a *main distribution*, having longer retention time and almost all the cells in memory came under this category, and (2) a *tail distribution*, which corresponds to a few memory cells with a shorter retention time [5]. For example with an average retention time of 6 s at 85°C retention time of normally distributed probability, retention time of the worst case is 100 ms at 5 sigma spread. This problematic *tail distribution* depends proportionally on the boron concentration of the p-well.

Hence, it was obvious that for the overall increase in the retention time, it is very important to reduce the tail distribution. Suggested methods include (1) reduction in the boron concentration of p-well which will reduce electric field of the depletion layer at storage node-p-n junctions, and (2) reduction of concentration of the deep level of thermionic field emission (TFE) current; TFE is a concept introduced for understanding of a relationship of *tail distribution* with boron concentration [6]. It is known that deep level TFE is related with interstitial silicon; hence, tail distribution can be improved by controlling the generation of point defects. Unless measures were taken to reduce the small number of problematic cells with shorter retention time, practical fabrication would become impossible.

Other studies have also been made for finding an explanation of leakage during tail modeling of the memory cell junction as a two-terminal p-n junction [101,102] as well as a three-terminal structure [103] taking into consideration the effect of gate electrode as well. It is concluded that GIDL also contributes to the leakage in weak cells in the tail. It is important, therefore, that GIDL current has to be controlled and reduced for improving data retention. It is confirmed in another study on an 80 nm RCAT technology [104] that GIDL current is a major component in determining the value of retention time. First, total leakage current is modeled as a combination of storage node leakage current and GIDL current. Using trap-assisted-tunneling model for the two current, maximum electric fields are evaluated as a function of bias conditions. The leakage model is then filled to reproduce tRET (data retention time) for different bias values of V_{DS}, V_{GS}, and V_{BS}, and the results confirm the contribution of GIDL as mentioned before.

Using retracted Si_3N_4 liner STI, a novel cell transistor is realized for improving the data retention time of high-density DRAMs in 0.15 μm technology. Use of retracted Si_3N_4 liner STI and optimized channel doping profile, from the point of view of local area electric field in the depletion region, reduces maximum electric field from channel to drain by about 15%. With reduction in maximum electric field, the junction leakage current of weak cells decreases and tail component of data retention improves [105].

Application of innovation in the fabrication process for improving DRAM retention time has further been used for sub-60 nm technology overcoming some limitations of existing devices. Recessed channel transistor, RCAT and SRCAT, discussed in Sections 4.2 and 4.2.1, respectively, have been employed extensively below 80 nm technologies to overcome the limitation of short-channel effect. However, they have higher body effect and sub-threshold slope in comparison to planar transistor and the effects are compounded due to the shrinking of feature size because of edge effect [106] of shallow trench isolation (STI). A new active isolation structure Lat Ex (lateral extended) has been adopted in an SRCAT to improve the data retention time. Figure 7.31 shows the schematic of the Lat Ex actively deployed in recessed channel structure, and it is compared with a conventional structure in terms of active area and S/D regions [107]. Junction leakage and GIDL current, both

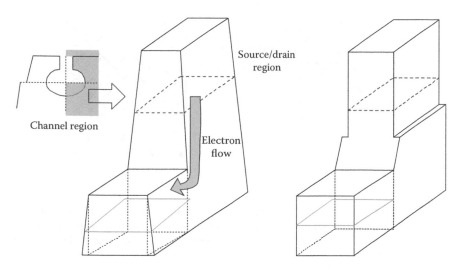

FIGURE 7.31

Schematic showing reduction in the area of channel depletion layer region and the area of S/D region in the Lat Ex active compared to conventional device at left. ("Lateral-Extended (Lat Ex.) Active for Improvement of Data Retention Time for Sub 60 nm DRAM Era", S. Lee et al., Sol. State-Device Research (ESSDRC), pp. 327–329, 2007.)

responsible for tail distribution of leakage current, are considerably reduced. One of the reasons for this is that band-to-band tunneling is reduced, affecting GIDL. Sub-threshold slope could be lowered up to 30% and the body effect is improved from 400 mV/V to 350 mV/V. As a result of these improvements in the characteristics of SRCAT, data retention characteristics become much better than a conventional SRCAT. It was expected that Lat Ex will be suitable even beyond 60 nm technology. Another attempt has been made employing RCAT and applying negatively biased off-state of the word line (NWL) to improve the sub-threshold characteristics of the cell transistor [108]. It was observed that the cumulative fail bit count (FBC) of the retention time *main distribution* decreases as NWL bias is strengthened. However, it was not so effective for the *tail distribution* because of its lesser effect on GIDL current. Therefore, a suitable NWL bias scheme and a magnitude of cell transistor threshold voltage are needed for full acceptability of the scheme.

It was suggested earlier that data retention time can also be improved by reducing vacancy-type defects in DRAM cell transistors because the amount of junction leakage current depends on it as well. The idea was pursued in a report indicating that the enhanced junction leakage current is due to triangular intrinsic stacking faults in the depletion layer of minority bits, through a trap-assisted tunneling [109]. Effort was made to control the defect growth through lattice strain, but success was insufficient and it was concluded that a small number of vacancy-type defects, like point, still remain [110]. Further study showed that reduction of silicon interstitials caused vacancy-type defect generation during annealing.

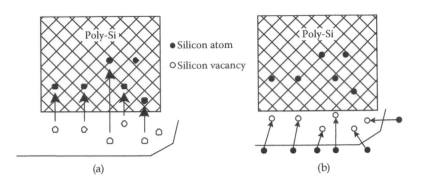

FIGURE 7.32
(a) Schematic diagram showing a possible reason for vacancy-type defects. (b) Under the reverse annealing condition, a large number of silicon interstitials are generated in the silicon substrate. (Redrawn from "Improvement of Data Retention Time Property by Reducing Vacancy-Type Defects in DRAM Cell Transistors", K. Okonogi et al., IEEE 44th Annual Int. Relia. Physics Symp., pp. 695–696, 2006.)

Hence, a *reverse annealing* was used to reduce such defects, which helps supply a large number of silicon atoms as illustrated in Figure 7.32. Data retention time of both majority and minority bits in a 0.11 μm technology DRAM was improved considerably. This report also gives a useful relationship between number of tail bits and the vacancy-type defect density [110]. Silicon atoms are absorbed in polysilicon during high-temperature annealing and remaining vacancies can generate vacancy-type defects, whereas in reverse annealing condition, interstitials are generated in the silicon substrate.

It is well established that tail component cells have an order of magnitude higher cell junction leakage current and GIDL current. The latter becomes especially serious in recessed-channel array transistors which are more likely to be used in sub-80 nm technology level. It is also known that the GIDL is primarily determined by the distribution of traps within a cell storage junction [111] and the current is generated by trap-assisted tunneling (TAT). These problematic cells with large leakage current are only a few parts per million and are generated during different fabrication processes such as high-power plasma etching, oxidation, annealing, and so on. Recent study has shown that the electric field at the gate overlapped source/drain region and the spatial location or energy levels of traps are the most important factors contributing toward higher leakage [112]. Only a small number of deep level traps primarily contribute the *tail distribution* and GIDL current dispersion due to TAT. If the deep traps are somehow excluded from high electric field region, and the number of traps is reduced, then the tail distribution cannot only be minimized, it can be eliminated, thereby increasing the data retention time corresponding to that of only *main distribution*.

For better understanding, three specimens of 1 Gbit DRAM at 100 nm, 60 nm, and 50 nm technology nodes were studied. The trap density per unit cell (Dit*) at Si/SiO_2 interface is estimated, which is then digitized. It gives an important result in that most cells realized in about 10 nm scale do not contain any traps; only a few have them. This means that the storage cells have defects or not; therefore, tail distribution shall not be present in the majority of the cells, which will have data retention times of several tens of seconds provided care is taken during the fabrication processes to minimize/eliminate reasons for the production of deep level traps especially in stronger electric field areas.

It is observed in a recent report that if passivation annealing is performed prior to the deposition of plasma nitride layer in DRAMs, it can improve junction leakage current considerably. It happens as crystalline defects are repaired due to sufficient generation of hydrogen when such a passivation annealing process is used [113]. In addition, variation in transistor threshold V_{th} is also reduced which enables a reduction in electric field and consequential reduction in leakage current. As a result DRAM data retention is improved further.

References

1. M. Aoki et al., "A 1.5V DRAM for Battery-Based Applications," *IEEE I.S.S.C. Conference*, pp. 238–239, 349, 1989.
2. Y. Nakagome et al., "A 1.5V Circuit Technology for 64 MB DRAMs," *Symp. VLSI Circuits*, pp. 17–18, 1990.
3. J.Y. Kim et al., "The Excellent Scalability of the RCAT (Recess Channel-Array-Transistor) Technology for Sub-70 nm DRAM Feature Size and Beyond," *Int. Symp. VLSI Techn.*, pp. 33–34, 2005.
4. K. Itoh, K. Sasaki, and Y. Nakagome, "Trends in Low-Power RAM Circuit Technologies," *Proc. IEEE*, Vol. 83, no. 4, pp. 524–543, 1995.
5. T. Hamamoto, S. Sugiura, and S. Swada, "On the Retention Time Distribution of Dynamic Random Access Memory (DRAM)," *IEEE Trans. Electr. Dev.*, Vol. 45, no. 6, pp. 1300–1309, 1998.
6. S.M. Ze, *Physics of Semiconductor Devices*, New York, Wiley, 1981, pp. 84–96.
7. D.J. Frank, "Power-Constrained CMOS Scaling Limits," *IBM J. Res. and Dev.*, Vol. 46, pp. 235–244, 2002.
8. Y. Nakagome et al., "Review and Future Prospects of Low-Voltage RAM Circuits," *IBM J. Res. and Dev.*, Vol. 47, no. 5/6, pp. 525–552, 2003.
9. T. Inukai and T. Hiramoto, "Suppression of Stand-By Tunnel Current in Ultra-thin Gate Oxide MOSFETs by Dual Oxide Thickness MTCMOS," *Int. Conf. S.S. Dev. and Mat.*, pp. 264–265, 1999.
10. K. Kimura et al., "Power Reduction Techniques in Megabit DRAMs," *IEEE J. Solid State Circ.*, Vol. SC-21, pp. 381–389, 1986.

11. Y. Watanabe et al., "Offset Compensating Bitline Sensing Scheme for High Density DRAMs," *J. Solid State Circuits*, Vol. 28, pp. 9–13, 1994.
12. S.H. Hong et al., "An Offset Cancellation Bit-Line Sensing Scheme for Low-Voltage DRAM Applications," *ISSCC Dig. Tech. Papers*, pp. 154–155, 2002.
13. T. Nagai et al., "A 17 ns 4-Mb CMOS DRAM," *J. Solid State Circuits*, Vol. 25, pp. 1538–1543, 1991.
14. J.Y. Sim et al., "A 1.0V 256 Mb SDRAM with Offset-Compensated Direct Sensing and Charge-Recycled Precharge Scheme," *ISSCC Dig. of Tech. Papers*, pp. 310–311, 2003.
15. S.S. Eaton et al., "A 100 ns 64 K Dynamic RAM Using Redundancy Techniques," *ISSCC*, pp. 84–85, 1981.
16. K. Itoh, "Trends in Megabit DRAM Circuit Design," *IEEE J. Solid State Circuits*, Vol. 25, no. 3, pp. 778–788, 1990.
17. K. Itoh et al., "An Experimental 1 Mb DRAM with On-Chip Voltage Limiter," *ISSCC*, pp. 282–283, 1984.
18. T. Sugibayashi et al., "A 30 ns 256 Mb DRAM with Multidivided Array Structure," *ISSCC Dig. Tech. Papers*, pp. 50–51, 1993.
19. K. Fujishima et al., "A 256K Dynamic RAM with Page-Nibble Mode," *IEEE J. Solid State Circuits*, Vol. SC-18, pp. 470–478, 1983.
20. T. Yoshihara et al., "A Twisted Bit Line Technique for Multi-Mb DRAMs," *ISSCC*, pp. 238–239, 1988.
21. H. Hidaka et al., "Twisted Bit-Line Architecture for Multi-Megabit DRAMs," *IEEE J. Solid State Circuits*, Vol. 24, pp. 21–27, 1989.
22. M. Aoki et al., "An Experimental 16-Mbit DRAM with Transposed Data-Line Structure," *ISSCC Dig. Tech. Papers*, pp. 250–251, 1988.
23. J.K. DeBrose, J.E. Lary, and E-J. Sprogis, "Signal Twist Layout and Method for Paired Line Conductors of Integrated Circuits," 1996 U.S. Patent, 5,534, 732.
24. M. Aoki et al., "A 1.5V DRAM for Battery-Based Applications," *Proc. Int. SS. Cir. Conf.*, pp. 238–239, 1989; and also *IEEE J.S.S.Cir.*, Vol. 24, pp. 1206–1212, 1989.
25. K. Kim, C. Hwang, and J. Lee, "DRAM Technology Perspective for Gigabit Era," *IEEE Trans. Elect. Dev.*, Vol. 45, pp. 598–608, 1998.
26. D.-S. Min and D.W. Langer, "Multiple Twisted Data Line Technique for Coupling Noise Reduction in Embedded DRAMs," *IEEE Custom Int. Cir, Conf.*, pp. 231–234, 1999.
27. D.-S. Min and D.W. Langer, "Multiple Twisted Data Line Techniques for Multigigabit DRAMs," *IEEE J. Solid State Circ.*, Vol. 34, pp. 856–865, 1999.
28. T. Yamagata et al., "Low Voltage Circuit Design for Battery-Operated and/or Giga DRAMs," *IEEE J. Solid State Circuits*, Vol. 30, pp. 1183–1188, 1995.
29. T. Mano et al., "Submicron VLSI Memory Circuits," *IEEE, ISSCC*, pp. 234–235, 311, 1983.
30. H. Tanaka et al., "Stabilization of Voltage Limiter Circuit for High-Density DRAM's Using Pole-Zero Compensation," *IEICE Trans. Electron.*, Vol. E 75-C, no. 11, pp. 1333–1343, 1992.
31. D. Chin et al., "An Experimental 16-Mbit DRAM with Reduced Peak Current Noise," *IEEE J. Solid State Circ.*, Vol. 24, pp. 1191–1197, 1989.
32. M. Horiguchi, "A Tunable CMOS-DRAM Voltage Limiter with Stabilized Feedback Amplifier," *IEEE J. Solid State Circ.*, Vol. 25, no. 5, pp. 1129–1135, 1990.
33. M. Horiguchi et al., "Dual-Operating-Voltage Scheme for a Single 5V 16 Mbit DRAM," *IEEE J. Solid State Circ.*, Vol. 23, pp. 612–617, 1988.

34. T. Furuyama et al., "A New On-Chip Voltage Converter for Sub-micrometer High-Density DRAM's," *IEEE J. Solid State Circ.*, Vol. SC-22, pp. 437–441, 1987.
35. M. Takada et al., "A 4 Mb DRAM with Half Internal Voltage Bit Line Precharge," *IEEE J. Solid State Circ.*, Vol. SC-21, pp. 612–617, 1986.
36. G. Kitsukawa et al., "A 1 Mbit Bic MOS DRAM Using Temperature-Compensated Circuit Techniques," *IEEE J. Solid State Circ.*, Vol. 24, pp. 597–602, 1989.
37. M. Akasura et al., "A 34 ns 256 Mb DRAM with Boosted Sense-Ground Scheme," *Proc. SSCC*, pp. 140–141, 1994.
38. D. Takashima et al., "Low Power On-Chip Supply Voltage Conversion Scheme for 1G/4G bit DRAMs," *Symp. on VLSI Circuits Digest of Tech. Papers*, pp. 114–115, 1992.
39. H. Tanaka et al., "A Precise On-Chip Voltage Generator for a Giga-Scale DRAM with a Negative Word-Line Scheme," *Symp. VLSI Circ. Dig. Tech. Papers*, pp. 94–95, 1998.
40. Y. Tsukikawa et al., "An Efficient Back-Bias Generator with Hybrid Pumping Circuit for 1.5V DRAMs," *Symp. VLSI Circ. Dig. Tech. Papers*, pp. 85–86, 1993.
41. M. Horiguchi et al., "Dual-Regulator Dual-Decoding-Trimmer DRAM Voltage Limiter for Burn-in Test," *IEEE J. Solid State Circ.*, Vol. 26, no. 11, pp. 1544–1549, 1991.
42. H. Tanaka et al., "Sub-1-µA Dynamic Reference Voltage Generator for Battery-Operated DRAM," *Symp. VLSI Circ. Dig. Tech. Papers*, pp. 87–88, 1993.
43. K. Sato et al., "A 4-Mb Pseudo SRAM Operating at 2.6 ± 1V with 3-µ,A Data Retention Current," *IEEE J. Solid State Circ.*, Vol. 26, pp. 1556–1562, 1991.
44. Y. Kagenishi et al., "Low Power Self Refresh Mode DRAM with Temperature Detecting Circuits," *VLSI Circ. Symp.*, pp. 43–44, 1993.
45. D-C. Choi et al., "Battery Operated 16 M DRAM with Post Package Programmable and Variable Self Refresh," *Symp. VLSI Circ. Dig. Tech. Papers*, pp. 83–84, 1994.
46. J. Nyathi and J. Delgado-Frias, "Self-Times Refreshing Approach for Dynamic Memories," *Proc. Annual Int. ASIC Conf.*, pp. 169–173, 1998.
47. Influent Corp., "Reducing Server Power Consumption by 20% with Pulsed Air Cooling," June 2009, http://www.influentmotion.com/Server White paper.pdf.
48. L. Minas and B. Ellison, "The Problem of Power Consumption in Server," *Intel Press Report*, 2009.
49. Micron, "TN-47-16 Designing for High Density DDR2 Memory Introduction," 2005.
50. JEDEC Committee JC-42-3, "JESD79-3D," Sept. 2009.
51. Y. Idei et al., "Dual-Period Self-Refresh Scheme for Low Power DRAMs with On-Chip PROM Mode Register," *IEEE J. Solid State Circ.*, Vol. 33, pp. 253–259, 1998.
52. S. Takase and N. Kushiyama, "A 1.6 Gb/s DRAM with Flexible Mapping Redundancy Technique and Additional Refresh Scheme," *Proc. Int. Sol. State Circ. Conf.*, pp. 410–411, 1999.
53. J. Kim and M. Papaefthymiou, "Dynamic Memory Design for Low Data-Retention Power," *Proc. PATMOS, 10th Int. Workshop*, pp. 207–216, 2000.
54. J. Kim and M. Papaefthymiou, "Block-Based Multiperiod Dynamic Memory Design for Low Data-Retention Period," *IEEE Trans. on VLSI Systems*, Vol. 11, pp. 1006–1018, 2003.
55. J. Stuecheli et al., "Elastic Refresh: Techniques to Mitigate Refresh Penalties in High Density Memory," *43rd IEEE/ACM Int. Symp. on Microarchitecture*, pp. 375–384, 2010.
56. M.J. Lee and K.W. Park, "A Mechanism for Dependence of Refresh Time on Data Pattern in DRAM," *IEEE Elect. Dev. Lett.*, Vol. 31, pp. 168–170, 2010.

57. S. Mutoh et al., "1-V Power Supply High-Speed Digital Circuit Technology with Multithreshold-Voltage CMOS," *IEEE J. Solid State Circuits*, Vol. 30, pp. 847–854, 1995.
58. S. Shigematsu et al., "A 1-V High-Speed MTCMOS Circuit Scheme for Power-Down Applications," *Symp. VLSI Circuit, Dig. Tech. Papers*, pp. 125–126, 1995.
59. H. Akamatsu et al., "A Low Power Data Holding Circuit with an Intermittent Power Supply Scheme for Sub-1V MT-CMOS LSIs," *Symp. VLSI Circuits, Dig. Tech. Papers*, pp. 14–15, 1996.
60. Z. Chen et al., "0.18 μm Dual V_t MOSFET Process and Energy-Delay Measurement," *Proc. IEDM*, pp. 851–854, 1996.
61. S. Heo and K. Asanovic. "Leakage-Based Domino Circuits for Dynamic Fine-Grain Leakage Reduction," *Symp. VLSI Circuits, Dig. Tech., Papers*, pp. 316–319, 2002.
62. S. Thompson et al., "Dual Threshold Voltages and Substrate Bias: Keys to High Performance, Low Power, 0.1 μm Logic Design," *Symp. VLSI Tech. Dig. Tech. Papers*, pp. 69–70, 1997.
63. H. Oyamatsu et al., "Design Methodology of Deep Submicron CMOS Devices for 1V Operation," *IEICE Trans. Electron.*, Vol. E79-C, pp. 1720–1724, 1996.
64. L. Wei et al., "Design and Optimization of Dual-Threshold Circuits for Low-Voltage Low-Power Applications," *IEEE Trans. VLSI Systems*, Vol. 7, pp. 16–24, 1999.
65. Zhamping Chen et al., "Estimation of Standby Leakage Power in CMOS Circuits Considering Accurate Modeling of Transistor Stacks," *ISPLED*, pp. 239–244, 1998.
66. S. Nareandra et al., "Scaling of Stack Effect and Its Application for Leakage Reduction," *ISPLED*, pp. 195–200, 2001.
67. K. Das and R. Brown, "Novel Ultra Low-Leakage Power Circuit Techniques and Design Algorithm in PD-SOI for Sub-1V Applications," *Proc. Internation. SOI Confer.*, pp. 88–90, 2002.
68. K. Das et al., "New Optimal Design Strategies and Analysis of Ultra-Low Leakage Circuits for Nano-Scale Technology," *Proc. ISLPED*, pp. 168–171, 2003.
69. International Technology Roadmap for Semiconductors, http://public.itrs.net/Files/2001.ITRS/Home.hlm, 2001 Edition.
70. R.M. Rao, J.F. Burns, and R.B. Brown, "Analysis and Optimization of Enhanced MTCMOS Scheme," *Proc. 17th International Conf. VLSI Design (VLSID'04)*, 2004.
71. J.C. Park and V.J. Mooney, "Sleepy Stack Leakage Reduction," *IEEE Trans. VLSI Systems*, Vol. 14, pp. 1250–1263, 2006.
72. J.T. Kao and A.P. Chandrakasan, "Dual-Threshold Voltage Techniques for Low-Power Digital Circuits," *IEEE J. Solid State Cir.*, Vol. 35, pp. 1009–1018, July 2000.
73. J. Kao, S. Narendra, and A. Chandrakasan, "MTCMOS Hierarchical Sizing Based on Mutual Exclusive Discharge Pattern," *ACM/IIEEE Design Automation Conf.*, pp. 495–500, June 1998.
74. T. Sakuta, W. Lee, and P. Balsara, "Delay Balanced Multipliers for Low Power/Low Voltage DSP Core," *IEEE Symp. Low Power Electronics*, pp. 36–37, 1995.
75. J. Kao, "Dual Threshold Voltage Domino Logic," *25th Eur. Solid State Circuit Conf.*, pp. 118–121, Sept. 1999.
76. G. Kitsukawa et al., "256 Mb DRAM Technologies for the File Applications," *IEEE J. Solid State Circ.*, Vol. 28, pp. 1105–1111, 1993.
77. T. Kawahara et al., "Subthreshold Current Reduction for Decoded-Driver by Self-Reverse Biasing," *IEEE J. Solid State Circuits*, Vol. 28, pp. 1136–1143, 1993.

78. Y. Nakagome et al., "Sub-1-V Swing Bus Architecture for Future Low Power VLSIs," *Symp. VLSI Circuits, Dig. Tech. Papers*, pp. 82–83, 1992.
79. H. Kawaguchi, K. Nose, and T. Sakurai, "A Super Cut-Off CMOS (SCCMOS) Scheme for 0.5V Supply Voltage with Pico Ampere Stand-by Current," *IEEE J. Solid State Circuits*, Vol. 35, pp. 1498–1501, 2000.
80. K. Seta et al., "50% Active-Power Saving without Speed Degradation Using Standby Power Reduction (SPR) Circuits," *ISSCC*, pp. 318–319, 1995.
81. T. Kuroda et al., "A 0.9V, 150 MHz, 10 mW, 4 mm², 2-D Discrete Cosine Transform Core Processor with Variable-Threshold-Voltage Scheme," *ISSCC Dig. Tech. Papers*, pp. 166–167, 1996.
82. H. Mizuno et al., "A 18-µA Standby Current 1.8V 200 MHz Microprocessor with Self-Substrate-Based Data Retention Mode," *IEEE J. Solid-State Cir.*, Vol. 34, pp. 1492–1500, 1999.
83. S.V. Kosonocky et al., "Enhanced Multi-Threshold (MTCMOS) Circuits Using Variable Well Bias," *Proc. ISPLED*, pp. 165–169, 2001.
84. M. Mizuno et al., "Elastic-V_T CMOS Circuit for Multiple On-Chip Power Control," *ISSCC*, pp. 300–301, 462, 1996.
85. K. Kumagi et al., "A Novel Powering-down Scheme for Low V_t CMOS Circuits," *Symp. VLSI Circuits, Dig. Tech. Papers*, pp. 44–45, 1998.
86. M.J. Lee and K.W. Park, "A Mechanism for Dependence of Refresh Time on Data Pattern in DRAM," *IEEE Elect. Dev. Lett.*, Vol. 31 (2), pp. 168–170, 2010.
87. M.J. Lee, K.W. Park, and J.H. Ahn, "Selective Negative Word Line Scheme for Improving Refresh," *Electronic Letters*, Vol. 47, No. 3, Feb., 2011.
88. K. Sato, et al., "A 20 ns Static Column 1 Mb DRAM in CMOS Technology," *IEEE, ISSCC Dig. Tech. Papers*, pp. 254–255, 1985.
89. D.S. Lee, Y.H. Jun, and B.S. Kong, "Simultaneous Reverse Body and Negative Word-Line Biasing Control for Leakage Reduction of DRAM," *IEEE J. Solid State Circ.*, Vol. 46, pp. 2396–2405, 2011.
90. M. Horiguchi et al., "Switched-Source-Impedance CMOS Circuit for Low Standby Subthreshold Current Giga-Scale LSI'S," *IEEE J. Solid State Circuits*, Vol. 28, pp. 1131–1135, 1993.
91. J.P. Halter and F.N. Najm, "A Gate-Level Leakage Power Reduction Method for Ultra-Low Power CMOS Circuits," *IEEE Custom. Integ. Cir. Conference*, pp. 475–478, 1997.
92. B.S. Deepaksubramaniam and A. Nunez, "Analysis of Subthreshold Leakage Reduction in CMOS Digital Circuits," *50th Midwest Symp. Circuits and Systems* (MWSCAS), pp. 1400–1404, 2007.
93. P. Ghafari, M. Anis, and M. Elmesry, "Impact of Technology Scaling on Leakage Reduction Techniques," *North East Workshop on Circuits and Systems, NEWCAS*, pp. 1405–1408, 2007.
94. A. Keshavarzi et al. "Technology Scaling Behavior of Optimum Reverse Body Bias for Standby Leakage Power Reduction in CMOS IC's," *ISLPED*, pp. 252–255, 1999.
95. K. Nose et al., "V_{TH}-Hopping Scheme to Reduce Subthreshold Leakage for Low-Power Processors," *IEEE J. Solid State Circuits*, Vol. 37, pp. 413–419, 2002.
96. S. Lee and T. Sakurai, "Run-Time Voltage Hopping for Low-Power Real-Time Systems," *IEEE/ACM Proc. Design Automation Conf.*, pp. 806–809, 2000.
97. K. Itoh, M. Horiguchi and M. Yamaoka, "Low-Voltage Limitations of Memory-Rich Nano Scale CMOS LSIs," *Proc. Eur. Solid State Circuits Conf.*, pp. 68–75, 2007.

98. S. Akiyama, et al., "Low-V_t Small-Offset Gated Preamplifier for Sub-1 V Gigabit DRAM Arrays," *IEEE Int. Solid State Cir. Conf., Dig. Tech. Papers*, pp. 142–143, 2009.

99. A. Kotabe, et al., "A 0.5 V Low-V_T CMOS Preamplifier for Low-Power and High-Speed Gigabit-DRAM Arrays," *IEEE J. Solid-State Cir.*, Vol. 45, pp. 2348–2355, 2010.

100. S. Tanakamaru and K. Takeuchi, "A 0.5 V Operation V_{TH} Loss Compensated DRAM Word-Line Booster Circuit for Ultra-Low Power VLSI Systems," *IEEE J. Solid-State Cir.*, Vol. 46, pp. 2406–2415, 2011.

101. A. Hairawa et al., "Local-Field-Enhancement Model of DRAM Retention Failure," IEDM Tech. Dig., pp. 157–160, 1998.

102. S. Ueno, Y. Irone, and M. Inuushi, "Impact of Two Trap-Related Leakage Mechanisms on the Tail Distribution of DRAM Retention Characteristics," IEDM Tech. Dig., pp. 37–40, 1999.

103. K. Saino et al., "Impact of Gate-Induced Drain Leakage Current on the Tail Distribution of DRAM Data Retention Time," IEDM Tech. Dig., pp. 837–840, 2000.

104. W-S. Lee et al., "Analysis on Data Retention Time of Nano-scale DRAM and Its Prediction by Indirectly Probing the Tail Cell Leakage Current," IEDM Tech. Dig., pp. 395–398, 2004.

105. J. Lee, D. Ha, and K. Kim, "Novel Cell Transistor Using Retracted Si_3N_4-Liner STI for the Improvement of Data Retention Time in Gigabit Density DRAM and Beyond," *IEEE. Trans. Electron Devices*, Vol. 48, pp. 1152–1157, 2001.

106. H. Fukutome et al., "Direct Measurement of Effects of Shallow-Trench Isolation on Carrier Profiles in Sub-50 nm N-MOSFETs," *Symp. on VLSI Tech.*, pp. 140–141, 2005.

107. S. Lee et al., "Lateral-Extended (Lat Ex.) Active for Improvement of Data Retention Time for Sub 60 nm DRAM Era," *Sol. State-Device Research (ESSDRC)*, pp. 327–329, 2007.

108. S. Park et al., "A Novel Method to Analyze and Design a NWL Scheme DRAM," *IEEE, 46th Annual Int. Relia. Physics Symp.*, pp. 701–702, 2008.

109. K. Okonogi et al., "Lattice Strain Design in W/WN/Poly-Si Gate DRAM for Improving Data Retention Time," IEDM Tech. Dig., pp. 65–68, 2004.

110. K. Okonogi et al., "Improvement of Data Retention Time Property by Reducing Vacancy-Type Defects in DRAM Cell Transistors," *IEEE 44th Annual Int. Relia. Physics Symp.*, pp. 695–696, 2006.

111. A. Bouhada, A. Tonhami, and S. Bakkali, "New Model of Gate-Induced Drain Current Density in an NMOS Transistor," *Microelectronic J.*, Vol. 29, pp. 813–816, 1998.

112. K. Kim and J. Lee, "A New Investigation of Data Retention Time in Truly Nanoscaled DRAMs," *IEEE Electron Dev. Lett.*, Vol. 30, pp. 846–848, 2009.

113. C.Y. Lee et al., "DRAM Data Retention and Cell Transistor Threshold Voltage Reliability Improved by Passivation Annealing Prior to the Deposition of Plasma Nitride Layer," *IEEE Trans. Dev. Mat. Reliability*, pp. 406–412, 2012.

8

Memory Peripheral Circuits

8.1 Introduction

DRAM organization has been described in Chapter 1. Among the constituents, the core occupies most of the chip area because of the large number of cells replicated in it, whereas peripherals are around it. The shape of the core is either square or rectangular, which makes the dimensions of the row and column decoders and sense amplifiers/drivers well matched with it. It results in proximity between the core and the decoders/sense amplifiers; therefore interconnections occupy less chip space and negligible delay occurs in signal transfer with the core. For the sake of reducing the core (cell) size and improving its performance, some of the performance standards set for a general logic design are compromised. The inherent structured nature of the core is helpful in controlling and minimizing noise generation, which allows relaxing the terms of robustness in comparison to general logic. Another important factor is the reduction in voltage swing at the bit lines during its (dis)charging, though output from the memory is at the full voltage levels. Peripheral circuits now play their role in randomly reaching the cell(s) for reading or writing through row and column buffers and decoders, whereas sense amplifiers help not only in attaining the full voltage level available at the output but also in refreshing the DRAM. The area consumed and the power dissipated by these peripheral circuits also become very important. In addition, propagation delay during access, refreshing, and outputting the signal become extremely important and are considerably affected by the peripherals also. Though the basic structure of row and column decoders has not changed with progressive generations, methods of application have changed with the modes of operation. Sense amplifiers have undergone major changes with the increase in DRAM density and reduction in the availability of sense signal and rise in noise levels due to various reasons. Another important component in the peripherals is the introduction of redundancy and error detection for enhancing the yield. This chapter introduces peripherals and the idea of redundancy.

8.2 Address Decoder Basics

DRAM operating modes are either asynchronous or synchronous. In synchronous mode all operations are controlled by a system clock, whereas in asynchronous DRAM, control signals $\overline{RAS}$ and $\overline{CAS}$ are used for reading, writing, refreshing, and so on. Most of the initial DRAMs were asynchronous; however, to increase data read, frequency synchronous modes were adopted. Irrespective of the operating mode, (1) address multiplexing is almost universal, mainly to reduce pin count and hence cost, and (2) random nature of the access of a cell still remains. In simplest form, all DRAM operations start with the lowering of $\overline{RAS}$ signal, as in this case the chip does not need chip select ($\overline{CE}$) signal, and activates row address buffers, row decoders, word line drivers, and bit-line sense amplifiers and then lowers $\overline{CAS}$ after a certain duration. One of the word lines gets selected and then some of the data from the selected word line is reached at the falling edge of the $\overline{CAS}$ signal through column address decoder.

8.2.1 Row Decoders

Input to a row decoder is an n-bit address and there are 2^n available outputs, of which only one output activates a row that contains the memory cell to be selected. In its simplest form, a 1 out of 2^n decoder is a collection of 2^n, logic gates each with 1^n. Each gate can be realized using an n-input NAND gate and an inverter; alternatively, it can be realized by an n-input NOR gate for each row. However, these simple realizations have many constraints. The layout of the wide (many) inputs gate must fit as per the pitch of the word line. In addition, NOR and NAND gates with three or more inputs suffer from large propagation delay which would add to the overall delay of the memory operation. Power consumption of the decoder is also an important consideration. Hence, the decoders are always realized by splitting the logic in terms of gates with two or three inputs only. The initial part of the logic, which decodes the address, is called a *predecoder*, and the later part, which provides the final output for the word line, is known as a *decoder*. Figure 8.1(a,b) show the basic schematic of a decoder and its implementation using six-input static CMOS NAND gates: a wide and slow gate. Figure 8.1(c) shows conversion of wider gate into a version of predecoder–decoder. Reduction in the number of inputs to the gates results in considerable improvements in the propagation delay and power consumption of the decoder. For an eight-input decoder, the expression for WL_0 can be written as

$$WL_0 = \frac{\overline{A_0}\overline{A_1}\ldots\ldots\ldots\overline{A_7}}{(A_0 + A_1)\ldots(A_6 + A_7)}$$

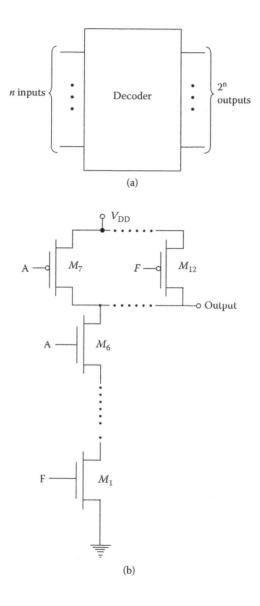

(a)

(b)

FIGURE 8.1
(a) An n-to-2n binary decoder. (b) Six-input NAND gate. (c) A possible predecoder and decoder configurations with 2 or 3 input gates.

One realization for this case can have two-bit predecoding, which requires both true and false bits. Output predecoders can then be combined in four-input NAND gates for final word line outputs as shown in Figure 8.2. If the predecoder is realized using complementary CMOS logic, number of transistor used in this decoder shall be $(256*8) + (4*4*4) + (4*4) = 2128$ in place of 4112 (including static inverters) if it was realized in a single stage. Four

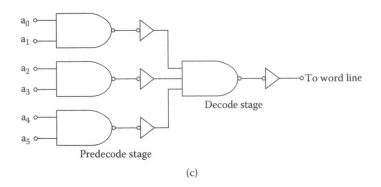

Predecode stage

Decode stage

To word line

(c)

FIGURE 8.1 (continued)

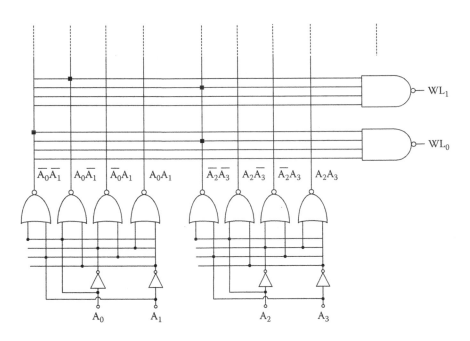

FIGURE 8.2
A NAND decoder using two-input predecoders.

input NAND decoders can further be converted in more than one logic level containing only two input gates and inverters. An additional advantage of using predecoders is that those decoder blocks which are not selected can be disabled through the addition of a select signal to each predecoder. This results in power saving from the unutilized decoder blocks. Word lines need sufficient capacity decoder drivers for charging and discharging [1], hence proper sizing of the transistors used in the gate becomes important.

8.2.1.1 Dynamic Decoders

Decoders can also be realized using dynamic logic. Basic forms of dynamic decoders are shown in Figure 8.3(a) and (b); these are 2-to-4 NOR decoder and 2-to-4 NAND decoder, respectively. Point of difference is that the NAND decoder shown is *active low* type in which selected word line is low, otherwise all lines are normally high. NAND decoders occupy smaller area and consume less power in comparison to NOR decoders but become very slow with increasing size. Hence, for larger decoders, a multilayer approach similar to the static decoder is used.

8.2.2 Column Decoders

Column decoder is essentially a 2^M-multiplexer, where M stands for the bits stored in a row (or bits in selected block row for a multi-block DRAM). During the read operation precharged bit lines are to discharge to the sense amplifiers, while during write a *low* operation bit line has to be driven low. Since read and write operations requirements are slightly different, different multiplexers are often used for the two operations for optimization, though a shared multiplexer is also used.

A simple but costly approach for realizing column decoders is based on pass transistor logic as shown in Figure 8.4(a). Control signals for the pass transistors are generated using a NOR-based predecoder. Advantage of this arrangement is that only one NMOS is on at a time which routes the selected column signal to the data output and makes it very fast. If more than one data is selected, as many transistors can route the output in parallel. However, the main disadvantage of the scheme is the requirement of $2M$ pass transistors for each bit line and $M*2^M$ transistors for the decoder circuit; the scheme becomes prohibitively large for large values of M. In fact, the number of transistors increases further for a shared read-write multiplexer in which instead of NMOS only, a transmission gate-parallel combination of an NMOS and a PMOS is used for the passage of full signal levels.

An alternative scheme using considerably fewer transistors based on binary selection of bits called *tree decoder* is shown in Figure 8.4(b). As column bit line drives the NMOS pass transistors, NOR decoder is not required, which reduces the total number of transistors considerably. As far as functionality is concerned, pass transistor network selects one out of every two bit lines at each stage. Major drawback of the scheme lies in presence of as many transistors in the data path as the number of address bits, which slows operation for large values of M.

Obviously, choice between the two main schemes depends on a number of factors, like chip area, speed requirement, and memory architecture. Hence, an optimum combination of the two schemes is used where a fraction of bits are predecoded, while the remaining bits are tree decoded.

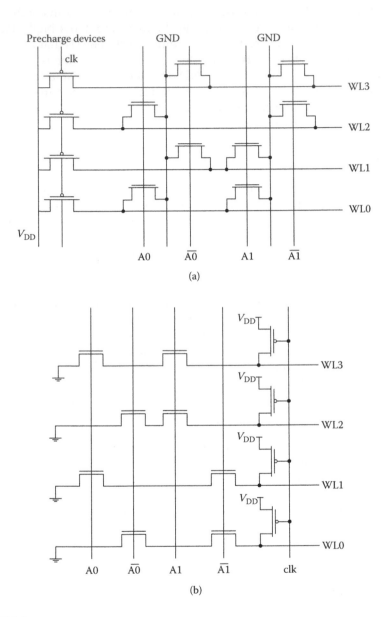

FIGURE 8.3
(a) Dynamic 2-to-4 NOR decoder. (b) A 2-to-4 MOS dynamic NAND decoder. This implementation assumes that all address signals are low during precharge.

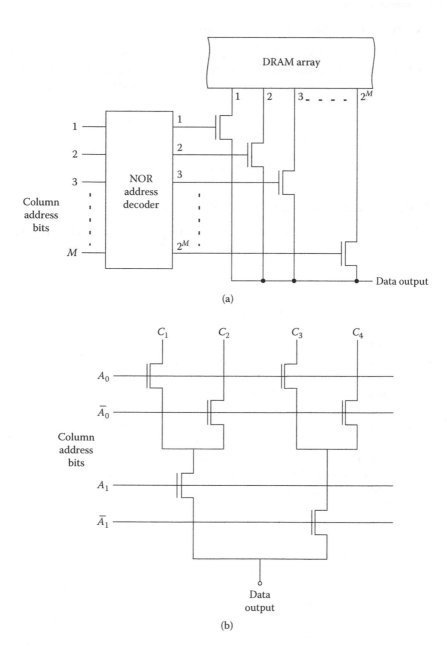

FIGURE 8.4

(a) Bit line (column) decoder arrangement using NOR address decoder and logic pass transistors. (b) Binary tree column decoder circuit for four bit lines.

8.3 Address Decoding Developments

At initial levels of DRAM density, the number of addresses and total number of I/O pins are small. With increased density, time multiplexing was introduced to reduce the pin count (hence the chip cost). Half the address bits were used to select one of the rows, followed by the application of the remaining half address bits to select a particular column. Separate application of row and column bits opened ways to some other methods of reading termed as *modes of operation* and different methods of address decoding. Some of the developments are briefly discussed in following sections.

8.3.1 Multiplexed Address Buffer and Predecoding

Use of predecoders was shown to result in a drop in the number of transistors required for row and column decoders. As the number of logic level is more than one, predecoding also reduces gate capacitance loading of row/column address drivers. However, for effective address multiplexing, the same address buffer, wherein the address bit information is stored, should be used for the rows and column. To achieve this objective, its address line is disconnected from the column address bus after latching the row address in the row decoder. Additional advantage comes in the form of reduced capacitive loading on the address driver during column access. Figure 8.5(a) shows a basic block form of predecoding with address buffers, where addresses are processed by one of four predecoders. The processing is preceded by storing bit information in an address buffer as shown. A CMOS DRAM address buffer is shown in Figure 8.5(b). A NAND input stage gets activated and samples address information through a CAB clock controlled by $\overline{RAS}/\overline{CAS}$ strobes. Bit information is then stored in the latch. A precharge clock $\overline{CABP}$ presets the address latch to high address state [2]. Additionally, buffers provide drivability and complimentary outputs.

During the 1980s most of the DRAM constituent components were converted from NMOS to CMOS. One of the main reasons was the need in power consumption reduction. In ref. [3] one such attempt was made while using CMOS row decoder in place of NMOS decoder. It was shown that in the CMOS only one decoder was discharging, which meant that capacitance involved was much less than in the NMOS decoder and the power dissipation in the decoder was reduced to only 4% compared to the NMOS version. Another scheme for the suppression of power consumption in CMOS word driver and decoders was given by Kitsukawa et al. [4]. The basic idea in the scheme was the reduction of sub-threshold current up to 3% of the conventional circuits by forcing self-reverse biasing of the driver transistors.

Keeping basic arrangement of buffering, predecoding, and final decoding, changes/improvements have been made in the row decoding with the

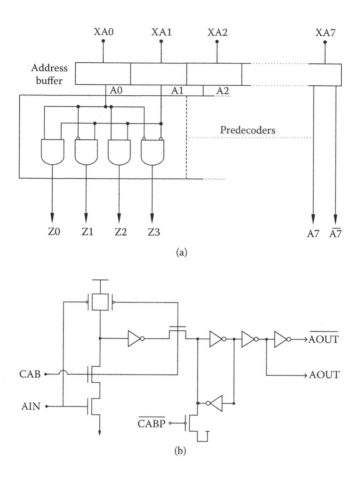

FIGURE 8.5

(a) CMOS dynamic RAM address predecoding scheme. (b) CMOS dynamic RAM address buffer circuit. (Redrawn from "A 70 ns High Density 64 K CMOS Dynamic RAM," R.J.C. Chwang et al., IEEE J.S.S. Circuits, Vol. SC-18, pp. 457–463, 1983.)

increasing memory density and changes in the architecture; however, most of these have been associated with SRAM [5,6]. These could well be used with DRAMs and hence a decoder circuit used with a 4.5 Mbit CMOS SRAM is discussed briefly, which is comparatively faster with 1.8 ns access time [7]. For comparison's sake, a conventional decoder scheme is shown in Figure 8.6(a) and a decoder using source-coupled logic (SCL) circuit, combined with reset circuits, is shown in Figure 8.6(b). In the conventional circuit, transistors turning on during evaluation phase with clock ϕ_p are larger than those turning on during precharging. However, it increases precharge delays, and hence large word line-signal pulse width resulting in long cycle time and high power dissipation. A reset signal in Figure 8.6(b) changes transistor aspect

Address input

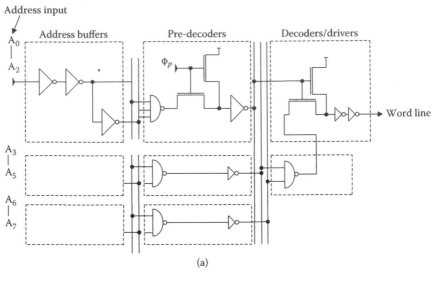

(a)

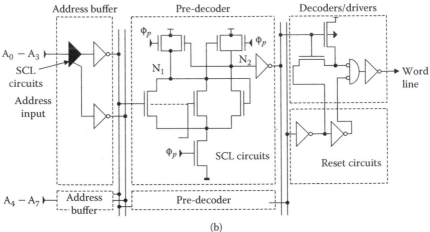

(b)

FIGURE 8.6

(a) A Conventional decoder. (b) Decoder using reset circuits and source-coupled-logic (SCL) circuits. (Modified from "A 1.8 ns Access, 550 MHz 4.5 Mb CMOS SRAM," H. Nimba et al., Proc. ISS. CC. Dig. Tech. Papers, paper SP 22.7, 1998.)

ratio which decreases the precharge time in the driver circuit. SCL circuit shown in Figure 8.6(b) provides OR and NOR output simultaneously. Such outputs are suitable for predecoders with large numbers of inputs, hence decoder with SCL circuit operates faster [7]. Combination of reset and SCL circuits reduced address input to word line signal delay by 32%.

8.3.2 Static Column Operation

Static column mode operation is similar to the page mode operations but without $\overline{CAS}$ strobe. Once a row is selected, output from columns is available as per the address as for static RAM. Figure 8.7(a) shows the timing diagram for such an operation [8]. Row addressing is same as that of a multiplex addressed DRAM beginning with the leading edge of $\overline{RAS}$ strobe. However, for the selection of column, only the address is to be selected and for further reading the next address is to be given without $\overline{CAS}$ strobe. Obviously, first data output depends on $\overline{RAS}$ access time like any conventional DRAM.

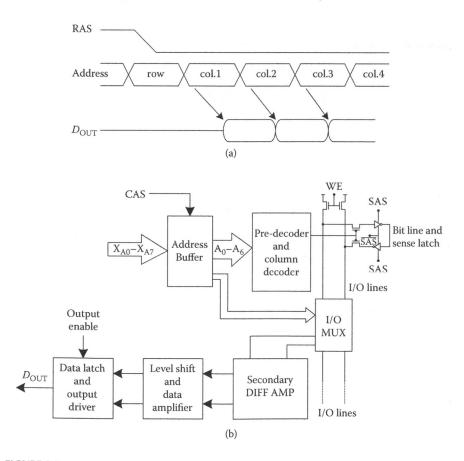

(a)

(b)

FIGURE 8.7

(a) Timing diagram of a static column DRAM. (Redrawn from "A 64 K DRAM with 35 ns Static Column Operation," F. Baba et al., IEEE J.S.S. Circuits, Vol. SC-18, pp. 447–451, 1983.) (b) A CMOS dynamic RAM static column design. (Redrawn from "A 1-Mbit CMOS Dynamic RAM with a Divided Bit line Matrix Architecture," R.T. Taylor and M.G. Johnson, IEEE J.S.S. Circuits, Vol. SC-20, pp. 894–902, 1985.)

For subsequent read outs, no pre-charging is needed which saves considerable time. Since $\overline{CAS}$ strobe is not used, operation becomes simpler and faster. Implementation slightly differs in different reports. Figure 8.7(b) shows an arrangement of static column decoder for 1 Mbit DRAM achieving 22 ns column cycle time. The CMOS sense amplifier behaves as static flip-flop once latched, and it is read like a static RAM cell. Sense amplifier is connected to differential I/O bus through NMOS read/write transistors. Static differential comparator feeds the static output buffer once it senses the small signal on the I/O lines. Static write circuit is also used for fast write operation [9].

8.4 DRAM Sense Amplifiers

An important figure of merit for a 1T1C DRAM cell is the charge transfer ratio given as $C_s/(C_s + C_{BL})$. Typical value of the charge transfer ratio lies between 0.05 and 0.2, and it becomes a figure of merit of the charge division which occurs when access transistor becomes on. Transfer of charge from storage capacitor makes the readout of the DRAM destructive. In addition, charge of C_s is lost quickly (in a few milliseconds) because of leakages. Therefore, the DRAM must be read continuously (disturbing the stored data) and small changes in the voltage of the bit line detected while reading are converted to the full logic value restored to the cell again. All this is done through the sense amplifier without which 1T1C DRAM cell will not function, in contrast to the SRAM case where the sense amplifier is used to speed up the read process, but it is not essential for the functionality.

8.4.1 Gated Flip-Flop Sense Amplifier

During initial stage of DRAM development, sense amplifiers (SAs) used gated flip-flop as shown in Figure 8.8 [10] which has the advantage of regeneration, and hence small initial unbalance quickly grows to full bias voltage. First, transistor T_c is turned on by clock ϕ_3, which connects both nodes of the flip-flop, and it selects dummy word lines on both sides as well. After establishing a precharge voltage on the two nodes, the flip-flop is switched off by switching ϕ_2, $\overline{\phi}_2$ and ϕ_3. For rewriting/refreshing, a storage cell and corresponding dummy word line on the opposite side are selected.

Depending upon the storage capacitor's state 1 or 0, small difference in voltage appears across the nodes of the flip-flop. To bring this small voltage to the full value, flip-flop is switched on by ϕ_2 and $\overline{\phi}_2$ and rewriting/refreshing is done simultaneously in the selected cell and finally the selected word line is deactivated. With basic approach remaining the same, one of the many

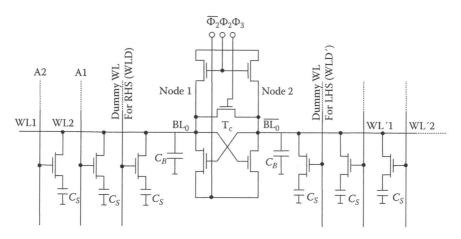

FIGURE 8.8
Sense/refresh circuit with neighboring dummy and storage cells. ("Storage Array and Sense/Refresh Circuit for Single Transistor Memory Cell," K.U. Stein et al., IEEE J. of Solid State Circuits, Vol. SC-7, pp. 336–340, 1972.)

variants of the basic gated flip-flop is given by Kuo et al. [11]. In this scheme the two bit lines were shorted together as before and a reference voltage level was written into the dummy cell structures. Once a cell is selected, two sides of the flip-flop were disconnected and data from the storage cell was then forced onto one side of the SA opposite to the side where reference level is supplied from the dummy cell onto the flip-flop node. As before, a voltage/charge differential appears across the two nodes of the SA. Load transistors are turned on, and both nodes rise higher, but regenerative action returns full logic back onto the cell very fast. However, a major practical problem with this arrangement still remained. The common mode potential at nodes continued to drift down due to junction leakage of the bit line to substrate. If sufficient time elapsed, charge differential became small enough for proper functioning. To avoid a possible problem from the floating bit line voltage, the circuit was modified by Foss and Harland [12] as shown in Figure 8.9. The two bit lines in this topology are clamped to a reference potential V_{REF}; the rest of the operation is similar to the previous cases. Once voltage/charge differential is set up, the sense clock ϕ_s grounds the common sources of the flip-flop and regenerative action starts. The load transistors are turned on by ϕ_L, and the low side rises to a level set by the ratio of the transistor geometries and the high side is pulled to $(V_{DD} - V_{th})$.

Another variation/modification of the gated flop or cross-coupled latch SA, which retains most of the features of earlier sense amplifiers, is shown in Figure 8.10(a), with three internally generated clocks ϕ_0, ϕ_1, and ϕ_2 and the chip enable $\overline{CE}$. Reference cell is precharged to ground and load transistors are turned on through ϕ_2 precharging both bit lines. First, transistor

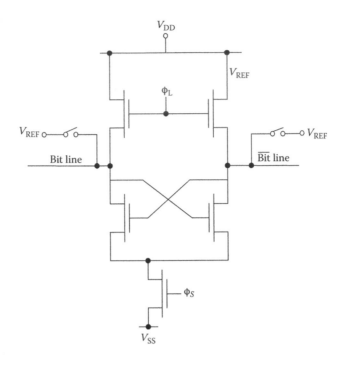

FIGURE 8.9
Sense amplifier given by Ross and Harland. (Redrawn from "Peripheral Circuits for One-Transistor Cell MOS RAM's," R.C. Foss and R. Harland, IEEE J. of Solid State Circuits, Vol. SC-10, pp. 225–261, 1975.)

T_3 is shut off and simultaneously, T_4 and T_5 are turned off through ϕ_2, so that the reference cell is isolated. Once a row is selected, the reference cell and the selected storage cell are connected to two sides of the SA. Soon T_9 is turned on and the signal is amplified on terminals B_1 and B_2 as shown in Figure 8.10(b). Sensitivity of the SA primarily depends on the threshold balance of the driver transistors T_1 and T_2, and the accuracy of the reference potential [13].

8.4.2 Charge-Transfer Sense Amplifier

In the gated flip-flop sense amplifiers, sense signal amplitude (Δv) goes on decreasing with the increase in the bit line capacitance as DRAM density increases, and/or with a decrease in the operating voltage V_{CC}. Along with other noises present, threshold imbalance in the cross-coupled pair of transistor also imposed a limitation on the minimum value of (Δv). Generally, a practical value of C_{BL}/C_S ranges between 8 to 10 for correctly reading data in the DRAM. In addition, analysis shows that set time of a latch circuit increases with reduced Δv or increased transistor threshold voltage imbalance [14]. It is therefore essential that for increased density/decreased operating voltage,

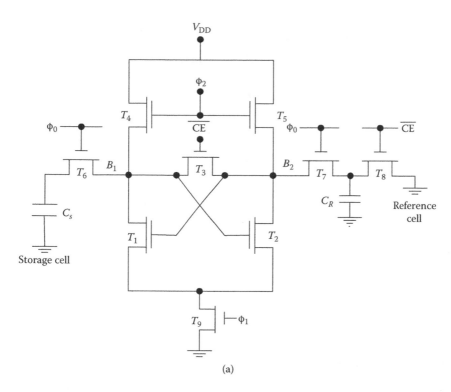

(a)

FIGURE 8.10
(a) Schematic of a sense amplifier showing only a single reference and storage cell, and (b) timing signals. ("A 16 384-Bit Dynamic RAM," C.N. Ahlquist et al., IEEE J. of Solid State Circuits, Vol. SC-11, No. 5, pp. 570–574, 1976.)

a sense amplifier should have better sensitivity than the conventional latch. One such method was promoted by Heller and others [15] in which sense signal is not the result of charge redistribution between the bit line capacitance and the storage capacitance, but it is preamplified using a charge-transfer method before feeding the sense signal to the latch. Bit line is then disconnected from the sense amplifier, making the sense signal independent of the bit line capacitance C_{BL}. Figure 8.11 shows the basics of a charge transfer preamplifier circuit and signal variations where charge of the memory cell capacitance C_S is transferred to a nearly equal size output capacitor C_o through the bit line. However, transistor T_2 isolates the highly capacitive bit line and the output capacitance C_o, and it can also preamplify the sense signal. Initially the word line is low which keeps C_s and C_{BL} disconnected. When clock ϕ_1 is asserted, output voltage rises quickly to *high* and forces T_1 to operate in the ohmic region. Since V_R makes T_2 always on, bit-line voltage is also raised to $(V_R - V_{th2})$. Transition of T_2 from saturation to cutoff takes place, gate voltage V_R is selected such that $V_R \le (V_D + V_{th2})$; hence, T_2 operates in saturation but moves to cutoff when bit line charges up to $(V_R - V_{th2})$. Transistor

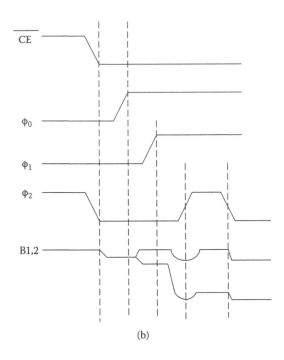

(b)

FIGURE 8.10 (continued)

T_1 is turned off by lowering ϕ_1 and the precharging state completes. Now the word line is raised and for a *high* stored on C_s bit-line voltage remains unchanged at $(V_R - V_{th2})$ with T_2 still off and output remains a *high*. However, for C_S voltage less than $(V_R - V_{th2})$, T_2 becomes momentarily on as bit line voltage dips by a small amount. Bit line recharges back to $(V_R - V_{th2})$ as it gets connected to the output and forces T_2 again to cut off, while potential of C_o drops by $\Delta V_0 = (V_R - V_{th2} - V_s)\, C_s/C_o$. Drop of charge on C_o is equal to the gain of charge on C_s and gain of the charge transfer preamplifier is $(C_{BL} + C_s)/C_o$. It is to be noted that the signal level change on C_o is independent of the bit line capacitance and threshold voltage of the charge transfer transistor.

The charge transfer circuit output is very well suited for connecting to the flip-flop for further amplification as it needs only two transistors and accepts differential input and write backs and refreshes the cell easily. As shown (within dotted rectangles) in Figure 8.12(a) charge transfer preamplifier circuit blocks, consisting of T_1, T_2 on the left and T_1', T_2' on the right, are connected to the latch devices T_3 and T_4. Transistor T_6 enables the latch when required and T_5 is the bit line equalizer as before. It is important to note that if capacitance at node 1 is higher than C_s, a small sense signal is preamplified into a larger signal.

Read/refresh and write cycle follows the sequence as mentioned below at the time instants cited in Figure 8.12(b). At instant t_1 precharging starts as discussed which is essential for read as well as write operation.

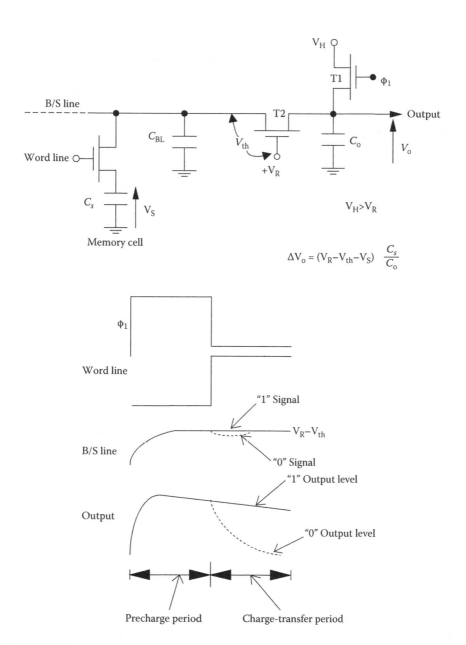

$$\Delta V_o = (V_R - V_{th} - V_S) \frac{C_s}{C_o}$$

FIGURE 8.11
Charge-transfer preamplifier circuit and signal variations. ("High-Sensitivity Charge-Transfer Sense Amplifier," L.G. Heller et al., IEEE J. of Solid-State Circuits, Vol. SC-11, pp. 596–601, 1976.)

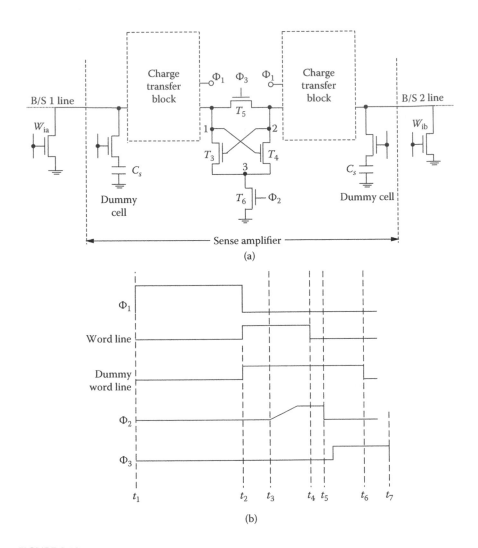

FIGURE 8.12
(a) Balanced charge transfer sense amplifier, and (b) signal variations. (Redrawn from High-Sensitivity Charge-Transfer Sense Amplifier," L.G. Heller et al., IEEE J. of Solid-State Circuits, Vol. SC-11, pp. 596–601, 1976.)

At instant t_2 charge transfer period begins. For a *low* stored cell, node 1 gains charge, which is lost by the cell, and left-side bit-line voltage is dropped. At the same time the dummy cell on the opposite side of the active cell is accessed, and it transfers half as much charge to node 2, dropping its potential nearly half as much as that at node 1. For the write cycle, instead of raising ϕ_2, data is written on bit line and latch proceeds to complete the writing and the word line is made low. By the time instant t_3 differential voltage between nodes 1 and 2 is established and T_6 is made on slowly, which pulls down node 3 which in turn makes transistor T_3 on, but at such a rate that potential at node 1 does not enable transistor T_4. Transistor T_2 works in linear region and left bit line reaches *low* level. At t_4 word line is made off, so C_s gets disconnected from the left bit line and *low* remains stored/refreshed.

At instant t_5, T_6 becomes off and T_5 becomes on through ϕ_2 and ϕ_3, respectively, and potential of the left bit line equalizes to that of the right side bit line to a middle level stored in the dummy cell. At instant t_6, the cycle ends; ready for next precharge and at t_7 nodes 1 and 2 become isolated.

A further modified and improved charge transfer circuit was given by Heller which increased preamplification speed through cross-coupling of isolation transistors of its earlier version [16].

DRAMs realized around 4 Kbit level used NMOS technology and supply voltage was 15 V to 12 V. Sensing was done at half V_{DD} either by shorting two half sections of the bit line or from a separate voltage regulator. Supply voltage being high, there was no problem of detecting deviations in bit line signal. After 4 Kbit level, NMOS DRAMs were standardized at $V_{DD} = 5$ V, which was even less than the earlier used half V_{DD}. With further increase in the DRAM density and reduced supply voltage of 5 V, its capacitance charge got reduced and half-V_{DD} bit line sensing faced limitation in its NMOS version resulting in reduced latching speed and the problem of static power consumption in low bit line side. Hence full V_{DD} bit line precharging came into practice. Even with the charge transfer technique, full V_{DD} bit line precharging was used. However with the change-over from NMOS to CMOS technology, major development took place with the application of half-V_{DD} bit line sensing with a host of advantages [17,18] like reduced peak current while sensing and precharging bit lines, reduced IR drop, reduced voltage bouncing noise due to wire inductance, reduced power for (dis)charging the bit lines, and so on; ref. [17] gives a good account of half V_{DD} precharging of bit lines and its important features. Some other bit line precharging techniques at 16 Mbit level have also been taken up [19–21] at other than ($V_{DD}/2$) levels.

DRAMs with half-V_{DD} precharging have now become almost universal. Usually they have their NMOSs located in GND-biased p-wells and PMOSs in V_{DD}-biased n-wells. Such a construction has certain limitations in spite of its inherent advantages especially when V_{DD} has to be decreased further for low-voltage operating systems. Threshold voltage of NMOS in N-channel sensing is increased due to the body effect arising out of effective *well-biasing* between GND-well and sense amplifier drive line. Body effect further aggravates the fluctuation of threshold voltage because of the limitations of the fabrication processes. The equalization current also gradually decreases, making it slower. Figure 8.13 shows one solution of the problem in the form of the well-synchronized sensing/equalization structure using a triple well, a modified structure of conventional scheme, in which voltage levels of

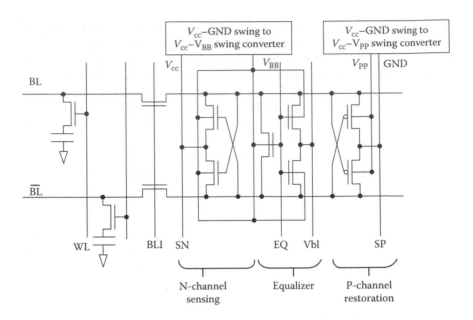

FIGURE 8.13
Well synchronized sensing/equalizing structure. (Adapted from "A Well-Synchronized Sensing/Equalizing Method for Sub-1.0 V Operating Advanced DRAMs," Ooishi et al., Symp. VLSI Circuit Tech. Papers, pp. 81–82, 1993.)

p-well and n-well are controllable independently by the control logic [22]. It makes the sense amplifier transistors free from body effect and consequential slowness of the DRAM. The circuit also has high sensitivity against channel length fluctuations, making it suitable for low voltage operations even below 1.0 V.

With DRAM density increasing beyond 4 Mbit/16 Mbit range, capacitance of the bit line increases even when a small part of it is activated (division of bit line is limited by the assembly technology on chip) and writing resistance from sense amplifier driving node to ground also increases in spite of using resistance reducing techniques. This larger characteristic time constant of the bit line necessitates much longer time for its signal development. Moreover, as large discharge current flows through the bit line and the sense amplifier driving transistor at the time of turning it on, large voltage drop takes place clamping sense amplifier driving node at nearly ground level and reducing effective gate-to-source voltage. In addition, sometimes there is a *weak column*, where for some reason initial sensing signal is smaller than for a normal column. In the weak column, signal development is very slow, which becomes a serious problem in high-density DRAM. A decoded source sense amplifier (DSSA) sensing is shown in Figure 8.14 that solves the two problems mentioned through decoding of each source and connecting it to

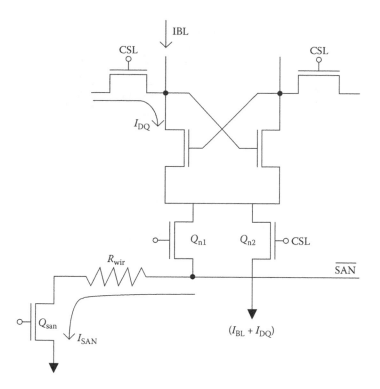

FIGURE 8.14
Sensing scheme of the decoded source sense amplifier showing only a selected column. (Redrawn from "Decoded-Source Sense Amplifier for High-Density DRAM's," J Okamura et al., IEEE J. of Solid State Circuits, Vol. 25, pp. 18–23, 1990.)

the sense amplifier driving node through the use of two additional transistors Q_{n1} and Q_{n2} only [23].

With the requirement of low voltage application, especially in mobile equipment employing multimedia systems, the sense signal from the memory cell became very small. It was reaching near the sense amplifier limits. The charge transfer sense amplifier technique of Section 8.4.2 which tried to overcome the limitation through preamplification was using full supply voltage precharging of the bit lines. However, with the established advantages of half-V_{CC} precharging it became imperative to apply it with the charge transfer sensing and Tsukude and others presented such a scheme in 1997 [24]. Earlier schemes of Heller [15,16] could not be directly applied with half-V_{CC} precharging as it required an excessively high sense amplifier precharge level (V_{SAP}). Principle of the proposed charge transfer presenting scheme (CTPS) is shown in Figure 8.15(a), where the transistor TG isolates the SA and the bit line and transfers the stored charge from the bit line to the SA node. As in the earlier charge transfer schemes, stored charge is first

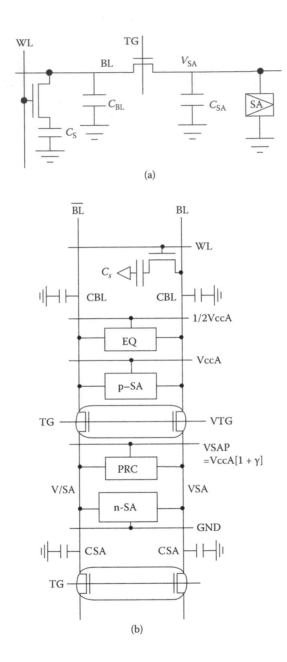

(a)

(b)

FIGURE 8.15
(a) Arrangement of charge transfer pre-sensing with 1/2Vcc BL-pre-charge. (Redrawn from "A 1.2 V to 3.3 V Wide Voltage-Range/Low-Power DRAM with a Charge-Transfer Presenting Scheme," M. Tsukude et al., IEEE J. of Solid-State Circuits, Vol. 32, pp. 1721–1727, 1997.) (b) Circuit organization of CTPS with 1/2Vcc BL precharge at VccA = 0.8 V. ("A 1.2 V to 3.3 V Wide Voltage-Range/Low-Power DRAM with a Charge-Transfer Presenting Scheme," M. Tsukude et al., IEEE J. of Solid-State Circuits, Vol. 32, pp. 1721–1727, 1997.)

transferred to the bit line and then to the SA node. Utilization of the half-V_{DD} bit line precharging CTPS with SA circuit arrangement for a DRAM cell is shown in Figure 8.15(b). For a 0.8 V array operation, bit line was precharged to 0.4 V and SA node was precharged to 1.6 V to satisfy the charge transfer voltage level conditions. During read/refresh operation, gate voltage of TG is to be selected for its proper operating node to transfer charge from bit line to SA node. As before, the whole operation is divided into charge transfer state and sensing state and is once again ready for the next precharging. A five times larger sense signal was easily obtained compared to the conventional sensing scheme and data transfer and sensing speed increased by 40% in 0.8 V array operation without the requirement of reduced threshold voltage in sense amplifier [24].

Well-synchronized sensing was used to suppress body effect in SA transistor, but it suffered from noise generation because of forward biasing between the bit line and the well before sensing. In a modified *charge-transfer well sensing scheme*, the problem was considerably reduced by having well potential less than half-V_{DD} [25]. Moreover, in this scheme the well under SA was not charged from the power supply, hence there was no power loss compared to the earlier well sensing [22] scheme. However, latching delay is larger than 10 ns in charge transfer well-sensing, which is not suitable for high speed applications.

Overdriving of the access transistor has been used quite effectively in DRAM cells. Overdriving of the SA has also been used by connecting external V_{DD} to it to enhance the data sensing speed while a lower internal array voltage is applied to maintain the stored voltage [26]. Sometimes initial voltage to the SA is greater than V_{DD} also, though for short duration, but it sends larger current creating difficulty in power supply design. Another low-voltage sense amplifier driver scheme was proposed to work at supply voltage below 0.9 V and was able to give higher speed using SA overdriving [27] using boost capacitors. Bit line was temporarily isolated from sense amplifier to reduce the capacitive load of the sense amplifier so as to reduce the column switching time. Though the isolation transistor drivers and the boost capacitor driver consume extra power, it was reduced using charge recycle technique. Bit line isolator (BI) transistors were temporarily closed during data sensing to reduce capacitive load on SA. For $V_{CC} = 0.9$ V, capacitors provided V_{pp} (= 1.8 V_{CC}) which drives BI and $V_{CC}/2$–precharge transistors. To reduce the power consumption, charge recycling effected a 28% reduction compared to the case without charge recycling, though overall power consumption was still more than 50% compared to conventional SA drivers [27]. It is not only the overdriving but also the method of overdriving of SA that is important for reducing the duration of sensing; Table 8.1 shows such a comparison. Higher density DRAM shall have impractically large sensing duration at reduced array voltage of 1.0 V or below, whereas conventional overdriving does not provide satisfactory performance. A distributed driver scheme along with meshed power line has shown improvement of 2.0 ns and

TABLE 8.1

Comparison of Sensing Duration for Non-Overdriven and Overdriven SAs
in Nanoseconds

Bit Line Swing (Volts)	DRAM Density, Technology Node (μm)	Conventional Non-Overdriven	Conventional Overdriven	Distributed Overdriven Scheme [28]
1.8	256 Mb (0.18)	5.6	4.8	3.0
1.6	256 Mb (0.18)			
	1 Gb (0.13)	6.2	4.9	3.2
1.4	1 Gb (0.13)	7.9	5.6	3.6
1.2	1 Gb (0.13)			
	4 Gb (0.09)	11.0	6.0	4.0
1.0	4 Gb (0.09)	Impractical	7.4	5.2
0.8	4 Gb (0.09)	Impractical	11.5	7.4

7.0 ns for 1 Gbit DRAM at 1.2 V of array voltage over a conventionally over-driven and non-overdriven SA [28].

8.4.3 Threshold-Voltage Mismatch Compensated Sense Amplifiers

With increased DRAM density, number of SAs also increased, occupying larger chip area as core of the SAs consisted of pairs of NMOS and PMOS transistors. Apart from the large number of SAs, use of planar transistors increases total chip area occupied. Advanced transistor structures, which are used in DRAM cores, are generally not used in sense amplifiers. Efforts are always on to see that overheads because of SAs are restricted, but a more serious problem encountered in the SAs is the generation of offset, forcing a minimum required sense signal. The difference between the sensing signal and offset voltage is critical as it affects the duration of signal restoration by the SA and DRAM cell refreshing period as well. Reasons for the generation of offset are different and the components responsible for it change their weightage at different density levels and the technologies used, as will be shown later, but one of the most significant reasons is the threshold voltage (V_{th}) mismatch between transistor pairs of the SAs. Even with the most advanced technology, threshold mismatch (ΔV_{th}) is unavoidable, though it is reducible, obviously at a higher manufacturing cost. It is, therefore, very important to use compensation/cancellation of the effect of ΔV_{th}, to have stable operation of the SAs. The problem was identified early and circuit techniques for V_{th} compensation were initiated even in 1979 [29]. The basic idea in such techniques was to change either gate voltage [29] or source voltage [30] to set gate source voltage in the transistor pairs equal to their own threshold voltage before reading the signal from a cell. Hierarchical data line architecture with direct sensing scheme was given by Kawahara et al. for V_{th} mismatch compensation in SAs along with reducing their area as well [31].

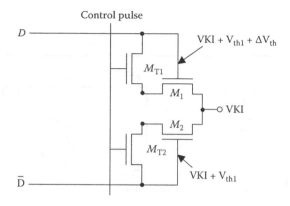

FIGURE 8.16

Schematic of sense amplifier operation ΔV_{th} compensation. (Redrawn from"A High-Speed, Small-Area, Threshold-Voltage-Mismatch Compensation Sense Amplifier for Gigabit-Scale DRAM arrays," T. Kawahara et al., IEEE J. Solid-State Circuits, Vol. 28, pp. 816–823, 1993.)

Figure 8.16 shows given V_{th} compensation scheme, where D and $\overline{D}$ is the sub-data-line pair which is different from global-data-line. Pair of transistors M_{T1} and M_{T2} is used for V_{th} compensation of the signal detection transistors M_1 and M_2. When transistors M_{T1} and M_{T2} are made on, these are converted as diodes for M_1 and M_2 while discharging data line capacitance corresponding to a level at its own V_{th} [31]. Normally, V_{th1} is higher (lower) than V_{th2} by ΔV_{th}, which needed cancellation. In this scheme, voltage in line D is set higher than $\overline{D}$ by ΔV_{th} when control pulse is on. Hence, original ΔV_{th} between M_1 and M_2 is cancelled.

Bit line direct sensing schemes have the advantage of fast signal development in comparison to the conventional cross-coupled sense amplifiers since they do not require any dynamic timing margin. However, static amplifiers require offset/imbalance compensation to maintain sensitivity. Several schemes have been given for compensating imbalance in dynamic cross-coupled SAs, which generally used bit line precharge level adjustment for the purpose [32,33]. To overcome their common drawback of slower operating speed, a direct bit line sensing scheme using a current mirror differential amplifier was proposed in 1999 [34], as shown in Figure 8.17. Thicker lines show the circuit part, which is used for read operation for the offset compensation scheme (OCS) in a simplified bit line SA. Switching is used in such a way that BL is automatically precharged to a level that compensates the offset, though accuracy of level shifting of BL also depends on the finite gain of the differential amplifier. The scheme worked satisfactorily at around 256 KB DRAM density level, but it could not be applied for manufacturing because of the large size of the compensating circuit, which could not fit in the pitch. A similar scheme was given by S. Hong and others which improved the data retention time of DRAM by nearly 2.4 times at an operating voltage of 1.5 V

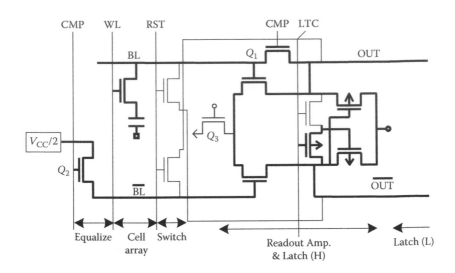

FIGURE 8.17
Schematic diagram of the OCS circuit. (Redrawn from "Offset Compensating Bit-Line Sensing Scheme for High Density DRAM's," Y. Watanabe, N. Nakamura and S. Watanabe, IEEE J. Solid-State Circuits, Vol. 29, pp. 9–13, 1994.)

and reduced required number of SAs as each SA supported larger number of cells [35]. Figure 8.18 shows the basic offset cancellation sense amplifier (OCSA) circuit. Normal read operation begins with $V_{DD}/2$ precharging, but simultaneously offset is canceled by forming current mirror on left-hand side and the sensed signal is allowed to develop with word line raised. After slight signal development, modification takes place in the form of signal *locking* using positive feedback with only M_8 on. Slightly enhanced bit line signal at the *lock* stage is able to overcome any remaining offset and signal restoration follows in usual manner.

The problem of having a differential amplifier and use of quite a few control switches, which could not be easily fabricated within a single pitch of a SA, still remained in the discussed schemes. Moreover, large current drawn by the differential amplifiers and delay caused by the offset cancellation process before activation of WL were the other drawbacks. Complication further arose because of the continued reduction in the DRAM core operating voltage, especially for mobile applications. For low voltages multistage amplification became imperative even with direct sensing. At the same time $V_{CC}/2$ precharge scheme even at low voltage was desirable, which necessitated the application of boosted voltage for bit line equalization. However, generation of boosted voltage required larger current for pump circuits. An offset compensated presensing (OCPS) scheme has been given which retains only BL and $\overline{BL}$ nodes in an SA pitch. Switching arrangement has also been simplified by using two separated amplifiers and only one of them provides offset cancellation at a time [36]. A charge recycle precharge (CRP) scheme was also

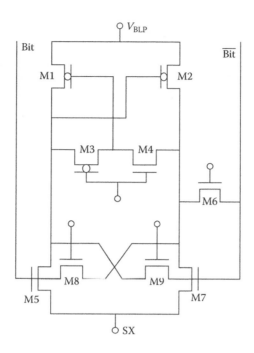

FIGURE 8.18
Offset-cancelation sense amplifier. (Redrawn from "Low-Voltage DRAM Sensing Scheme with Offset-Cancellation Sense Amplifier," S. Hong et al., IEEE J. Solid-State Circuits, Vol. 37, pp. 1356–1360, 2002.)

employed where core voltage was 1.0 V, boosted bit-line equalization voltage was 1.2 V, and WL voltage was 2.7 V. Overall chip size overhead due to the use of OCPS was 3% and at $V_{CC} = 1.0$ V, access time of 25 ns could be achieved.

8.4.4 Low-Voltage Charge-Transferred Presensing

Due to the nonscalable nature of the threshold voltage of the access transistor, its drivability is reduced at low DRAM core voltage. For any up-voltage conversion, pumping circuits consume power. Available charge-transferred presenting schemes (CTPS) [15,16,24] did improve performance at low voltage but suffered from constraints. Biasing should be within close limits so that isolation transistor(s) remain in saturation, and ratio of storage capacitance with bit line capacitance also affects the operating mode of the isolation transistor. A new scheme eliminates any constraint on the bias levels of the conventional CTPS [37]. Figure 8.19 shows basics of the proposed CTPS scheme. Transistor N_7 and N_8 are turned on after the initial charge sharing phase, which allows initial presenting on account of capacitance difference between C_{BL} and C_{SBL}. Regenerative action due to positive feedback in the transistor N_9 network takes place. Normal sensing by P_1 and P_2 follows, which is preceded by the pulling down of one of the bit lines. Presence of

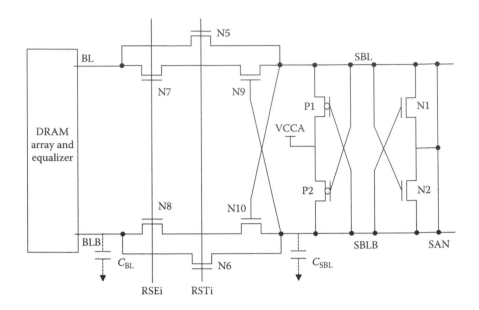

FIGURE 8.19
Circuit diagram of the charge transfer presenting scheme (CTPS). (Redrawn from "Charge-Transferred Presenting, Negatively Precharged Word-Line, and Temperature-Insensitive-Up Schemes for Low-Voltage DRAMs," J.Y. Sim et al., IEEE J. Solid-State Circuits, Vol. 39, pp. 694–703, 2004.)

positive feedback eliminates any chance of polarity change between SBL and SBLB. If the threshold voltage of the access transistor is less than half of array supply voltage, then there is no constraint on bias voltage for circuit reliability [37]. In conventional CTPS schemes BL flipping can occur if column line is selected early during the sensing process; therefore, to avoid it, selection of column line is delayed by about 3 ns. In the proposed scheme, presence of positive feedback resulted in the sensing of this time delay at array voltage of 1.3 V, without any time delay.

Trends for DRAM supply voltage indicated it to be reduced to 1.0 V by 2010; therefore working with data line voltage of 0.5 V was becoming essential [38]. At this reduced voltage, sense signal available from bit lines is very near to the sensing limit and sensing speed also becomes slow. Charge transfer schemes tried to overcome this problem, as discussed earlier, at a relatively larger voltage. A boosted charged transfer preamplifier scheme (BCTP) tried to solve the problem for a 0.5 V DRAM and aimed to provide higher speed as well [39]. However, some fine reports by Hitachi have served well in this direction [40–43], and Figure 8.20 shows the development. First among such schemes in Figure 8.20 (without the block in dotted lines) is a low V_{th} gated preamplifier (LGA) in parallel with a conventional high V_{th} SA. Because of the low value of V_{th} in the LGA, it quickly amplifies the small bit line differential signal (v_{so}) even with $V_{DD}/2$ array precharging. The amplification is

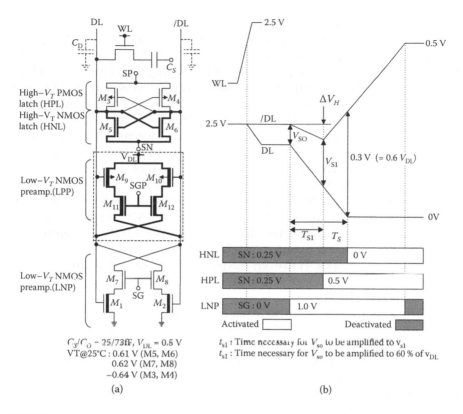

FIGURE 8.20
(a) Schematic diagram of a sense amplifier having low VT preamplifier and high VT CMOS latch (SANP), modified to CPA and for asymmetrical cross-coupled SA (ASA). (b) Operational sequence of the SANP. ("Asymmetric Cross-Coupled Sense Amplifier for Small-Sized 0.5 V Gigabit-DRAM Arrays," A. Kotabe et al., IEEE Asin S.S. Circ. Conf., 2010.)

achieved by turning transistors M_7 and M_8 on though SG and bit line differential voltage value is reached above the offset voltage of the high V_{th} SA (v_{s1}). Full restoration of voltage level is done by the high V_{th} conventional SA and the SG level turns off the low V_{th} preamplifier to reduce leakage. At 70 nm technology a 128 Mbit DRAM was fabricated using LGA which has a read access time of 14.3 ns at read current of 75 mA and row access time of 16.4 ns at an array voltage of 0.9 V. Application of LGA for gigabit DRAMs had certain limitations. For example, threshold voltage of the high-V_{th} NMOS should be less than 0.61 V and threshold voltage of the high V_{th} PMOS should be greater than −0.64 V so that leakage current does not increase above 100 μA for 128 SAs. Secondly, with data line voltage at 0.5 V, and with optimum threshold voltage of low V_{th} NMOS of 0.3 V, achievable minimum sensing time t_s is 15.6 ns. One of the reasons t_s could not be reduced was the voltage drop at the high-level data line (ΔV_H) as it increased the total charge supplied for the data line by high V_{th} PMOS.

To improve upon the LGA, a CMOS low V_{th} preamplifier (CPA) was developed in which low V_{th} PMOS cross couple (M9, M10) and PMOSs (M11, M12) are added to it as shown in Figure 8.20 (block in dotted line now included) [41,42]. To achieve both better speed and data sensing with low power, a low-V_{th} (NMOS and PMOS) activation scheme was used in which both old and newly added preamplifiers get activated simultaneously. It results in reduced value of ΔV_H, thereby increasing the sensing speed. The rest of the operation is similar to the LGA. In this scheme sensing time was reduced to 6 ns (62% shorter than for LGA) and writing time was reduced to 16.3 s (72% less than a high V_{th} CMOS latch only). Data line charging current also reduced by 26% with decreases in data line voltage from 0.8 V to 0.5 V.

Addition of low V_{th} PMOS preamplifier improved the operation, but it also increased chip area; an overall increase of 9% for 128 Mbit bank. Hence, further modification was done in which two cross couples of CPA are eliminated which not only reduced required area but provided faster operation as well as low data leakage [43]. However, the modified simpler structure, as such, called asymmetric cross-coupled SA (ASA) was slow in sensing and writing due to high V_{th} PMOS latch and large leakage from low V_{th} NMOS preamplifier. The problem was solved by using an overdrive of common source (SP) and reduction in the threshold of PMOS. Power dissipation increased due to increased overdrive voltage, hence a new overdrive of the high V_{th} PMOS with transfer-gate clocking is used which sped sense and restore operations. An adaptive leakage control of the preamplifier was also used to considerably reduce leakage and maintain safe limits. Figure 8.20 shows the structure (all components shown with thick lines deleted) of a column circuit using ASA for high-speed operation and small leakage in active standby mode.

8.5 Error Checking and Correction

Alpha particle-induced soft error is a serious problem in high-density DRAMs. Apart from steps taken for its prevention, error checking and correction (ECC) circuits are extremely important. A 4 Mbit DRAM fabricated in 0.8 μm CMOS technology has employed concurrent 16-bit ECC(C-ECC) along with duplex bit line architecture (DBA) [44]. The DBA architecture provides multiple bit internal operations and with C-ECC multiple-bit error checking and correction could become possible without expanding memory.

In the conventional bidirectional parity code method, vertical and horizontal cell data are processed and checked [45]. However, method becomes complex for multiple data output. On the other hand, in C-ECC, all V-directional data is checked concurrently with simplified process [44]. The C-ECC consumes less chip area (~7.5% of whole area) than the conventional ECC.

Improvement of nearly 10^{10} was obtained in the soft error rate (SER) in the 16-bit concurrent ECC.

Other than the horizontal/vertical (HV) parity code, built-in ECC circuits also use Hamming code. The Hamming code circuit checks all data bits on an ECC data group simultaneously. Therefore, it has higher error-correcting frequency and provides better reliability. Built-in circuits can use formation of ECC data groups through subarrays. For example, a 256 Kbit DRAM used a short ECC code with eight data bits and four check bits [46]. In a similar scheme, an HV parity ECC data group is formed with data associated with an identical word line in a subarray. Parity checker monitors both horizontal and vertical groups where selected data resides.

Though chip area used is less than first scheme, it incurs a penalty of 5 ns in access time [47]. Limitations of larger chip area consumption or access time delay in different implementations are improved in another Hamming code ECC technique using eight data bits and four check bits. ECC data group is formed with cells that are selected by an identical word line and single error bit is found by calculating the syndrome S using parity checker matrix [48]. The built-in Hamming code ECC combined with redundancy could reduce SER about 100 times less than using HV parity code ECC.

8.6 On-Chip Redundancy Techniques and ECC

As early as 1967, spare rows were used to replace defective ones to get improved yield and enhanced reliability [49]. Soon spare columns were also used along with spare rows [50]. A separate area was allocated to accommodate spare word lines, bit lines, decoders, and a few additional reset lines as shown in Figure 8.21. Decoders in the separate area are different and have two output states. In one state the decoder does not respond to any address, but in the second state an address can be attributed to it, which is the address of the faulty row/column being replaced. However, it is extremely important to find the faulty row/column correctly and get it replaced by the spare line quickly. Different approaches have been used for storing the information about the faulty lines and for the implementation of the redundancy.

In the initial level analysis, it was observed that faulty bits were either due to gross imperfections or due to random defects in the photolithography, processing, and materials [51]. Random defects were further classified as (1) in the correctable area where a cell or word/bit line only was faulty and it could be replaced by the spare line, and (2) in an uncorrectable part due to which most of the chip fails. Uncorrectable or fatal defects can occur in supply lines, clock wires, or address lines. Obviously chip productivity highly depends on uncorrectable defect-susceptible area. In a typical example, for a four-critical-defects-per-4-Kbit array, effective yield was shown to be less

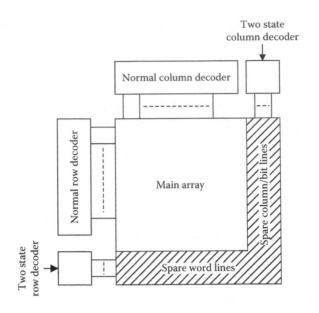

FIGURE 8.21
Spare word lines, bit lines, and decoders for improving yield. (Redrawn from "Multiple Word/
Bit line Redundancy for Semiconductor Memories," S.E. Schuster, IEEE. J. Solid State Circuits,
Vol. SC-13, pp. 698–703, 1978.)

than 2%. However, it increased to more than 70% with the addition of seven
word and seven bit spare lines [51].

Laser fusing redundancy circuits have been used in many applications for
improving yield without speed degradation and with minimum number of
links [52], but fusing involved a mechanically difficult process. Alternatively
an increase in the number of spare word/bit lines has been used to reduce
the required number of fuses, which occupied as much as 25% of the array
area [53]. Comparatively fewer spare rows and column lines have been used
while increasing yield five times and requiring fewer number of fuses to be
blown [54]. In this report spare memory cells form two logical groups of four
rows and two groups of two spare columns. This array architecture provided
maximum efficiency in removing random defects. Laser-blown polysilicon
fuses have been used at 1 Mbit DRAM density level, which could provide
81% repairable die area through redundancy [55].

Use of redundant circuits and ECC was implemented on a 50 ns 16 Mbit
DRAM providing excellent fault tolerance [56]. The 16 Mbit chip was divided
in four independent quadrants, each one having its bit redundancy and data
steering, word redundancy system, ECC circuitry, and SRAM. Each quad-
rant contained four array blocks with 1024 word lines. Each WL contained
eight ECC words, where each word had 128 data bits and 9 check bits. With

16 bits for the redundant bit lines, combined with 8 ECC words each word line comprised of 1112 bits.

An optimum odd-weight Hamming code was used with 128 data bits and 9 check bits [57] for double-error detection/single-error correction. Single cell error could be effectively corrected using the ECC circuit, but, for more than one bit failure, correction needed redundancy circuits. For 64 data blocks, each contained 2048 ECC words and two redundant bit lines. Combined effect of ECC and redundant circuits showed dramatic improvements. To obtain an expected yield of 50%, without the use of ECC circuits, no more than 28 random single bit failures were allowed, whereas the number of permissible failures increased to 428 and 2725, respectively, for the two cases when only ECC was used and when ECC was combined with redundancy circuits.

8.7 Redundancy Schemes for High-Density DRAMs

Use of spare column and bit lines was very effective in yield and reliability improvement up to 64–256 Kbit DRAM range, though the simple arrangement faced a few constraints:

1. Degradation in raw yield forced deployment of larger numbers of spare row/column lines and spare decoders. Resultant chip area overhead became excessively large.

2. Defects/faults in the redundancy overhead area also became significant.

3. With increase in DRAM density, memory array division started both without and with hierarchical word line. Arrangement of redundancy which worked in single core became prohibitively large in multi-divided array.

4. Because of the connecting wire strips becoming thin with increased DRAM density, chances of lines open circuiting or short circuiting with the supply or ground lines increased. It caused DC characteristic faults, especially increase in the standby current (I_{SB}) fault.

It became necessary to improve redundancy techniques without which the main advantage of the DRAM, lowest cost per bit, was likely to be under threat. In early stages of DRAM production, an estimation of yield was as given in Table 8.2 [4]. It was based on the assumed effective defect density D of 2/cm^2 [58] and increase of chip area 1.5× with each DRAM generation. Out of the total chip area, memory cells, decoders, word drivers, and sense

TABLE 8.2

Estimated Yield Improvement Range through Line
and Subarray Replacement with D = 2/Cm² and Chip
Area 1.5× per Generation

DRAM density	16 Mb	64 Mb	256 Mb	1 Gb
		Yield (%)		
Without redundancy	12–6	6–~1%	nil	nil
Line replacement	62–50	50–37	37–25	25–10
Subarray replacement	82–75	75–65	65–55	55–40

amplifiers were assumed to take up nearly 80% and the remaining area was left for the other peripherals. It was observed that the then used spare line redundancy resulted in only about half of the yield at 256 Mbit level compared to that at 16 Mbit level of DRAMs.

8.7.1 Subarray Replacement Redundancy for I_{SB} Faults

Yield of gigabit DRAMs is mainly affected by DC characteristic faults, like short circuit between a word line (at ground level during standby) and a bit line (precharge voltage-$V_{CC}/2$) causing flow of excessively large standby current (I_{SB}). Replacement of defective word/bit line will remove the fault, but large I_{SB} current would continue to flow, which needs to be minimized/stopped. In one scheme I_{SB} was limited to ~15 µA/short circuit using a current limiter through the precharge supply line [59].

In a similar approach DC current is cut off by a power switch controlled by a fuse [60]. However, a preferred scheme is given by Kitsukawa and others for I_{SB} fault correction through on-chip replacement of a defective subarray [4]. Figure 8.22 shows an example of an I_{SB} fault where bit line at precharge voltage ($V_{CC}/2$) is shown to be short circuited with the grounded word line, and also shows a subarray replacement scheme for its correction. For implementing the scheme, each subarray contains power switches for the precharge voltage ($V_{CC}/2$) and the cell plate voltage V_{PL}. The subarray also needs timing signal and a fuse ROM to control the time signal gate. Power switches of the defective subarray are turned off and those for the spare array are turned on; hence, the faulty I_{SB} current path is cut off. With the increase in the number of subarrays in high-density DRAMs, the scheme finds favor.

8.7.2 Flexible Redundancy Technique

In conventional redundancy applications in DRAMs without cell array division, defective word line addresses are programmed in the address comparators (ACs) and compared with the input address. The number of comparators is equal to the number of spare word lines (say L); hence, only L defective

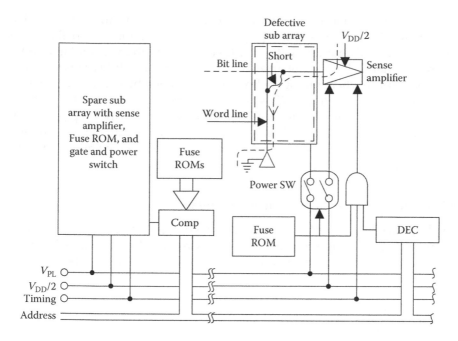

FIGURE 8.22
Subarray having I_{SB} fault replacement redundancy technique. (Redrawn from "256 Mb DRAM Circuit Technologies for File Applications," G. Kitsukawa et al., IEEE J.S.C. Circuits, Vol. 28, pp. 1105–1113, 1993.)

word lines can be replaced [54,61]. With increasing DRAM density the cell array is divided many times (M); then inter-subarray (SAR) replacement, in which defective line in one array should not be replaced by spare line in another SAR as it increases memory array control complexity considerably. The following two approaches have been suggested [58]:

1. Simultaneous replacement scheme in which ACs equal L in a subarray and each AC compares only the intra-SAR address signals and output is sent to all SARs. This means that to replace one defective line (L-1), other functional lines with the same intra-SAR address are also replaced. Hence, it requires a larger number of spare lines increasing wasteful chip area, and the possibility of a defect in the spare line, which is replacing a faulty line, also increases.

2. In a similar scheme, called individual replacement, every spare line in each SAR has its own ACs. Therefore, the number of ACs increases and equals the product $L*M$, but has some advantages over scheme 1. above. Statistically, required value of L for as many defects becomes small and only one line is replaced by the spare line, avoiding wasteful chip area. However, more area is taken up by the ACs.

Problems and limitations in the two schemes were minimized in a flexible intra-SAR replacement scheme [58]. The ACs are connected to the spare lines through OR gates, so that each AC compares both intra- and inter-SAR address signals. This kind of arrangement provides flexibility between ACs and spare lines, where one spare line can be reached by several ACs; hence, the number of ACs required (C) is also flexible as per the relationship $L \leq C \leq L^*M/m$, where m is the number of SARs in which spare lines replace faulty lines simultaneously. It was observed that the flexible redundancy technique was more effective at higher density where the yield became mainly dependent on fatal defects discussed in Section 8.7.1.

8.7.3 Inter-Subarray Replacement Redundancy

For high-density DRAMs chances of clustered defects increase. For such repairs in intra-SAR replacement techniques, required spare lines L in a SAR must be more than or equal to the defective lines, which means larger value for L or larger chip overhead.

At least two inter-SAR replacement redundancy techniques shown in Figure 8.23 have been employed in which defective lines could be repaired by the spare line in any SAR. In the distributed-spare line approach, shown in Figure 8.23(a), each SAR has spare line, but these lines can replace a faulty line in any other SAR as well. Hence, a cluster of L^*M defects can be repaired in any SAR, with L being the average number of defective lines (not maximum) in a SAR. Number of ACs equals L^*M, though it can be reduced a bit [62].

Concentrated-spare line approach shown in Figure 8.23(b) is based on a separate SAR whose lines are used as spare lines and can replace any defective line [59,63]. The remaining SARs do not have any spare lines. Number of lines in the spare SAR is flexible (L'); obviously, L' defects clustered anywhere can be rectified. The number of ACs used equal L' and since the L' number is flexible, use of comparators becomes economical. However, a separate SAR needs an extra decoder, sense amplifier, and so on. Access time penalty is also a little more than intra-SAR replacement technique because an activated SAR may be changed as per requirement instead of an activated line only.

8.7.4 DRAM Architecture–Based Redundancy Techniques

Redundancy area unit (RAU) is a memory area comprising either rows or columns that can be repaired, if faulty, by a single spare line. Obvious a system having larger RAU needs fewer spare lines. At higher DRAM density level, multibank architecture is now common, but because of more than one bank being active simultaneously, spare elements cannot be shared. More spares are now required, forcing RAU to be small, even when all spare lines in a bank might not be used. Since high-speed multibit column access is needed, a defect replacement from a spare line located at a large distance physically adds to the access delay which is another reason for having smaller RAU

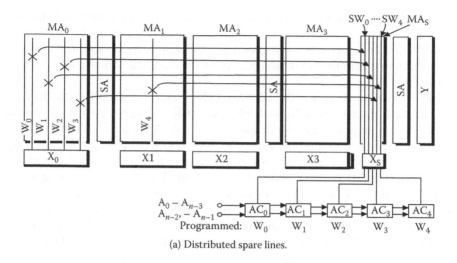

(a) Distributed spare lines.

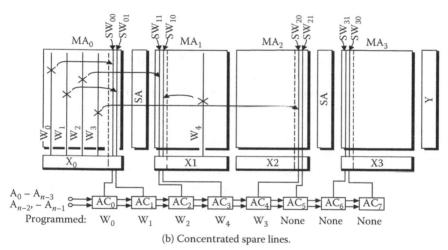

(b) Concentrated spare lines.

FIGURE 8.23
Inter-subarray replacement redundancy techniques. ("Redundancy Techniques for High-Density RAMs," M. Horiguchi et al., Int. Conf. Innovative Systems in Silicon, pp. 22–29, 1997.)

in DRAMs. Flexible redundancy technique of Section 8.7.2 did introduce a flexible relation between fuse set and spare element, but it was subjected to certain constraints and chip area efficiency was less, which was improved in another technique called *flexible mapping redundancy* (FMR) [64]. The main feature of the scheme is that instead of maintaining an equal number of spare lines and fuse sets with one-to-one correspondence between them, a different approach is adopted. In a flexible correspondence, which is obtained by mapping fuses in a fuse set, numbers of fuse sets that store the addresses of the defects are less than the numbers of spare elements. The mapping data that are stored in a fuse set are based on the defect distribution information. Both row and column redundancies are effectively used. For a 16-bank DRAM, 13% of chip area saving was obtained in comparison to a conventional scheme and the total area occupied by the FMR scheme was 5.2% of the chip.

For increasing bandwidth, DRAMs often used an architecture having hierarchical data lines with multiple I/Os [65]. In such structures column redundancy acquires greater importance [66]. A flexible column redundancy circuit for such architecture has been reported involving I/O shifting [67]. Fuse mapping algorithm use minimizes redundancy circuit overhead. Proposed flexible column redundancy circuit has RAU containing eight segments and two redundant segments at each end of the eight segments. In each of the eight segments there are 64 column select lines (CSLs) and four I/O pairs and each redundant column has four redundant CSLs and four I/O pairs. The scheme works in such a way that a maximum of eight columns can be repaired for eight segments. A control circuit for shift switches and the mapping table was also given which determines the state of the switches. Number of fuses used in the scheme was fewer than those used in the FMR scheme [64] mentioned above.

A dual CSL column redundancy scheme (DCCR) has two symmetrical I/O buses in one I/O block, which translates into an action in which one CSL and one spare column line selection (SCSL) can be activated simultaneously on repairing access. Hence, in high-speed DRAMs with multiple bit prefetch structures, the proposed scheme repairs defective bits by single units [68]. In a four-bit prefetch, RAU of the DCCR becomes as large as eight times compared to a conventional column redundancy scheme. From this brief discussion in this section, it is obvious that architecture of the DRAM has great influence on the method of application of redundancy, and needs to be pursued for future benefits.

References

1. J.M. Rabaey, A. Chandrakasan, and B. Nikalic, *Digital Integrated Circuits—A Design Perspective*, Pearson Edition, 2003.
2. R.J.C. Chwang et al., "A 70 ns High Density 64 K CMOS Dynamic RAM," *IEEE J.S.S. Circuits*, Vol. SC-18, pp. 457–463, 1983.
3. K. Kimura et al., "Power Reduction Techniques in Megabit DRAM's," *IEEE J.S.S. Circuits*, Vol. SC-21, pp. 382–388, 1986.
4. G. Kitsukawa et al., "256 Mb DRAM Circuit Technologies for File Applications," *IEEE J.S.C. Circuits*, Vol. 28, pp. 1105–1113, 1993.
5. J.J. Barnes et al., "Circuit Techniques for a 25 ns 16 K x 1 SRAM Using Address-Transition Detection," *IEEE J.S.C. Circuits*, Vol. SC-19, pp. 455–461, 1984.

6. L.F. Childs and R.T. Hirase, "A 18 ns 4K x 4 CMOS SRAM," *IEEE J.S.S. Circuits*, Vol. SC-19, pp. 454–551, 1984.
7. H. Nimbu et al., "A 1.8 ns Access, 550 MHz 4.5 Mb CMOS SRAM," *Proc. ISS. CC. Dig. Tech. Papers*, paper SP 22.7, 1998.
8. F. Baba et al. "A 64 K DRAM with 35 ns Static Column Operation," *IEEE J.S.S. Circuits*, Vol. SC-18, pp. 447–451, 1983.
9. R.T. Taylor and M.G. Johnson, "A 1-Mbit CMOS Dynamic RAM with a Divided Bit Line Matrix Architecture," *IEEE J.S.S. Circuits*, Vol. SC-20, pp. 894–902, 1985.
10. K.U. Stein et al., "Storage Array and Sense/Refresh Circuit for Single Transistor Memory Cell, " *IEEE J. of Solid State Circuits*, Vol. SC-7, pp. 336–340, 1972.
11. C. Kuo et al., "Sense Amplifier Design Is Key to the One-Transistor Cell in 4096 Bit RAM," *Electronics*, Vol. 46, pp. 116–121, 1973.
12. R.C. Foss and R. Harland, "Peripheral Circuits for One-Transistor Cell MOS RAM's," *IEEE J. of Solid State Circuits*, Vol. SC-10, pp. 225–261, 1975.
13. C.N. Ahlquist et al., "A 16384-Bit Dynamic RAM," *IEEE J. of Solid State Circuits*, Vol. SC-11, no. 5, pp. 570–574, 1976.
14. W.T. Lynch and H.J. Boll, "Optimization of the Latching Pulse for Dynamic Flip-Flop Sensors," *IEEE J. of Solid State Circuits*, Vol. SC-9, pp. 49–55, 1974.
15. L.G. Heller et al., "High-Sensitivity Charge-Transfer Sense Amplifier," *IEEE J. of Solid-State Circuits*, Vol. SC-11, pp. 596–601, 1976.
16. L.G. Heller, "Cross-Coupled Charge-Transfer Sense Amplifier," *IEEE, ISSCC. Digest of Tech. Papers*, pp. 20–21, 1979.
17. N.C.C. Lu and H.H. Chao, "Half-V_{DD} Bit-Line Sensing Schemes in CMOS DRAMs," *IEEE J. of Solid State Circuits*, Vol. SC-19, pp. 451–454, 1984.
18. S. Fujii et al., "A 50 μA Standby 1 Mx 1/256 x 4 CMOS DRAM," *IEEE J. of Solid State Circuits*, Vol. SC-21, pp. 627–634, 1986.
19. T. Mano et al., "Circuit Techniques for 16 Mb DRAMs," *ISSCC Dig. Tech. Papers*, pp. 22–23, Feb. 1987.
20. D. Chin and W. Hwang, "Quarter-V_{DD} Sensing Schemes in CMOS DRAMs," *IBM Tech. Disc. Bull.*, pp. 3750–3751, Jan. 1987.
21. S.H. Dhong et al., "High-Speed Sensing Scheme for CMOS DRAMs," *IEEE J. of Solid State Circuits*, Vol. 23, pp. 34–40, 1988.
22. T. Ooishi et al., "A Well-Synchronized Sensing/Equalizing Method for Sub-1.0 V Operating Advanced DRAMs," *Symp. VLSI Circuit Tech. Papers*, pp. 81–82, 1993.
23. J. Okamura et al., "Decoded-Source Sense Amplifier for High-Density DRAM's," *IEEE J. of Solid State Circuits*, Vol. 25, pp. 18–23, 1990.
24. M. Tsukude et al., "A 1.2 V to 3.3 V Wide Voltage-Range/Low-Power DRAM with a Charge-Transfer Presenting Scheme," *IEEE J. of Solid State Circuits*, Vol. 32, pp. 1721–1727, 1997.
25. T. Yamagata et al., "Circuit Design Techniques for Low-Voltage Operating and/or Giga-Scale DRAMs," *ISSCC, Dig. of Tech. Papers*, pp. 113–114, 1995.
26. M. Nakamura et al., "A 29 ns 64 MB DRAM with Hierarchical Array Architecture," *ISSCC, Dig. of Tech. Papers*, pp. 246–247, 1995.
27. K. Gotoh et al., "A 0.9 V Sense-Amplifier Driver for High-Speed Gb-Scale DRAMs," *IEEE Symp. VLSI Circuits, Dig. of Tech. Papers*, pp. 108–109, 1996.
28. T. Takashahi et al., "A Multi-Gigabit DRAM Technology with 6 F^2 Open-Bit-Line Cell Distributed Over-Driven Sensing and Stacked-Flash Fuse," *IEEE Intern. S.S. Circ. Conf.*, pp. 465–468, 2001.

29. S. Suzuki et al., "Threshold Difference Compensated Sense Amplifier," *IEEE J. Solid State Circuits*, Vol. SC-14, pp. 1066–1070, 1979.
30. T. Mano et al., "Submicron VLSI Memory Circuit," *ISSCC Dig. Tech. Papers*, pp. 234–235, 1983.
31. T. Kawahara et al., "A High-Speed, Small-Area, Threshold-Voltage-Mismatch Compensation Sense Amplifier for Gigabit-Scale DRAM Arrays," *IEEE J. Solid State Circuits*, Vol. 28, pp. 816–823, 1993.
32. S. Suzuki and M. Hirata, "Threshold Difference Compensated Sense Amplifier," *IEEE J. Solid State Circuits*, Vol. SC-15, pp. 1066–1070, 1979.
33. T. Furuyama, S. Saito, and S. Fuiji, "A New Sense Amplifier Technique for VLSI Dynamic RAM's," *IEDM Tech. Dig.*, pp. 44–47, 1981.
34. Y. Watanabe, N. Nakamura, and S. Watanabe, "Offset Compensating Bit-Line Sensing Scheme for High Density DRAM's," *IEEE J. Solid State Circuits*, Vol. 29, pp. 9–13, 1994.
35. S. Hong et al., "Low-Voltage DRAM Sensing Scheme with Offset-Cancellation Sense Amplifier," *IEEE J. Solid State Circuits*, Vol. 37, pp. 1356–1360, 2002.
36. J.Y. Sim et al., "A 1.0 V 256 Mb SDRAM with Offset-Compensated Direct Sensing and Charge-Recycled Precharge Schemes," *IEEE Inter. S.S. Circuits Conf. Paper*, 17.6, 2003.
37. J.Y. Sim et al., "Charge-Transferred Presenting, Negatively Precharged Word-Line, and Temperature-Insensitive Up Schemes for Low-Voltage DRAMs," *IEEE J. Solid State Circuits*, Vol. 39, pp. 694–703, 2004.
38. K. Itoh, "Adaptive Circuits for the 0.5 V Nanoscale CMOS Era," *IEEE Int. S.S.C. Conf. Dig. Tech. Papers*, pp. 14–17, 2009.
39. H.C. Chow and C.L. Hsich, "A 0.5 V High Speed DRAM Charge Transfer Sense Amplifier," *50th Midwest Symp., MWCAS*, pp. 1293–1296, 2007.
40. S. Akiyama et al., "Low V_t Small-Offset Gated Preamplifier for Sub-IV Gigabit DRAM Arrays," *IEEE Int. S.S. Circuits Conf.*, pp. 142–144, 2009.
41. A. Kotabe et al., "CMOS Low-V_T Preamplifier for 0.5-V Gigabit-DRAM Array," *IEEE Asian S.S. Cir. Conf.*, pp. 213–216, 2009.
42. A. Kotabe et al., "0.5 V Low V_T CMOS Preamplifier for Low-Power and High-Speed Gigabit-DRAM Array," *IEEE J. S.S. Circ.*, Vol. 45, pp. 2348–2355, 2010.
43. A. Kotabe et al., "Asymmetric Cross-Coupled Sense Amplifier for Small-Sized 0.5 V Gigabit-DRAM Arrays," *IEEE Asian S.S. Circ. Conf.*, pp. 1–4, 2010.
44. T. Yamada et al., "A 4 Mbit DRAM with 16-bit Concurrent ECC," *IEEE J.S.S. Cir.*, Vol. 23, pp. 20–26, 1988.
45. J. Yamada et al., "A Submicron VLSI Memory with a 4b-at-a-Time Built-in ECC Circuit," *Proc. ISSCC Dig. Tech. Papers*, pp. 104–105, 1984.
46. Micron Technology Inc., Boise, ID, MT1256/MT4064 data sheet, 1984.
47. T. Yamada, "Selector-Line Merged Built-in ECC Technique for DRAM's," *IEEE J.S.S. Cir.*, Vol. SC-22, pp. 868–873, 1987.
48. K. Furutani et al., "A Built-In Hamming Code ECC Circuit for DRAM's," *IEEE J.S.S. Cir.*, Vol. 24, pp. 50–56, 1989.
49. E. Tammaru and J.B. Angell, "Redundancy for VLSI Field Enhancement," *IEEE J. Solid State Circuits*, Vol. SC-2, pp. 172–182, 1967.
50. A. Chen, "Redundancy in VLSI Memory Array," *IEEE J. Solid State Circuits*, Vol. SC-4, pp. 291–293, 1969.
51. S.E. Schuster, "Multiple Word/Bit Line Redundancy for Semiconductor Memories," *IEEE J. Solid State Circuits*, Vol. SC-13, pp. 698–703, 1978.

52. R.P. Cenker et al., "A Fault Tolerant 64K Dynamic RAM," *ISSCC Dig. Tech. Papers*, pp. 150–151, 1979.

53. V.G. McKenny, "A 5V 64K EPROM Utilizing Redundant Circuitry," *ISSCC Dig. Tech. Papers*, pp. 146–147, 1980.

54. S.S. Eaton et al., "A 100 ns 64 K Dynamic RAM Using Redundancy Techniques," *IEEE Int. S.S. Cir. Conf.*, pp. 84–85, 1981.

55. R.T. Taylor and M.G. Johnson, "A 1 Mbit CMOS Dynamic RAM with a Divided Bit Line Matrix Architecture," *IEEE J. S.S. Circuits*, Vol. SC-20, pp. 894–902, 1985.

56. H.L. Kalter et al., "A 50 ns 16-Mb DRAM with a 10-ns Data Rate and On-Chip ECC, *IEEE J.S.S. Circ.*, Vol. 25, pp. 1118–1128, 1990.

57. J.A. Field, "A High-Speed On-Chip System Using Modified Hamming Code," IBM General Technology Division, VT Rep. TR 19.90496, 1990.

58. M. Horiguchi et al., "A Flexible Redundancy Technique for High-Density DRAM's," *IEEE J.S.S. Circ.*, Vol. 26, pp. 12–17, 1990.

59. T. Kirihata et al., "Fault-Tolerant Designs for 256 Mb DRAM," *IEEE J. S.S. Circ.*, Vol. 31, pp. 558–566, 1996.

60. K. Furutani et al., "A Board Level Parallel Test and Short Circuit Failure Repair Circuit for High-Density Low Power DRAMs," *Symp. VLSI Circuits, Dig. Tech. Papers*, pp. 70–71, 1996.

61. K. Shimohigashi et al., "Redundancy Techniques for Dynamic RAMs," *Proc. 14 Conf. S.S. Devices*, pp. 63–67, 1982.

62. M. Horiguchi, "Redundancy Technique for High-Density DRAMs," *Proc. IEEE Int. Conf. Innovative Systems in Silicon*, pp. 22–29, 1997.

63. K. Ishibashi et al., "A 12.5 ns 16 Mb CMOS SRAM with Common-Centroid-Geometry-Layout Sense Amplifiers," *IEEE J.S.S. Circ.*, Vol. 29, pp. 411–418, 1994.

64. S. Takase et al., "A 1.6-GByte/s DRAM with Flexible Mapping Redundancy Technique and Additional Refresh Scheme," *IEEE J.S.S. Circ.*, Vol. 34, pp. 1600–1606, 1999.

65. C. Kim et al., "A 2.5 V, 72-Mbit, 2.0 Gbytes/s Packet-Based DRAM with a 1.0 Gb/s/Pin Interface," *IEEE J.S.S. Circ.*, Vol. 34, pp. 645–652, 1999.

66. Namekawa et al., "Dynamically Shift-Switched Data Line Redundancy Suitable for DRAM Macro with Wide Data Bus," *Symp. VLSI Circ.*, pp. 149–152, 1999.

67. Y-W. Jeon, Y-H. Jun, and S. Kim, "Column Redundancy Scheme for Multiple I/O DRAM Using Mapping Table," *Electronic Letters*, Vol. 36, pp. 940–942, 2000.

68. J-G. Lee et al., "A New Column Redundancy Scheme for Yield Improvement of High Speed DRAMs with Multiple Bit Pre-fetch Structure," *Symp. VLSI Circ. Dig. Tech. Papers*, pp. 69–70, 2001.

Index